Astronomical Photometry

Astronomical Photometry

Arne A. Henden
Systems & Applied Sciences Corp.
and
Dept. of Astronomy
Indiana University

and

Ronald H. Kaitchuck
Dept. of Astronomy
Indiana University

VAN NOSTRAND REINHOLD COMPANY
NEW YORK   CINCINNATI   TORONTO   LONDON   MELBOURNE

Copyright © 1982 by Van Nostrand Reinhold Company Inc.

Library of Congress Catalog Card Number: 81-11444
ISBN: 0-442-23647-6

All rights reserved. No part of this work covered by the copyright hereon may be reproduced or used in any form or by any means—graphic, electronic, or mechanical, including photocopying, recording, taping, or information storage and retrieval systems—without permission of the publisher.

Manufactured in the United States of America

Published by Van Nostrand Reinhold Company Inc.
135 West 50th Street, New York, N.Y. 10020

Van Nostrand Reinhold Limited
1410 Birchmount Road
Scarborough, Ontario MIP 2E7, Canada

Van Nostrand Reinhold Australia Pty. Ltd.
17 Queen Street
Mitcham, Victoria 3132, Australia

Van Nostrand Reinhold Company Limited
Molly Millars Lane
Wokingham, Berkshire, England

15 14 13 12 11 10 9 8 7 6 5 4 2 1

Library of Congress Cataloging in Publication Data

Henden, Arne A.
  Astronomical photometry.

  Includes index.
  1. Photometry, Astronomical.   I. Kaitchuck,
Ronald H.   II. Title.
QB135.H44   522'.62   81-11444
ISBN 0-442-23647-6   AACR2

PREFACE

Most people who do astronomical photometry have had to learn the hard way. Books for the newcomer to this field are almost totally lacking. We had to learn by word-of-mouth and searching libraries for what few references we could find.

The situation improved markedly after we began our graduate studies in astronomy at Indiana University. We then had access to professional astronomers with many years of experience in photometry. Indiana has a very good astronomical library, and the copy machine was used heavily by both of us. Nevertheless, information was still gleaned in a piecemeal fashion. It became obvious to us that, as tedious as our educational process had been, it must be a frustrating experience for those with more limited reference resources. We were also aware of the many "tricks" we had learned which somehow never found their way into print. With this in mind, we set out to write a reference text both to spare the beginner some of the hardships and mistakes we encountered, and to encourage others to share in the satisfaction of doing meaningful research.

Our basic approach was to create a self-contained book that could be used by the interested amateur with little or no college background, and by the astronomy major who is new to photometry. By self-contained, we mean the inclusion of sections on observational techniques, construction, and reference material such as standard stars. In addition, we added substantial theoretical background material. The more esoteric material was placed in appendices at the back of the book, thereby retaining the beginning level throughout the bulk of the manual yet providing heavier reading for the most advanced student.

Photoelectric photometry is a relatively small field of science, and therefore does not have the large commercial suppliers of instrumentation. We have tried wherever possible to indicate sources of equipment, not to recommend any particular brand but to indicate starting points for any equipment selection procedure. Any implied endorsement is unintentional.

Similarly, we advocate certain techniques in both the data acquisition and reduction. There are as many methods in photoelectric astronomy as there are observers and we will certainly have made some arguable statements. We have tried to only present techniques that we have used and found successful.

This book would not have been possible without the dedicated help of Professor Martin S. Burkhead, who instructed us in observational techniques and acquainted us with Indiana University facilities, and Professor R. Kent Honeycutt, who provided much of our theoretical background knowledge by course material and stimulating conversations. Both these professors, Russ Genet and Bob Cornett have proofread much of the text for which we are very grateful. We would also like to thank Thomas L. Mullikin for writing the section on occultation techniques. Our wives contributed more time and effort into proofreading and correcting than we would like to admit!

We hope that reading this book will instill in you the excitement and satisfaction that we have found in astronomical photometry. Good luck!

Arne A. Henden
Ronald H. Kaitchuck

CONTENTS

Preface / v

1. AN INTRODUCTION TO ASTRONOMICAL PHOTOMETRY / 1

1.1 An Invitation / 1
1.2 The History of Photometry / 5
1.3 A Typical Photometer / 9
1.4 The Telescope / 11
1.5 Light Detectors / 13
   a. Photomultiplier Tubes / 13
   b. PIN Photodiodes / 18
1.6 What Happens at the Telescope / 23
1.7 Instrumental Magnitudes and Colors / 25
1.8 Atmospheric Extinction Corrections / 28
1.9 Transforming to a Standard System / 29
1.10 Other Sources on Photoelectric Photometry / 30

2. PHOTOMETRIC SYSTEMS / 33

2.1 Properties of the UBV System / 34
2.2 The UBV Transformation Equations / 37
2.3 The Morgan-Keenan Spectral Classification System / 38
2.4 The M-K System and UBV Photometry / 42
*2.5 Absolute Calibration / 50
2.6 Differential Photometry / 52
2.7 Other Photometric Systems / 54
   a. The Infrared Extension of the UBV System / 55
   b. The Strömgren Four-Color System / 55
   c. Narrow-Band $H\beta$ Photometry / 57

3. STATISTICS / 60

3.1 Kinds of Errors / 60
3.2 Mean and Median / 61
3.3 Dispersion and Standard Deviation / 64
3.4 Rejection of Data / 66
3.5 Linear Least Squares / 68
   *a. Derivation of Linear Least Squares / 69
   b. Equations for Linear Least Squares / 70
3.6 Interpolation and Extrapolation / 73
   a. Exact Interpolation / 74
   b. Smoothed Interpolation / 76
   c. Extrapolation / 77
3.7 Signal-to-Noise Ratio / 77
3.8 Sources on Statistics / 78

4. DATA REDUCTION / 80

4.1 A Data-Reduction Overview / 80
4.2 Dead-Time Correction / 81
4.3 Calculation of Instrumental Magnitudes and Colors / 85
4.4 Extinction Corrections / 86
   a. Air Mass Calculations / 86
   b. First-order Extinction / 88
   *c. Second-order Extinction / 90
4.5 Zero-Point Values / 91
4.6 Standard Magnitudes and Colors / 92
4.7 Transformation Coefficients / 93
4.8 Differential Photometry / 95
*4.9 The (U-B) Problem / 98

5. OBSERVATIONAL CALCULATIONS / 101

5.1 Calculators and Computers / 101
5.2 Atmospheric Refraction and Dispersion / 104
   a. Calculating Refraction / 104
   b. Effect of Refraction on Air Mass / 106
   c. Differential Refraction / 107
5.3 Time / 108
   a. Solar Time / 108
   b. Universal Time / 109
   c. Sidereal Time / 110

d. Julian Date / 112
*e. Heliocentric Julian Date / 113
5.4 Precession of Coordinates / 116
5.5 Altitude and Azimuth / 119
*a. Derivation of Equations / 119
b. General Considerations / 122

6. CONSTRUCTING THE PHOTOMETER HEAD / 124

6.1 The Optical Layout / 124
6.2 The Photomultiplier Tube and Its Housing / 128
6.3 Filters / 134
6.4 Diaphragms / 138
6.5 A Simple Photometer Head Design / 141
6.6 Electronic Construction / 147
6.7 High-Voltage Power Supply / 149
 a. Batteries / 149
 b. Filtered Supply / 150
 c. RF Oscillator / 153
 d. Setup and Operation / 155
6.8 Reference Light Sources / 155
6.9 Specialized Photometer Designs / 157
 a. A Professional Single-beam Photometer / 157
 b. Chopping Photometers / 159
 c. Dual-beam Photometers / 161
 d. Multifilter Photometers / 163

7. PULSE-COUNTING ELECTRONICS / 167

7.1 Pulse Amplifiers and Discriminators / 167
7.2 A Practical Pulse Amplifier and Discriminator / 170
7.3 Pulse Counters / 172
7.4 A General-Purpose Pulse Counter / 173
7.5 A Microprocessor Pulse Counter / 178
7.6 Pulse Generators / 181
7.7 Setup and Operation / 182

8. DC ELECTRONICS / 184

8.1 Operational Amplifiers / 185
8.2 An Op-Amp DC Amplifier / 188
8.3 Chart Recorders and Meters / 193

x CONTENTS

   8.4 Voltage-to-Frequency Converters / 195
   8.5 Constant Current Sources / 196
   8.6 Calibration and Operation / 197

9. PRACTICAL OBSERVING TECHNIQUES / 202

   9.1 Finding Charts / 202
      a. Available Positional Atlases / 203
      b. Available Photographic Atlases / 204
      c. Preparation of Finding Charts / 205
      d. Published Finding Charts / 206
   9.2 Comparison Stars / 207
      a. Selection of Comparison Stars / 208
      b. Use of Comparison Stars / 209
   9.3 Individual Measurements of a Single Star / 210
      a. Pulse-counting Measurements / 210
      b. DC Photometry / 213
      c. Differential Photometry / 216
      d. Faint Sources / 218
   9.4 Diaphragm Selection / 220
      *a. The Optical System / 220
      *b. Stellar Profiles / 223
      c. Practical Considerations / 224
      d. Background Removal / 226
      e. Aperture Calibration / 227
   9.5 Extinction Notes / 228
   9.6 Light of the Night Sky / 229
   9.7 Your First Night at the Telescope / 231

10. APPLICATIONS OF PHOTOELECTRIC PHOTOMETRY / 238

   10.1 Photometric Sequences / 238
   10.2 Monitoring Flare Stars / 240
   10.3 Occultation Photometry / 245
   10.4 Intrinsic Variables / 248
      a. Short-period Variables / 249
      b. Medium-period Variables / 250
      c. Long-period Variables / 254
      d. The Eggen Paper Series / 254
   10.5 Eclipsing Binaries / 256

10.6 Solar System Objects / 270
10.7 Extragalactic Photometry / 272
10.8 Publication of Data / 273

APPENDICES / 279

A. First-Order Extinction Stars / 279
B. Second-Order Extinction Pairs / 286
C. UBV Standard Field Stars / 290
D. Johnson UBV Standard Clusters / 297
   D.1 Pleiades / 298
   D.2 Praesepe / 298
   D.3 IC 4665 / 302
E. North Polar Sequence Stars / 305
F. Dead-Time Example / 308
G. Extinction Example / 311
   G.1 Extinction Correction for Differential Photometry / 311
   G.2 Extinction Correction for "All-Sky" Photometry / 313
   G.3 Second-Order Extinction Coefficients / 320
H. Transformation Coefficients Examples / 322
   H.1 DC Example / 322
   H.2 Pulse-counting Example / 327
I. Useful FORTRAN Subroutines / 335
   I.1 Dead-Time Correction for Pulse-Counting Method / 336
   I.2 Calculating Julian Date from UT Date / 336
   I.3 General Method for Coordinate Precession / 337
   I.4 Linear Regression (Least Squares) Method / 338
   I.5 Linear Regression (Least Squares) Method Using the UBV Transformation Equations / 339
   I.6 Calculating Sidereal Time / 340
   I.7 Calculating Cartesian Coordinates for 1950.0 / 341
J. The Light Radiation from Stars / 342
   J.1 Intensity, Flux, and Luminosity / 342
   J.2 Blackbody Radiators / 349
   J.3 Atmospheric Extinction Corrections / 351
   J.4 Transforming to the Standard System / 355
K. Advanced Statistics / 358
   K.1 Statistical Distributions / 358
   K.2 Propagation of Errors / 361
   K.3 Multivariate Least Squares / 363

K.4 Signal-to-Noise Ratio / 366
    a. Detective Quantum Efficiency / 367
    b. Regimes of Noise Dominance / 370
K.5 Theoretical Differences Between DC and Pulse-Counting Techniques / 372
    a. Pulse Height Distribution / 372
    b. Effect of Weighting Events on the DQE / 373
K.6 Practical Pulse-DC Comparison / 377
K.7 Theoretical S/N Comparison of a Photodiode and a Photomultiplier Tube / 378

INDEX / 389

Astronomical Photometry

CHAPTER 1
AN INTRODUCTION TO ASTRONOMICAL PHOTOMETRY

1.1 AN INVITATION

In the direction of the constellation Lyra, at a distance of 26 light years, there is a star called Vega. Unknown to the ancients who named this star, its surface temperature is almost twice that of the sun and each square centimeter of its surface radiates over 175,000 watts in the visible portion of the spectrum. This is roughly 100 times the power of all the electric lights in a typical home, radiating from a spot a little smaller than a postage stamp. After traveling for 26 years, the light from Vega reaches the neighborhood of the sun diluted by a factor of 10^{-39}. Of this remaining light, approximately 20 percent is lost by absorption in passing through the earth's atmosphere. Approximately 30 percent is lost by scattering and absorption in the optics of a telescope. A 25-centimeter (10-inch) diameter telescope pointed at Vega will collect only one half-billionth of a watt at its focus. Of this, only a fraction is actually detected by a modern photoelectric detector. This incredibly small amount of energy corresponds to one of the brightest stars in the night sky. The amazing thing is that the stars can be seen at all! Perhaps even more amazing is that starlight can be accurately measured by a device which can be constructed at a cost of a few hundred dollars. Such is the nature of astronomical photometry.

The photometry of stars is of fundamental importance to astronomy. It gives the astronomer a direct measurement of the energy output of stars at several wavelengths and thus sets constraints on the models of stellar structure. The color of stars, as determined by measurements at two different spectral regions, leads to information on the star's temperature. Sometimes these same measurements are used as a probe of interstellar dust. Photometry is often needed to establish a star's distance and size. The Hertzsprung-Russell diagram, the key to understanding stellar evolution, is based on photometry and spectroscopy.

Finally, many stars are variable in their light output either due to internal changes or to an occasional eclipse by a binary partner. In both cases, the light curves obtained by photometry lead to important information about the structure and character of the stars. The photometry of stars, especially at several wavelengths, is one of the most important observational techniques in astronomy.

Astronomy differs from other modern sciences in an important aspect. Vast commercial laboratories or university facilities are not necessary to undertake important research. An amateur astronomer with a modest telescope or persons with access to a small college observatory equipped with a simple photoelectric photometer can make *a valuable and a needed contribution to science.* The number of stars, galaxies, and nebulae vastly outnumber the professional astronomers. As an example, there are less than 3000 professional astronomers in the United States and yet there are over 25,000 catalogued variable stars. On any given night, almost all of them go unobserved! Furthermore, an observer with a large telescope will concentrate on faint objects for which a large instrument was intended. Very little time is given to the brighter stars even though they are no better understood than the faint ones! A small telescope is well suited for these objects and even a simple photometer can produce first-class results in the hands of a careful observer. This book is, in part, an invitation and a guide to the amateur astronomer or persons with access to small college observatories to share in the satisfaction of astronomical research through photoelectric photometry. In Chapter 10, research projects for a small telescope are discussed. However, to give the reader a "feel" for what can be done, we cite two examples now. Figure 1.1 shows a light curve for the short-period eclipsing binary V566 Ophiuchi. This light curve was collected over several nights using a homemade photometer on a 30-centimeter (12-inch) telescope. One result of this study was the discovery of a change in the orbital period, the first such change seen for this binary in 13 years. These changes are believed to be related to mass transfer, from one of the two stars to the other, which in turn is related to changes in the stars. Thus, indirectly, photometry allows stellar evolution to be seen!

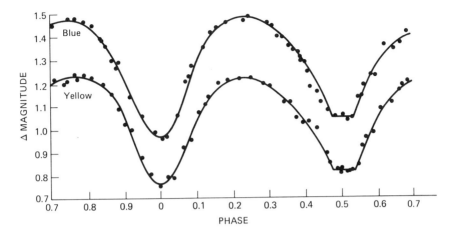

Figure 1.1 Light curve of an eclipsing binary.

Figure 1.2 shows the light curve of the double-mode cepheid TU Cassiopeiae taken with a 40-centimeter (16-inch) telescope. The scatter in the light curve is not due to poor photometry, but rather to the beating of two pulsations, with different periods, that are occurring in this star. Careful determination of both periods allows theorists to determine its mass, and the temporal variations in the light curve amplitude give constraints on its evolutionary behavior. Monitoring this type of star over long periods of time is essential, but difficult for the professional astronomer to do without preventing colleagues from using the telescope for their research.

This book is also directed toward a second type of reader. Often undergraduate or graduate students in astronomy are faced with the prospect of beginning their research only to discover that the "how-to"of photometry is lacking in textbooks. Much of the necessary information, such as lists and finding charts of standard stars, is scattered throughout the literature. Hopefully, this book will go a long way toward solving this problem. This reader will be more interested in the observing techniques and data reduction and less interested in construction details than the amateur astronomer. In like manner, there are theoretical sections that will be of less interest to the amateur. These sections have been either marked by an asterisk or placed in the appendices. The amateur astronomer may read or skip these sections as

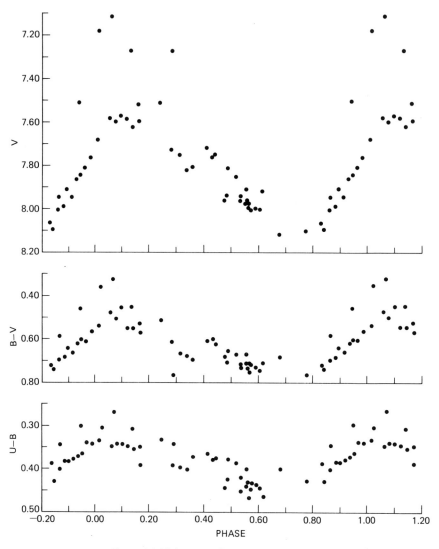

Figure 1.2 Light curve of a double-mode cepheid.

they are not mandatory for the construction and successful use of a photometer.

Overall, this book is intended to be a thorough guide from theory through practical circuits and construction hints to worked examples of data reduction. As a first step, we discuss the history of photometry and then consider a layout of a photoelectric photometer.

1.2 THE HISTORY OF PHOTOMETRY

A person need not own a telescope or a photometer to know that stars differ greatly in apparent brightness. It is therefore not surprising that the first attempt to categorize stars predates the telescope and was based solely on the human eye. Over 2000 years ago, the Greek astronomer Hipparchus divided the naked eye stars into six brightness classes. He produced a catalog of over 1000 stars ranking them by "magnitudes" one through six, from the brightest to the dimmest. In about A.D. 180, Claudius Ptolemy extended the work of Hipparchus, and from that time, the magnitude system became part of astronomical tradition. In 1856, N. R. Pogson confirmed Herchel's earlier discovery that a first magnitude star produces roughly 100 times the light flux[†] of a sixth magnitude star. The magnitude system had been based on the human eye, which has a nonlinear response to light. The eye is designed to suppress differences in brightness. It is this feature of the eye which allows it to go from a darkened room into broad daylight without damage. A photomultiplier tube or a television camera, which responds linearly, cannot handle such a change without precautionary steps. It is this same feature which makes the eye a poor discriminator of small brightness differences and the photomultiplier tube a good one. Pogson decided to redefine the magnitude scale so that a difference of five magnitudes was *exactly* a factor of 100 in light flux. The light flux ratio for a one-magnitude difference is $100^{1/5}$ or $10^{2/5}$ or 2.512. This definition is often referred to as a *Pogson scale*. The flux ratio for a two magnitude difference is $(10^{2/5})^2$ and a three magnitude difference is $(10^{2/5})^3$ and so on. In general,

$$F_1/F_2 = (10^{2/5})^{m_2-m_1} \tag{1.1}$$

where F_1, F_2 and m_1, m_2 refer to the fluxes and magnitudes of two stars.
This can be rewritten as

$$\log(F_1/F_2) = \tfrac{2}{5}(m_2 - m_1) \tag{1.2}$$

or

$$m_1 - m_2 = -2.5 \log F_1/F_2. \tag{1.3}$$

[†]See Appendix J for a discussion of flux, intensity, luminosity, and blackbody radiation.

Note that the 2.5 is exact and not 2.512 rounded off. Equation 1.1 tells us that the eye responds in such a way that equal magnitude *differences* correspond to equal flux *ratios*. Pogson made his new magnitude scale roughly agree with the old one by defining the stars Aldebran and Altair as having a magnitude of 1.0.

The human eye can generally interpolate the brightness of one star relative to nearby comparisons to about 0.2 magnitude. This is an acceptable error for certain programs such as the monitoring of long-period, large-amplitude variable stars. Because of the speed of measurement, visual photometry can be performed in sky conditions unsuitable for other forms of measurement. However, many problems exist with visual photometry, not the least of which are systematic errors such as color sensitivity differences between observers, difficulty in extrapolating to fainter stars, and lack of accuracy. The latter can be reduced somewhat by mechanical means introduced in the nineteenth century, so that the light from a variable artificial star visible to the observer can be adjusted to the same brightness as the object being measured. This type of photometer was invented by Zöllner and reduced the error to about 0.1 magnitude. A brief description of this device can be found in Miczaika and Sinton.[1]

Photography was quickly applied to photometry by Bond[2] and others at Harvard in the 1850s. The density and size of the image seemed to be directly related to the brightness of the star. However, the magnitudes determined by the photographic plate are not, in general, the same as those determined by the eye. The visual magnitudes are determined in the yellow-green portion of the spectrum where the sensitivity of the eye reaches a peak. The peak sensitivity of the basic photographic emulsion is in the blue portion of the spectrum. Magnitudes determined by this method are called *blue* or *photographic magnitudes*. The more recent panchromatic photographic plates can yield results which roughly agree with visual magnitudes by placing a yellow filter in front of the film. Magnitudes obtained in such manner are referred to as *photovisual*. Photographic photometry quickly showed that the old visual scale was not accurate enough for photographic work. What was needed was a new system, based on photographic photometry, and defined by a large number of standard stars. The unknown magnitude of a star could then be found by comparing it to the standards and applying Equation 1.1, where the flux ratio is determined by the image densities on the film. Because the brightest stars are not always well positioned for observation, a group of standard stars was defined in the vicinity of the north celestial pole. For a Northern Hemisphere observer, these stars would always be above the horizon. The group became known as the *North Polar Sequence*. Their magnitudes were defined so that the brightest stars in the sky would still be close to the photovisual magnitude of one. This system became known as the *International System*. At Mount Wilson Observatory, stars in 139 selected regions of the sky were established as secondary standards by comparison with the North Polar Sequence. Some of these stars were as faint as nineteenth magnitude. However, a large departure from Pogson's scale had occurred for the fainter stars because the nonlinearity of the photographic plate was not properly taken into account. Photographic photometry is still in use today, but primarily as a method of interpolating between nearby comparison stars, giving an error of about 0.02 magnitude. Photography offers a permanent record with a vast multiplexing advantage: thousands of images are recorded at one time.

Because of the difficulties inherent in visual and photographic methods, the application of the photoelectric method of measuring starlight in the late 1800s ushered in a new era in astronomy. Most early work such as that of Minchin[3] used selenium *photoconductive* cells which changed their resistance upon exposure to light. These cells are similar to the photocells found in some modern cameras. A constant voltage source was applied to the cell and the resulting variable current was measured with a galvanometer (a very sensitive current indicator.) A galvanometer is not used very often at present, primarily because of its bulk and its difficult calibration and operation. Joel Stebbins and F. C. Brown[4] were the first to use the selenium cell in the United States. Stebbins and his students were involved in most of the later development of photoelectric photometry (see Kron[5] for more details).

Some of the major disadvantages of the selenium cell were its low sensitivity (only bright stars and the moon were measured), narrow spectral response, and lack of commercial availability. Each cell had to be made individually, and it often took dozens of trials to produce a sensitive cell. Even so, in 1910 Stebbins[6] published a light curve of Algol of far greater precision than ever before, showing for the first time the shallow secondary eclipse that had eluded visual observers.

The discovery of the *photoelectric cell* in 1911 promised more sensitive measurements. These cells were similar to a tube-type diode using sodium, potassium, or other alkaline electrodes. A voltage of approximately 300 volts was applied, and when the cell was exposed to light, electrons liberated by the photoelectric process created a small current. This response was *linear,* that is, a source twice as bright gave twice the current. Schultz,[7] working with Stebbins, used the photoelectric cell to record light from Arcturus and Capella. Similar systems were being developed in Europe by Guthnick[8] and Rosenberg.[9] For many years, the problems associated with selenium cells plagued the newer design. Commercial photoelectric cells were not available until the 1930s. Galvanometers hung directly on the telescope and had to be kept level. The limit of detectability for the photoelectric cell-galvanometer combination was about a seventh magnitude star for a 40-centimeter (16-inch) telescope. The reader is referred to Stebbins[10] for details about these early measurements.

The electronic amplifier was introduced into astronomy by Whitford,[11] stepping up the feeble photocurrents to the point where less expensive meters and, more importantly, chart recorders could be used. At the same time, however, tube thermionic noise and amplifier instabilities were now problems and became the limiting component of a photoelectric system. The late 1920s and the 1930s also saw the advent of wide-band filters and the increasing adoption of the photoelectric photometer.

The invention of the electron multiplier tube or photomultiplier in the late 1930s was an important advance for astronomy. This tube is essentially a photocell with the addition of several cascaded secondary electron stages which allow noiseless amplification of the photocurrent. Whitford and Kron[12] used a prototype photomultiplier for automatic guiding. RCA introduced the 931 photomultiplier just prior to World War II and the 1P21 during the war. Kron[13] was the first to use these tubes for astronomical purposes. With the prototype tubes and a galvanometer, eleventh magnitude stars were measured on the Lick 36-inch refractor.

It became clear with the development of photoelectric techniques that the North Polar Sequence had not been established with enough accuracy. The new photoelectric magnitude systems are now defined by the choice of filters, photomultiplier tube and a network of standard stars. The definition of these systems are taken up in detail in Chapter 2.

Recent years have seen improvements on existing photometric systems, but no major changes. Various filter combinations and newer photocathode materials extending measurements from the near-ultraviolet

AN INTRODUCTION TO ASTRONOMICAL PHOTOMETRY 9

to the near-infrared are being used. Less noisy amplifiers and pulse counting techniques have been developed to retrieve the feeble pulsed current. Innovative new designs that are in the prototype stage at this writing promise a bright future for the photoelectric measurement of starlight.

1.3 A TYPICAL PHOTOMETER

The heart of any photometer is the light detector. This device is explained in detail in Section 1.5. For now, it is sufficient to say it is a device that produces an electric current which is proportional to the light flux striking its surface. The output of the detector must be amplified before it can be measured and recorded by a device such as a strip-chart recorder. The detector is mounted in an enclosure on the telescope called the *head*, which allows only the light from a selected star to reach the light-sensitive element. Figure 1.3 shows the principal components of a photoelectric photometer, when the light detector is a photomultiplier tube. The telescope shown is a Cassegrain type, but any type may be used. The components enclosed by a dashed line are contained in the head, with its relative size exaggerated for clarity.

The first component is a circular diaphragm whose function is to exclude all light except that coming from a small area of sky surround-

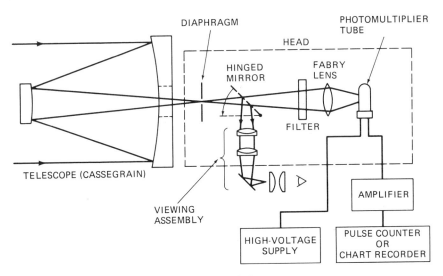

Figure 1.3 A typical photometer.

ing the star under study. The sky background between the stars is not totally dark for a number of reasons, not the least of which is city light scattered by dust particles in the atmosphere. Some of this background light also enters the diaphragm. The telescope must be offset from the star in order to make a separate measurement of the sky background, which can then be subtracted from the stellar measurement. The size of a stellar image at the focal plane of the telescope will vary with atmospheric conditions. Some nights it may seem nearly pinpoint in size while on other nights atmospheric turbulence may enlarge the image greatly. For this reason, a slide containing apertures of various sizes replaces the single diaphragm. To keep the effects of the sky background at a minimum, it is advantageous to use a small diaphragm. On the other hand, this puts great demands on the telescope's clock drive to track accurately for the duration of the measurement. On any given night, a few minutes of trial and error are necessary to determine the best diaphragm choice.

The next component is a diaphragm viewing assembly. This consists of a movable mirror, two lenses, and an eyepiece. Its purpose is to allow the astronomer to view the star in the diaphragm to achieve proper centering. When the mirror is swung into the light path, the diverging light cone is directed toward the first lens. The focal length of this lens is equal to its distance from the diaphragm (which is at the focal point of the telescope). This makes the light rays parallel after passing through the lens. The second lens is a small telescope objective that refocuses the light. The eyepiece gives a magnified view of the diaphragm. Once the star is centered, the mirror is swung out of the way and the light passes through the filter. As with the diaphragm, this is part of a slide assembly that allows different filters to be selected. The choice of filters is dictated by the spectral regions to be measured and is discussed in Chapter 6.

The next component is the *Fabry lens*. This simple lens is very important. Its purpose is to keep the light from the star projected on the same spot on the detector despite any motions the star may have in the diaphragm because of clock drive errors or atmospheric turbulence. This is necessary because no photocathode can be made with uniform light sensitivity across its surface. Without the Fabry lens, small variations in the star's position would cause false variations in the measurements. The focal length of this lens is chosen so that it projects an image of the primary mirror, illuminated by the light of the star, on the detector.

The final component in the photometer head is the photomultiplier tube. It is usually housed in its own subcompartment with a dark slide so that it can be made light-tight from the rest of the head. The tube is surrounded by a magnetic shield that prevents external fields from deviating the paths of the electrons and hence changing the output of the tube. Details on the construction of the photometer head are discussed in Chapter 6.

1.4 THE TELESCOPE

Before the reader rushes out to buy parts and start construction of that shiny new photometer, there is a very important practical consideration to be tackled. *Take a good, hard look at your telescope.* Most amateur-built telescopes, and even those commercially made for amateurs, are not directly suitable for photometry. The problem is usually not optical, but rather mechanical. These telescopes are seldom designed to carry the weight of a photometer head at the focal plane. Even the simplest head containing an uncooled detector weighs in the neighborhood of 4.5 kilograms (10 pounds). The telescope should be capable of being rebalanced to carry this load and the clock drive must still be capable of tracking smoothly. Furthermore, the mount must be sturdy enough so that small gusts of wind do not shake the telescope and move the star out of the diaphragm.

If your telescope has a portable mount, there should be some provision for attaining an equatorial alignment to better than 1°. There are several techniques of alignment that have been discussed in the literature.[14,15,16,17] The clock drive must have sufficient accuracy to keep a star centered in a diaphragm long enough to make a measurement. Typically, this means 5 minutes when using a diaphragm size of 20 arc seconds. Many clock drive systems have difficulty doing this. It is not uncommon for amateur drive systems to suffer from periodic tracking error. This is because of cutting errors in making the worm gear, and results in the telescope oscillating between tracking too slowly and too fast. The cure is to use a large worm gear of good quality. It is essential to have slow-motion controls on both axes. It is nearly impossible to center a star in a small diaphragm by hand. For right ascension, the slow-motion control can be the standard variable frequency drive corrector in common use today. The mechanical declination slow-motion controls supplied by most telescope manufacturers are far too coarse.

The declination motion should be as slow as the right ascension slow motion. It may be possible to gear down an existing system that is too coarse. An especially convenient method is to motorize the declination motion and then operate both axes by pushbuttons in a single hand control.

There are some requirements of the optical system as well. First of all, a large F-ratio is preferred. A small F-ratio produces a light cone that diverges very rapidly inside the head. This means that the components must be placed uncomfortably close together near the focal point. Photometers have been placed on telescopes with F-ratios as small as five. However, an F-ratio of eight or larger is recommended. A large F-ratio has a second advantage. It is highly desirable that the angular diameter of a diaphragm on the sky be kept as small as possible. This reduces the sky background light that enters the photometer. With a short F-ratio telescope, this becomes difficult since the diaphragm holes cannot be drilled small enough with a conventional drill press.

Another important consideration is the location of the focal point. Some telescopes are designed so that the prime focus never extends outside the drawtube. However, the diaphragm must be placed at the prime focus (see Figure 1.3). It may be necessary to move an optical element in the telescope to accomplish this.

Finally, the choice of the optical system itself is important. Refracting telescopes have very serious disadvantages. The glass of the objective lens does not transmit ultraviolet light. Hence, the U magnitude of the UBV system cannot be measured. Note that this problem also applies to Schmidt-Cassegrain telescopes (like the Celestron) though to a lesser extent, since the lens is very thin. A second problem with refractors is chromatic aberration. No matter how well the lens is made, not all wavelengths have a common focal point. The modern achromatic lens minimizes this effect, but perfect correction is not possible. When the diaphragm is at the focal point of blue light, some of the red light is excluded from the photometer, because the red light cone is too wide to pass through the diaphragm. The only solution is to use very large diaphragms that allow a large amount of sky background light to reach the detector. This makes the measurement of faint stars very difficult because the detector sees more sky background light than star light.

Thus, the Newtonian and Cassegrain telescopes are preferred. However, there is still a potential problem. Most small reflecting telescopes come with mirrors which have been overcoated with silicon monoxide.

As the coating ages, it converts to silicon dioxide, which does not transmit ultraviolet light as well. The solution is to keep the overcoating always fresh, or not to overcoat the mirrors, or simply plan not to do any ultraviolet measurements.

We recommend that you modify or improve your telescope before you spend a very frustrating night of attempting photometry with an inadequate telescope. Lest we end this section on too negative a note, it should be emphasized that these modifications are well worth the effort and will result in a much better telescope.

1.5 LIGHT DETECTORS

Since the late 1940s, the most commonly used light detector in astronomy has been the photomultiplier tube. However, a solid-state detector known as the photodiode may well become important in the near future. We discuss each of these devices in turn in this section.

1.5a Photomultiplier Tubes

The key to the operation of the photomultiplier tube is the *photoelectric effect,* discovered in 1887 by Heinrich Hertz. He found that when light struck a metal surface, electrons were released, with the number of electrons released each second being directly proportional to the light intensity. The photoelectric effect is perfectly linear in this regard. The kinetic energy of the released electrons depends on the frequency of the light source and not on its brightness. For a given metal, there is a certain minimum frequency below which no electrons are released no matter how intense the light source may be.

The explanation of the photoelectric effect was given by Albert Einstein in 1905 for which he was later awarded the Nobel Prize. He pictured light as a stream of energy "bullets" or *photons,* each containing an amount of energy directly proportional to the frequency and inversely proportional to the wavelength of the light. Because electrons are bound to the metal by electrical forces, a certain minimum energy is required to free an electron. When an electron absorbs a photon, it gains the photon's energy. However, unless the frequency is above a certain value, the energy is insufficient for the electron to escape the metal. For frequencies higher than this threshold value, the electron can escape and any excess energy above the threshold becomes the kinetic energy of the electron. For all frequencies above the threshold value, the number of electrons released is directly proportional to the number of photons striking the metal surface.

There are other ways of releasing electrons from a metal surface which are also of interest. *Thermionic emission* is essentially the same as the photoelectric effect except that the energy that releases the electrons comes from heating of the metal rather than from light. *Secondary emission* is the release of electrons because of the transfer of kinetic energy from particles that hit the metal surface. Finally, *field emission* is the removal of electrons from the metal by a strong external electric field. All of the above effects come into play in a photomultiplier tube.

Most photomultiplier (PM) tubes are about the size of the old-fashioned vacuum tubes used in radios and televisions. The components of the tube are contained by a glass envelope in a partial vacuum, so that the electrons can travel freely without colliding with air molecules. Figure 6.2 shows a photograph of an RCA 1P21 PM tube. The heart of the tube is the metal surface that releases the photoelectrons. Since this surface is at a large negative voltage with respect to ground, it is called the *photocathode*.

Photocathodes are not constructed of simple, common metals but rather a combination of metals (antimony and cesium in the case of the 1P21). The metals are chosen to give the desired spectral response and light sensitivity. For a typical photocathode material, the quantum efficiency is about 10 percent. (Of every 100 incident photons, only 10 will be successful in releasing a photoelectron. The energy from the remaining 90 photons is absorbed by the metal and dissipated in other ways.) The current produced by the photoelectrons is very weak and difficult to measure even for bright stars. For this reason, the early use of photocells met with limited success.

The PM tube differs from the photocell in that the PM tube amplifies this current internally. In order to accomplish this, the photoelectrons released by the photocathode are attracted to another metal surface by an electric potential. This metal surface is called a *dynode* and in the 1P21 it is at a potential 100 V less negative than the photocathode. As a result, this dynode looks positive compared to the photocathode. Photoelectrons are accelerated toward its surface, and the impact of each releases about five more electrons by the process of secondary emission. These electrons are in turn accelerated toward another dynode that is 100 V less negative than the previous dynode. Once again, the process of secondary emission releases about five electrons per incident electron.

This process is then repeated at other dynodes. The 1P21 has nine dynodes, so for each photoelectron emitted at the photocathode there are 5^9 or two million electrons emitted at the last dynode. This tube is said to have an internal gain of two million. These electrons are then collected at a final metal surface, called the *anode*, from which they flow through a wire to the external electronics.

Figure 1.4 shows the arrangement of the photocathode, dynodes, and anode inside a 1P21. The arrows show the paths of the electrons (for simplicity, not all the electron paths are shown). There are other PM tube designs and Figure 1.4 shows, schematically, the "venetian blind," and the "box-and-grid" types. The 1P21 is called a "squirrel-cage" design. Note that the 1P21 is a "side-window" design while the others pictured in Figure 1.4 are examples of "end-on" tubes.

The current amplification produced by the dynode chain is an extremely important characteristic of the PM tube. This amplification is essentially noise-free. Unlike the early photocells, far less external amplification is required. As a result, the external amplifier noise is relatively unimportant.

While the amplification process of a PM tube is noise-free, there are, unfortunately, noise sources within the tube. Noise is defined as any output current that is not the result of light striking the photocathode. With the PM tube sitting in total darkness, with the high voltage on, there is a so-called "dark current" which is produced by the tube. This current is a result of electrons released at the dynodes by thermionic and field emission. Even at room temperature, the dynodes are warm

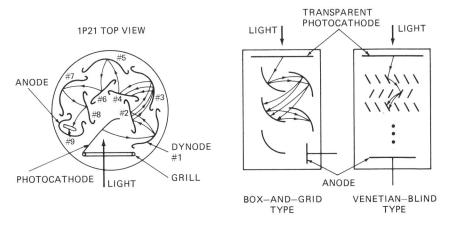

Figure 1.4 Photomultiplier tube designs.

enough for an electron to be released occasionally. When this happens, the electron is accelerated and amplified by the remaining dynode chain.

The obvious solution to large dark currents is to reduce the temperature of the tube. Most professional astronomers cool the PM tube with dry ice, almost totally eliminating thermionic emission. The amateur astronomer need not exert this much effort as an uncooled tube is still very useful. The only problem is that very faint stars are difficult to detect because the current they produce at the anode may be as small or smaller than the dark current. There is little that can be done to eliminate field emission because the tube must contain strong electric fields. However, in practice this noise source is very small compared to thermionic emission.

The current that leaves the anode is still very weak and requires amplification before it can be easily measured. There are two general ways to accomplish this. Because each photoelectron produces a burst of electrons at the anode, a pulse amplifier can be used to amplify each burst and convert it to a voltage pulse that can be counted electronically. The number of pulses counted in a given time interval is a measure of the number of photons that strike the photocathode in the same time interval. (We use the terms *pulse counting* and *photon counting* interchangeably whenever referring to the technique of counting individual photoelectron pulses caused by an incident photon on the photocathode of a photomultiplier tube.) The second technique is to use a DC amplifier and to smooth the bursts to look like a continuous current. This current is amplified and measured by a meter or a strip-chart recorder. Both techniques are discussed in detail in Chapters 7 and 8.

No photocathode material releases the same number of electrons at all wavelengths even when the light source is equally bright at all wavelengths. The spectral response of a photomultiplier is an important characteristic to know. Figure 1.5 shows the spectral responses of a few types of photocathode materials. The most common in astronomical use today is the cesium antimonide (Sb-Cs) surface, used in the first mass-produced photomultiplier, the RCA 931A. The RCA 1P21 is the successor to this tube and is used to define the *UBV* photometric system. The spectral response of this surface is labeled "S-4" in Figure 1.5 (the "S" numbers refer to different spectral responses). The light sensitivity peaks near 4000 Å, cutting off at the blue end near 3000 Å while on the red end there is a long tail to 6000 Å. Individual tubes vary and some-

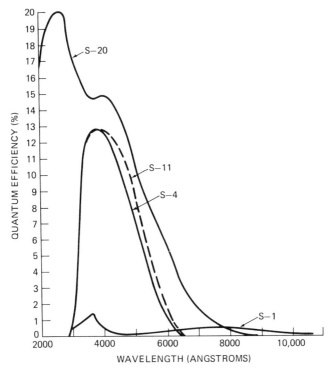

Figure 1.5 Photocathode spectral response.

times the response goes beyond 7000 Å, producing a problem when blue or ultraviolet filters are used. These filters transmit some light in the red and the tube detects red light passed by these filters. For red stars, this "red leak" can cause an error in a blue magnitude of a few percent. This problem is discussed later.

In Figure 1.4, two types of photocathodes are illustrated. The 1P21 is a side-window device. The light strikes the front surface of the photocathode and electrons are released from the same front surface. This is called an opaque photocathode. With a semitransparent photocathode, used with "end-on" PM tubes, the light strikes the front surface and the electrons are released from the *back* surface. These two types of cathodes, even if made of the same material, have a slightly different spectral response. The semitransparent photocathodes tend to be more red-sensitive. For this reason, a semitransparent photocathode made of Sb-Cs is designated S-11, not S-4.

Another important photocathode material is the so-called "tri-alkali" designated as S-20. Figure 1.5 shows that this material covers much the same spectral range as the S-4 but with much useful sensitivity in the near-infrared. This material is extremely sensitive in blue light and has a quantum efficiency of 20 percent. By contrast, the Ag-O-Cs, S-1 surface has a very low quantum efficiency of a few tenths of a percent. However, it has an extremely broad response up to 11,000 Å. There is a wide dip in its sensitivity centered at 4700 Å. The advantage of the S-1 photocathode is that a single tube can be used to measure from the blue to the infrared. The disadvantage, of course, is that you are limited to fairly bright objects.

Noise is a problem with PM tubes designed for infrared work. Infrared photons carry very little energy, which means the photocathode must be made of a material in which the electrons are bound very loosely. Unfortunately, this means they are very easy to release thermally. Hence all infrared tubes are cooled with dry ice, and detectors for the far-infrared are cooled to an even lower temperature with liquid nitrogen or liquid helium.

1.5b PIN Photodiodes

To date, very little experimentation with photodiodes for astronomical photometry has been published. However, these devices look very promising.[18,19,20,21,22,23] A well-designed photodiode photometer is now commercially available from Optec, Inc.[24] To understand the photodiode, we will review the operation of an ordinary diode briefly. More complete explanations can be found in most elementary electronics texts.

In an isolated atom, electrons are confined to orbits about the nucleus, which correspond to sharply defined energy levels. When atoms are linked in a crystal of a solid, the energy level structure is quite different. In a simplified view, the energy levels become two distinct bands. The lower band "contains" all the electrons (at least at very low temperatures) while the upper band is empty. There is a gap between the two energy bands that represents energy states unavailable to the electrons. If an electron somehow receives sufficient energy to reach the upper band, it can move freely through the crystal, unattached to any one atom. An external electric field easily can cause these electrons to move. For this reason, the upper band is called the *conduction band*. The electrons in the lower band are involved in the chemical bonds to neighboring atoms in the crystal, so this band is called the *valence band*. For solids that are insulators, the gap between the two bands is very large. It is very unlikely that an electron from the valence band will receive enough energy to promote it to the conduction band. Therefore, insulators are poor conductors of electric current. Likewise, a conductor is a material in which the two bands merge and electrons can easily move into the conduction band. Semiconductors have a small gap between energy bands. Germanium (Ge) and silicon (Si) are the two most commonly used semiconductor materials for making diodes and transistors.

Semiconductors without impurities are called *intrinsic semiconductors*. They have a rather low conductivity, but not as low as an insulator. If a semiconductor is "doped" with impurities, its conductivity can be increased markedly. Si and Ge each have four valence electrons per atom, which are used in bonding to four adjacent atoms when making a crystal. The process of doping involves replacing a few of these atoms with atoms that have one fewer (three) or one more (five) valence electrons. Suppose a Ge crystal is doped with arsenic, which has five valence electrons. Four of these electrons are used to bind the atom in the crystal with four neighboring Ge atoms. However, the fifth electron is loosely bound with an energy just below the conduction band. This electron cannot be in the valence band since this band is "full." A small amount of energy promotes this electron into the conduction band. Thus, doping has greatly increased the electrical conductivity. The impurity atom, arsenic, in this case is referred to as a *donor* because it supplied the extra electron. A semiconductor doped in this way is referred to as an *n-type* because a negative charge was donated.

The conduction of the crystal is also increased if it is doped with atoms that have only three valence electrons. These atoms are one short of completing their bonds with neighboring atoms. Thus, a "hole" exists. This atom has an unfilled energy level and a nearby valence electron can move into this location. Of course, this electron leaves a hole behind. In this way, it is possible for holes to migrate through the crystal. This impurity is labeled an *acceptor* because it accepts a valence electron from elsewhere in the crystal. Acceptor-doped crystals are referred to as *p-type* semiconductors because the current carriers are holes, or a lack of electrons, which look positive by comparison to the electrons.

A diode is made by bringing p-type and n-type material together. The

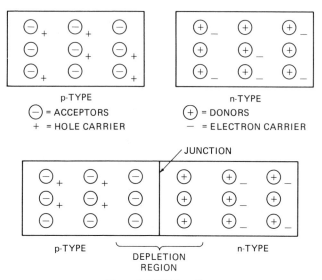

Figure 1.6 P-N junction.

surface of contact is referred to as the *junction*. Holes from the p-side and electrons from the n-side diffuse across the junction until an equilibrium is reached. The result is a region on either side of the junction where there are no charge carriers because the electrons and holes have combined to annihilate each other. This is called the *depletion region,* as shown in Figure 1.6. An electrostatic field is produced across the junction because the p-side now has an excess of electrons filling the holes and the n-side now has an excess of holes because it has lost electrons to the p-side. This results in a situation where an electron from the n-side is not likely to cross the junction because it is faced with an excess of electrons on the p-side that repel it. Similarly, a hole from the p-side will no longer cross to the n-side. If an external electric circuit is connected to the diode, such that the n-side is connected to a positive potential and the p-side to a negative potential (called *reversed biased*), no current will flow. This is because the external potential only increases the potential difference across the depletion layer. If the contacts are reversed, the current from the external circuit will tend to neutralize the charge difference across the junction. The potential difference drops and current flows. It is in this manner that alternating current can be converted into direct current, because during only one half of the cycle, when the diode is *forward biased,* will current be allowed to flow.

Normally, when electronics texts discuss the operation of a diode, as we have done above, they fail to mention one additional "complication." A graphic illustration of this is seen in the following experiment. Go to your local electronics store and find a glass-encapsulated diode. With a knife, scrape off the black paint that coats the glass. Connect a voltmeter capable of reading a few tenths of a volt across the diode. Shine a bright light on the diode and watch the meter. The added "complication" is just what makes diodes interesting to astronomers; they are highly light-sensitive. Light energy absorbed at the p-n junction raises an electron from the valence band to the conduction band. Such an electron is repelled by the p-side of the depletion region and attracted by the n-side. The opposite is true for the hole left behind. This process, when repeated over and over as light continues to strike the junction, results in the voltage detected in the above experiment. A diode used to measure light in this manner is said to be used in the *photovoltaic mode*.

In practice, diodes designed for light detection are constructed differently from ordinary diodes. In the so-called PIN photodiode a p-type layer (P) is separated from the n-type layer (N) by an intrinsic layer (I). The light is absorbed in the intrinsic layer, creating an electron-hole pair. The hole is attracted to the n-material and the electron to the p-material after drifting through the I layer. The function of the I layer is to reduce noise current produced by such effects as electron-hole pairs created by thermal processes. Nevertheless, this is still the major source of noise in a photodiode.

There are numerous advantages to a photodiode as a detector in a photometer. One advantage is seen in Figure 1.7, which shows that a blue-enhanced photodiode is an efficient detector from the ultraviolet to the infrared. Furthermore, the quantum efficiency of the photodiode is much better than the photomultiplier, reaching 90 percent in the near-infrared. Even though an S-1 photomultiplier can also span this range of wavelengths, it has a quantum efficiency of only a few tenths of a percent. Compared to photomultiplier tubes, photodiodes are also less expensive, much smaller, and do not require a high-voltage supply. It would thus appear that the professional astronomers should rush to replace the photomultipliers with photodiodes. The reason this has not occurred is that the photomultiplier tube still has one very important advantage. The dynode chain of the photomultiplier yields an internal current amplification (gain) of about 10^6. This is not the case for the photodiode. Therefore, the external electronics must amplify an addi-

22 ASTRONOMICAL PHOTOMETRY

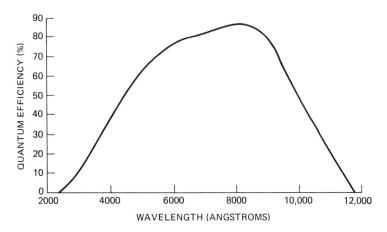

Figure 1.7 Approximate quantum efficiency of a blue-enhanced photodiode.

tional factor of 10^6 and this introduces noise. For a photodiode to be competitive with a photomultiplier tube, a well-designed amplifier is required and both the photodiode and the amplifier should be cooled. The internal gain of a photodiode is unity, which means that pulse-counting techniques cannot be used, so that DC photometry is required. In Appendix K, it is shown that DC is inferior to pulse counting when it comes to measuring faint stars, but for bright stars the photodiode works well. This fact, combined with its convenience, makes the photodiode a detector to be seriously considered. Also in Appendix K, a theoretical comparison of the photodiode and the photomultiplier is presented in order to help one decide on the best light detector for one's observing program and budget.

The size of the active area (light-sensitive area) of a photodiode should be kept fairly small to minimize the noise introduced by thermally produced electron-hole pairs. Thus, unlike the photomultiplier, the photodiode is placed at the focus of the telescope. This necessitates some design changes in the photometer head. Figure 1.8 illustrates the optical layout schematically when a photodiode is used. The first difference is that no diaphragms are used. The light-sensitive area of the photodiode is so small (typically 0.5 millimeter across) that it acts as its own diaphragm. It is not possible to place a viewing eyepiece behind the diaphragm. Instead, the eyepiece must be placed in front of the photodiode and equipped with a cross hair for centering the star on the photodiode. The placement of the photodiode as shown eliminates the need

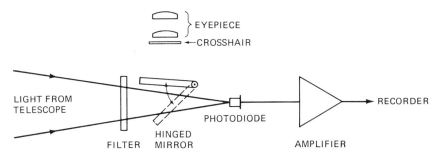

Figure 1.8 Photodiode photometer.

for a Fabry lens. The spectral response of the photodiode necessitates some special considerations when choosing filters. These are discussed in Chapter 6.

1.6 WHAT HAPPENS AT THE TELESCOPE

In later chapters, we discuss observing techniques and data reduction in detail. However, for the benefit of the novice, we now outline the observing procedure and define some terms. The actual observing pattern depends on the goal of the project and the form in which the final data are needed. In general, one of two techniques is followed. The simplest observing scheme, and the one highly recommended to the beginner, is *differential photometry*. In addition to its simplicity, it is the most accurate technique for measuring small variations in brightness. This technique is widely used on variable stars, especially short-period variables and eclipsing-binary systems.

In differential photometry, a second star of nearly the same color and brightness as the variable star is used as a *comparison star*. This star should be as near to the variable as possible, preferably within one degree. This allows the observer to switch rapidly between the two stars. Another extremely important reason for choosing a nearby comparison star is that the extinction correction (Section 1.8) can often be ignored, because both stars are seen through nearly identical atmospheric layers. All changes in the variable star are determined as magnitude differences between it and the comparison star. It is important that the comparison star be measured frequently because the altitude of these objects is continuously changing throughout the night. This type of photometry can be extremely accurate (0.005 magnitude) and is highly recommended where atmospheric conditions can be quite variable, such as the midwestern United States. Any star that meets the criteria can be a comparison star. However, it is a good idea to pick a second star, called the check star, as a test of the nonvariability of the comparison star. The check star need be measured only occasionally during the night.

The observational procedure is to alternate between the variable and comparison stars, measuring them a few times in each filter. A "measurement" consists of centering the star in the diaphragm and then moving the flip mirror out of the light path so that the light can fall on the detector. You then record the meter reading on your amplifier along with the time. If you are using pulse-counting equipment, you record the number displayed on your counter. If you are very lucky, your microcomputer can record it for you! Once this has been done for each filter, the star is moved out of the diaphragm and the sky background is recorded through each filter. This is necessary since the measurements of the star really include the star and the sky background.

The magnitude differences between the variable and comparison stars in each filter can then be calculated using the expression

$$m_x - m_c = -2.5 \log (d_x/d_c) \quad (1.4)$$

where d_x and d_c represent the measurement of the variable and the comparison stars minus sky background, respectively. If different amplifier gains were used for the two stars, this must also be included. An advantage of differential photometry is that no calibration to the standard photometric system is necessary for many projects. The disadvantage is that your magnitude differences will not be exactly the same as those measured on the standard system. However, if you are using the specified detector and filters, and have matched the color of the comparison and variable stars, your results will not differ very much (see Section 2.6). A further disadvantage is that your final results will be in *differences*. You will not be able to specify the actual magnitudes or colors of the variable star unless you standardize the comparison star. However, these results are good enough for many projects such as determining light curve shapes or the times of minimum light of an eclipsing binary.

The second technique is the most general and commonly used by professional astronomers. It is also the most demanding on the quality of the sky conditions. In this scheme, numerous program stars, located in many different places in the sky, are to be measured to determine their magnitudes and colors. As before, each star and its sky background are measured through all filters. However, because each star is observed at a different altitude above the horizon, each is seen through a slightly different thickness of the earth's atmosphere. Therefore, observations must also be made of another set of stars of known magnitudes and colors to determine the atmospheric extinction corrections. Finally, a set of standard stars must be observed to determine the transformation coefficients so the measurements of the program stars can be transformed into magnitudes and colors of a standard system, such as the *UBV* system. This procedure often involves less observing time than it would appear at first glance. This is because it is often possible to use some of the same observations to determine the extinction corrections and the transformation coefficients. Furthermore, the transformation coefficients need only be determined occasionally. Details of the procedures are treated in Chapters 4 and 9.

1.7 INSTRUMENTAL MAGNITUDES AND COLORS

It would appear to the beginner that the determination of a star's magnitude is fairly simple and, furthermore, that magnitude can be simply related to the star's light flux. Unfortunately, the latter is far from true. To see this more clearly, we rewrite Equation 1.3 as

$$m_1 = m_2 - 2.5 \log F_1 + 2.5 \log F_2. \tag{1.5}$$

Suppose star "2" is a reference star of magnitude zero and star "1" is the unknown. Then

$$m_1 = q - 2.5 \log F_1 \tag{1.6}$$

where q is a constant. Since there is now only one star the subscript "1" can be dropped in favor of lambda (λ) to remind us that the magnitude depends on the wavelength of observation. Thus,

$$m_\lambda = q_\lambda - 2.5 \log F_\lambda. \tag{1.7}$$

Again, this equation only seems to verify the simple relationship between magnitude and flux. However, the above equation refers to the *observed* flux. The observed flux is related to the actual flux in a very complicated way. The problems can be broken into two groups: (1) extinction because of absorption or scattering of the stellar radiation on its way to the detector and (2) the departure of the detecting instrument from one with ideal characteristics. We now discuss these two problems in turn.

There are two sources of absorption of the stellar flux: interstellar absorption because of dust and absorption within the earth's atmosphere. The former is generally neglected for published observations, but the latter is usually taken into account. The earth's atmosphere does not transmit all wavelengths freely. For example, ultraviolet light is heavily absorbed. Human life can be thankful for that! Observatories at higher elevations have less of the absorbing material above them, while others located near large bodies of water have more water vapor above them. In addition, the atmosphere scatters blue light much more than red light.

Not all telescopes transmit light in the same manner and this can be a function of wavelength. For instance, glass absorbs ultraviolet light heavily, and various aluminum and silver coatings have different wavelength dependences of the reflectivity. Also, in practice it is not possible to measure the flux from a star at one wavelength. Any filter transmits light over an interval of wavelengths. Despite the best efforts of the manufacturers, no two filters or light detectors can be made with exactly the same wavelength characteristics. As a result, no two observatories measure the same observed flux for a given star.

A calibration process is necessary to enable instruments to yield the same results. The observed flux, F_λ, is related to the actual stellar flux, F^*_λ, outside the earth's atmosphere, by

$$F_\lambda = \int_0^\infty \phi_A(\lambda)\phi_T(\lambda)\phi_F(\lambda)\phi_D(\lambda) F^*_\lambda d\lambda$$

where

$\phi_A(\lambda)$ = fractional transmission of the earth's atmosphere
$\phi_T(\lambda)$ = fractional transmission of the telescope
$\phi_F(\lambda)$ = fractional transmission of the filter
$\phi_D(\lambda)$ = efficiency of the detector (1.0 corresponds to 100 percent).

This expression can be very complicated and the many factors are usually poorly known. It is for this reason that stellar fluxes are very difficult to measure accurately. Fortunately, the determination of stellar magnitudes does not require a knowledge of most of these factors, except in an indirect manner. The magnitude scheme requires only that certain stars be *defined* to have certain magnitudes, so that magnitudes of other stars can be determined from observed fluxes that are corrected only for atmospheric absorption. This is why the seemingly awkward magnitude system has survived so long.

The only remaining problem is to account for the individual differences among telescope, filter, and detector combinations. This is where the set of standard stars comes into use. By observing a set of known stars, it is possible for each observatory to determine the necessary transformation coefficients to transform their instrumental magnitudes to the common standard system.

In practice, a star is not measured in flux units. The detector produces an electrical output that is directly proportional to the observed stellar flux. In DC photometry, the amplified output current of the detector is measured, while in pulse-counting techniques the number of counts per second is recorded. In either case, the recorded quantity is only proportional to the observed flux. Symbolically,

$$F_\lambda = Kd_\lambda \tag{1.8}$$

where d_λ is the practical measurement (i.e., current or counts per second), and K is the constant of proportionality. Equation 1.7 can be written as

$$m_\lambda = q_\lambda - 2.5 \log K - 2.5 \log d_\lambda \tag{1.9}$$

or

$$m_\lambda = q'_\lambda - 2.5 \log d_\lambda \tag{1.10}$$

This then relates the actual measurement, d_λ, to the instrumental zero point constant q'_λ, and to the *instrumental magnitude,* m_λ. The *color index* of a star is defined as the magnitude difference between two different spectral regions. If the subscripts 1 and 2 refer to these two regions, then a color index is defined as

$$m_{\lambda 1} - m_{\lambda 2} = q'_{\lambda 1} - q'_{\lambda 2} - 2.5 \log d_{\lambda 1} + 2.5 \log d_{\lambda 2} \quad (1.11)$$

or

$$m_{\lambda 1} - m_{\lambda 2} = q_{\lambda 12} - 2.5 \log (d_{\lambda 1}/d_{\lambda 2}) \quad (1.12)$$

where the zero point constants have been collected into a single term, $q_{\lambda 12}$. Again the quantity $(m_{\lambda 1} - m_{\lambda 2})$ is in the instrumental system. The transformation from the instrumental system to the standard system is discussed shortly. Before that transformation can be made, it is necessary to correct for the absorption effects of the earth's atmosphere.

1.8 ATMOSPHERIC EXTINCTION CORRECTIONS

Even on the clearest of nights, the stars are dimmed significantly by absorption and scattering of their light by the earth's atmosphere. The amount of light loss depends on the height of the star above the horizon, the wavelength of observation and the current atmospheric conditions. Because of this complex behavior, the measured magnitudes and color indices are corrected to a location "above the earth's atmosphere." In other words, they are corrected to give the same values an observer in space would measure. In this way, measurements by two different observatories can be effectively compared.

A measured magnitude, m_λ, is corrected to the magnitude that would be measured above the earth's atmosphere, $m_{\lambda 0}$, by the following equation,

$$m_{\lambda 0} = m_\lambda - (k'_\lambda + k''_\lambda c) X, \quad (1.13)$$

where k'_λ is called the *principal extinction coefficient* and k''_λ is the *second-order extinction coeffcient*. This second-order term is often small enough to be ignored in practice. Here c is the observed color index and

AN INTRODUCTION TO ASTRONOMICAL PHOTOMETRY 29

X is called the *air mass*. At the zenith, X is 1.00 and it grows larger as the altitude above the horizon decreases. To a good approximation,

$$X = \sec z, \quad (1.14)$$

where z is the zenith distance (90° − altitude) of the star.

Just as the sun grows red in color as it sets, the atmospheric extinction process affects the color indices of stars. A measured color index, c, is transformed to a color index as seen from above the earth's atmosphere, c_0, by the following expression:

$$c_0 = c - k'_c X - k''_c X c. \quad (1.15)$$

as above, k'_c and k''_c represent the principal and second-order extinction coefficients, respectively. The subscript c is a reminder that the value of the coefficient depends on the two wavelength regions measured. That is to say, the extinction coefficient for a color index based on a blue and a yellow filter is not the same as that based on a yellow and red filter. The extinction coefficients, k'_λ, k''_λ, k'_c and k''_c are determined observationally. The details of this technique will be discussed in Chapter 4. The derivation of the above extinction equations can be found in Appendix J.

1.9 TRANSFORMING TO A STANDARD SYSTEM

A system of magnitudes and colors, such as the *UBV* system, is defined by a set of standard stars measured by a particular detector and filter set. In order for observers at different observatories to be able to compare observations, the observations must be transformed from the instrumental systems (which are all different) to a standard system. It is important for the observers to match the equipment used to define the system of standard stars as closely as possible. However, no two filter sets or detectors are exactly the same. Hence, it is necessary for all observers to measure the standard stars in order to determine how to transform their observations to the standard system.

A derivation of the transformation equations can be found in Appendix J. Only the results are be stated here. Once the observed magnitude

30 ASTRONOMICAL PHOTOMETRY

has been corrected for atmospheric extinction, it can be transformed to a standardized magnitude (M_λ) by

$$M_\lambda = m_{\lambda 0} + \beta_\lambda C + \gamma_\lambda \qquad (1.16)$$

where C is the standard color index of the star, β_λ and γ_λ are the color coefficient and zero-point constant, respectively, of the instrument. The standardized color index is given by

$$C = \delta c_0 + \gamma_c \qquad (1.17)$$

where c_0 is the observed color index which has been corrected for atmospheric extinction. Again, δ is a color coefficient and γ_c is a zero-point constant. These coefficients and zero-point constants are determined for each photometer system by the observation of standard stars. The details of this are taken up in Chapter 4.

1.10 OTHER SOURCES ON PHOTOELECTRIC PHOTOMETRY

There are several sources relating to photoelectric photometry that are available in good astronomical libraries. Some of these are obscure and are difficult to locate. Most of the references listed below are out of print or are sections of expensive texts. However, if you are interested in more detail than can be found in this text, we recommend looking at those references available in your area.

- Irwin, J. B., ed. 1953. *Proceedings of the National Science Foundation Astronomical Photoelectric Conference.* Flagstaff, Arizona: Lowell Observatory. This book has considerable detail on sky conditions and site selections for observatories.
- Wood, F. B., ed. 1953. *Astronomical Photoelectric Photometry.* Washington, D.C.: AAAS. This is the proceedings of a symposium on December 31, 1951. Contains many references of early photometry and describes DC, AC, and pulse-counting techniques as practiced at that time.
- Hiltner, W. A., ed. 1962. *Astronomical Techniques.* Chicago: Univ. of Chicago Press. Three chapters of this book are of particular interest: Lallemand ("Photomultipliers"), Johnson ("Photoelectric Pho-

tometers and Amplifiers"), and Hardie ("Photoelectric Reductions").
- Whitford, A. E. 1962. "Photoelectric Techniques." In *Handbuch der Physik*. Berlin: Springer-Verlag Co. Edited by S. Flugge, p. 240. This chapter is a well-rounded description of photomultiplier tube photometry.
- Wood, F. B. 1963. *Photoelectric Astronomy for Amateurs.* New York: Macmillan. This text is low level and understandable, but is incomplete and contains out of date circuitry.
- *AAVSO*, 1967. *Manual for Astronomical Photoelectric Photometry.* Cambridge: AAVSO. The AAVSO has tried to produce a short manual to start observers on photometry. A new edition is forthcoming.
- Golay, M. 1974. *Introduction to Astronomical Photometry.* Holland: D. Reidel. For complete theoretical descriptions of wide-band photometry, this text is hard to beat. Requires extensive mathematics and astronomy background.
- Young, A. T. 1974. In *Methods of Experimental Physics: Astrophysics* **vol. 12A**. Edited by N. Carleton. New York: Academic Press. This is extremely complete in the problems arising in photometry and should be required reading.

In addition, some professional observatories have their own small manuals that can be obtained directly from them.

Amateur and professional astronomers interested in photometry are strongly encouraged to join the International Amateur-Professional Photoelectric Photometry (IAPPP) association. The goal of this group is to foster communication on the practical aspects of photometry. This is accomplished through annual IAPPP symposia and the IAPPP Communications. Interested persons should contact either of the following people:

> Dr. Douglas S. Hall
> Dyer Observatory
> Vanderbilt University
> Nashville, TN 37235

> Mr. Russell M. Genet
> Fairborn Observatory
> 1247 Folk Road
> Fairborn, OH 45324

Amateur astronomers are encouraged to coordinate their photometric observing programs with those of other amateurs by contacting one of the following organizations.

American Association of Variable Star Observers (AAVSO)
187 Concord Ave.
Cambridge, MA. 02138
U. S. A.

Royal Astronomical Society of New Zealand
Variable Star Section
P. O. Box 3093 Greerton
Tauranga, New Zealand

REFERENCES

1. Miczaika, C. R., and Sinton, W. M. 1961. *Tools of the Astronomer*. Cambridge, Mass.: Harvard Univ. Press, p. 156.
2. Bond, W. C. 1850. *Annals of the Harvard College Observatory*, I, 1, CXLIX.
3. Minchin, G. M. 1895. *Proc. Roy. Soc.* **58**, 142.
4. Stebbins, J., and Brown, F. C. 1907. *Ap. J.* **26**, 326.
5. Kron, G. E. 1966. *Pub. A.S.P.* **78**, 214.
6. Stebbins, J. 1910. *Ap. J.* **32**, 185.
7. Schultz, W. F. 1913. *AP. J.* **38**, 187.
8. Guthnick, P. 1913. *Ast. Nach.* **196**, 357.
9. Rosenberg, H. 1913. *Viert. der Ast. Gesell.* **48**, 210.
10. Stebbins, J. 1928. *Pub. Washburn Obs.* XV, 1.
11. Whitford, A. E. 1932. *Ap. J.* **76**, 213.
12. Whitford, A. E., and Kron, G. E., 1937. *Rev. Sci. Inst.* **8**, 78.
13. Kron, G. E. 1946, *Ap. J.* **103**, 326.
14. Davis, F. W., Jr. 1973. *Griffith Observer* (May), 8.
15. Custer, C. P. 1973. *Sky and Tel.* **46**, 329.
16. Souther, B. L. 1978. *Sky and Tel.* **55**, 78.
17. Souther, B. L. 1978. *Sky and Tel.* **55**, 173.
18. De Lara, E., Chavarria, K. C., Johnson, H. L. and Moreno, R. 1977. *Revistia Mexicana de Astron. y Astrof.* **2**, 65.
19. Schumann, J. D., 1977. In *Astronomical Applications of Image Detectors with Linear Response*, I. A. U. Colloquium No. 40, 31-1.
20. Fisher, R. 1968. *Appl Optics* **7**, 1079.
21. Masek, N. L. 1976. *South. Stars* **26**, 175.
22. Corney, A. C. 1976. *South. Stars* **26**, 177.
23. McFaul, T. G. 1979. *J. AAVSO* **8**, 64.
24. Optec Inc., 119 Smith, Lowell, MI 49331.

Chapter 2
Photometric Systems

The basic goal of astronomical photometry sounds simple enough: to measure the light flux from a celestial object. So it would seem that simply placing a light detector at the focus of a telescope is all that is needed. The problem begins when different observers using different light detectors and telescopes try to compare or combine their data. Even though they may have been observing the same star at exactly the same time, their measurements will not necessarily be the same. This difference is due to the different spectral response of the telescope and detector. To take an extreme example, suppose a detector is mostly sensitive to blue light while a second is mostly sensitive to red. Stars are not equally bright at all wavelengths so the two detectors cannot possibly give the same results for the same star.

The obvious first step toward a uniform data set would be to have all observers use the same kind of detector. It would also be extremely valuable to isolate and measure certain portions of the spectrum containing features that indicate physical conditions of the star. This can be achieved by using a detector with a broad spectral response with individual spectral regions isolated by filters transmitting only a limited wavelength interval to the detector. Every observer should match the detector and filters as closely as possible to a common system. However, even this is not enough to yield strict uniformity as it is impossible to manufacture identical light detectors and filters. Thus, a third and final component is necessary: standard stars. Observations of the same, non-variable (hopefully!) stars, of known magnitudes and colors, will allow each observer to determine his own coefficients for Equations 1.16 and 1.17. It is then possible to measure any unknown star and use these equations to transform the results to a common photometric system.

This is how a photometric system is defined: by specifying the detector, filters, and a set of standard stars.

Photometric systems can be broken into three rough categories based on the size of the wavelength intervals transmitted by their filters. Wide-band systems (such as the *UBV* system) have filter widths of about 900 Å, while intermediate-band filter widths are about 200 Å. Narrow-band systems are used to isolate and measure a single spectral line and may have widths of 30 Å or less. While the narrow-band systems give very specific spectral information, they transmit only a small fraction of the light of the star. Unless a large telescope is used, their use is limited to very bright stars. A discussion of various photometric systems can be found in Golay.[1]

At the end of this chapter, an intermediate-band and a narrow-band system are considered. However, in what follows, and throughout the remainder of this book, the wide-band *UBV* system will be discussed. By adopting a single system, specific examples of observing techniques and data reduction can be used, avoiding a very general discussion that would be of much less benefit to the beginner. However, many of those procedures can be applied to any system. The choice of the *UBV* system as "the system" for this book is based on a number of considerations. The *UBV* system has become popular among astronomers, and there exists a considerable data base of *UBV* observations in the literature. Being a wide-band system makes it especially suitable for users of small telescopes. The photomultiplier tube and filters used to define the system are readily available and relatively inexpensive. There is also a fairly extensive set of standard stars. The reader should not assume that the choice of the *UBV* system for this book means that the other systems are less important or yield measurements that tell us less about the stars. As you will see later, the *UBV* system is not a perfect system, and for some research projects, other systems are preferred.

2.1 PROPERTIES OF THE *UBV* SYSTEM

The *UBV* system was defined and established by H. L. Johnson and W. W. Morgan.[2,3] They desired to establish a photoelectric system that would yield results comparable to the yellow and blue magnitudes of the International System (see Section 1.2), to have a third color for better discrimination of stellar attributes, and to be closely tied to the Morgan-Keenan (M-K) spectral classification system. The *UBV* system

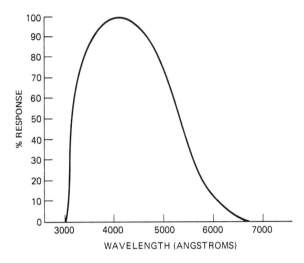

Figure 2.1 Typical response function of a 1P21 photomultiplier tube.

was developed around the RCA 1P21 photomultiplier tube and three broad-band filters that give a visual magnitude (V), a blue magnitude (B), and an ultraviolet magnitude (U). The response function of the 1P21 is shown in Figure 2.1 and the transmission curves of the filters are shown in Figure 2.2.

The V filter is yellow with a peak transmission around 5500 Å. This filter was chosen so that the V magnitude is almost identical to the pho-

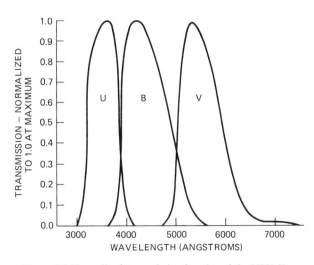

Figure 2.2 Normalized transmission function of the UBV filters.

tovisual magnitude of the International System. The long wavelength cutoff is determined by the response of the 1P21 and not the filter. The blue (B) filter is centered around 4300 Å but has some transmission over most of the sensitivity range of the 1P21. The B magnitude corresponds well with the earlier blue photographic magnitudes. This filter actually consists of two: a blue filter and an ultraviolet blocking filter. This latter filter prevents the B magnitude from being affected by the Balmer discontinuity, which is discussed later. The U filter is centered on 3500 Å and has two problems. This filter has a red "leak," that is, it transmits some light in the near infrared. This red light must be blocked by a second filter or the red leak must be measured and subtracted from the U measurement. The second problem is that the short wavelength cutoff is not set by either the filter or the photomultiplier, but instead by the earth's atmospheric ultraviolet transmission. This is a function of the observatory's altitude and can be variable depending on atmospheric conditions. Thus the UBV system is not totally filter-defined.

The UBV standard stars were measured by Johnson's original photometer without any transformation. In other words, except for some additive constants, the UBV system is the instrumental system of that photometer. The zero points of the color indices, $(B - V)$ and $(U - B)$, are defined by six A0 V stars. These stars are α Lyr, γ UMa, 109 Vir, α Crb, γ Oph, and HR 3314. The average color index of these stars is defined to be zero, that is

$$(B - V) = (U - B) = 0.$$

The system was originally defined with 10 primary standard stars. Just 10 stars, spaced over the entire sky, is an insufficient number to allow other observatories to calibrate their photometers. Johnson and Morgan established a more extensive list of secondary standards that are closely tied to the ten primary stars. Appendix C lists these stars. Secondary standards were also established within three open-star clusters. These clusters are especially valuable for UBV calibration since the uncertainties in atmospheric extinction are less important because of the proximity of the stars. The names of these clusters along with finder charts are found in Appendix D.

2.2 THE UBV TRANSFORMATION EQUATIONS

Through the observations of standard stars, an observer can take instrumental measurements of program stars and transform them to the standard *UBV* system. In Chapter 1, the transformation equations are presented in a general form to be applied to any photometric system. It is customary to change the symbols used in the transformation equations to indicate the use of the *UBV* system. Equation 1.10 is replaced by

$$v = -2.5 \log d_v \quad (2.1)$$
$$b = -2.5 \log d_b \quad (2.2)$$
$$u = -2.5 \log d_u \quad (2.3)$$

where v, b, u and d_v, d_b, d_u represent the instrumental magnitudes and measurements through the V, B, and U filters, respectively. The constants, q', have been dropped because they can be "absorbed" by the zero-point constant in the transformation equations. Equation 1.12 is replaced by

$$(b - v) = -2.5 \log d_b/d_v \quad (2.4)$$
$$(u - b) = -2.5 \log d_u/d_b \quad (2.5)$$

The lower case u, b, v refer to the instrumental system while U, B, V refer to the standard system. The magnitude and colors corrected for atmospheric extinction, Equations 1.13 and 1.15 become

$$v_0 = v - k'_v X \quad (2.6)$$
$$(b - v)_0 = (b - v)(1 - k''_{bv} X) - k'_{bv} X \quad (2.7)$$
$$(u - b)_0 = (u - b) - k'_{ub} X \quad (2.8)$$

In the *UBV* system, k''_{ub} is defined to be zero (more about this later), and experience has shown that k''_v is very small so it is not included in Equation 2.6. Equations 1.16 and 1.17 become

$$V = v_0 + \epsilon(B - V) + \zeta_v \quad (2.9)$$
$$(B - V) = \mu(b - v)_0 + \zeta_{bv} \quad (2.10)$$
$$(U - B) = \psi(u - b)_0 + \zeta_{ub} \quad (2.11)$$

where ϵ, μ, ψ are the transformation coefficients and ζ_v, ζ_{bv}, ζ_{ub} are the zero-point constants. These six values are found by observations of the standard stars in Appendices C and D. The details of this calibration are given in Chapter 4.

2.3 THE MORGAN-KEENAN SPECTRAL CLASSIFICATION SYSTEM

Spectral classification is a very important topic in stellar astronomy and the reader can find an elementary review of this topic in the books by Abell,[4] Swihart,[5] and Smith and Jacobs,[6] to name but a few. A more advanced discussion is given by Keenan.[7] A brief review is given here because of the close relationship of the *UBV* system to the Morgan-Keenan (M-K) spectral classification system.

The first large-scale classification of stellar spectra began in the 1920s at Harvard College Observatory and became known as the *Henry Draper Catalog*. Over 400,000 stellar spectra were classified. At first, the stars were broken into a few groups based on the strength of the hydrogen absorption lines. The groups were designated A through P, from strongest to the weakest lines. In time, it became clear that some of these classes did not exist but were a product of poor quality spectrograms. Furthermore, simply arranging the spectrograms so the hydrogen lines varied from strong to weak did not produce a continuous and logical pattern in the remaining lines. Consequently, some classes were dropped and the remainder were rearranged. The result was a scrambled alphabetic sequence (O, B, A, F, G, K, M) but a logical and continuous variation in strength of all spectral lines. Better quality spectrograms have led to the development of 10 subclasses indicated by a number (zero through nine) following the letter. The sun is designated as a G2 while Vega is an A0 star. Figure 2.3 shows several spectra of main sequence stars. At the bottom of the figure, a typical filter plus photomultiplier tube response function for the *UBV* system is shown. It is customary to refer to stars near the beginning of the sequence as *early type* and those near the end as *late type*. That is, an A0 star is an earlier type than an F5, and a K0 is earlier than a K5.

We now know that the spectral sequence is an ordering by stellar surface temperature. For instance, O stars are approximately 50,000 K while M stars are 3000 K. The changing pattern of spectral lines in Figure 2.3 is a direct result of the change in the stellar temperature. An O type star shows few lines because most atoms are totally ionized.

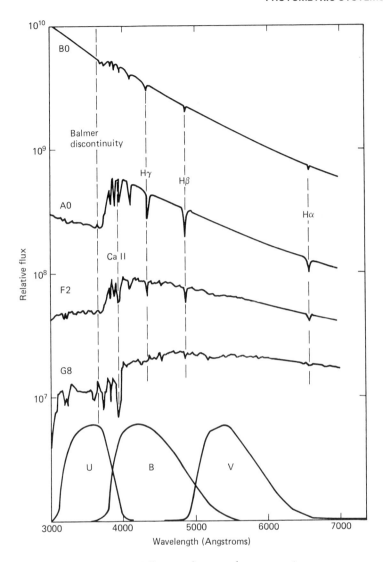

Figure 2.3 Spectra of some main sequence stars.

However, lines of He II (singly ionized helium) are fairly strong and are sometimes seen in emission. As one progresses towards the B class, the He II lines grow weaker and He I (neutral, un-ionized helium) and hydrogen lines grow stronger. By class B2, the He I lines dominate the spectrum. Hydrogen and ionized metal[†] lines grow stronger until early type A. Hydrogen and many ionized metals reach maximum strength at A0. By late A and into early F, the ionized metal lines grow while hydrogen lines decrease rapidly. Through classes F to G the spectral lines of Ca II strengthen reaching a peak at G2. Neutral metals continue to gain in strength as their ionized counterparts disappear. By late K, molecular bands appear and neutral metal lines dominate. The hydrogen lines are essentially gone and the calcium lines are still strong. By late K and into M, the bands of titanium oxide become prominent. Lines of the neutral metals are still stronger.

As stated earlier, the V filter was chosen to match the old visual magnitudes and the B filter to match the photographic magnitudes. The U filter was chosen to measure a spectral feature. In Figure 2.3, it is easy to see that the hydrogen lines dominate the early spectral types. The spacing between these lines becomes closer and closer until at the Balmer limit they merge and the absorption becomes continuous. Therefore, at the Balmer limit (3647 Å) there is a sharp drop in the continuum level, called the *Balmer discontinuity*. Figure 2.3 also shows that the U filter straddles this discontinuity. Thus, the $(U - B)$ color index is sensitive to the strength of the discontinuity, which in turn is a function of the star's spectral type. Note that the effective wavelength of observation through the U filter depends on the strength of the Balmer discontinuity. If the discontinuity is strong, very little light is received shortward of 3647 Å. The light measured through the U filter is that which passes through the "red wing" of the filter, longward of 3647 Å. Thus, we are effectively looking at a wavelength that is longer than the middle of the filter bandpass. On the other hand, a star that has a very weak discontinuity supplies light roughly equally across the bandpass of the filter. Then the effective wavelength of observation is near the center of the bandpass. An important consequence of this effect is that the second-order atmospheric extinction coefficient for $U - B$ has a complicated behavior with spectral type. That is to say, unlike the behavior of

---

[†]Astronomers use the term "metal" in a very different sense than do chemists. The term is used to designate any element other than hydrogen or helium.

k''_{bv}, k''_{ub} does not vary smoothly with spectral type, but rather shows a double sawtoothed variation from type O to M. To avoid the time-consuming process of correcting k''_{ub}, Johnson and Morgan defined it to be zero. Since second-order terms are small, the error introduced by this definition is of the order of 0.03 in $U - B$. A more detailed discussion of this problem can be found in Section 4.9.

The Harvard System is a one-dimensional classification scheme. However, it was realized rather early that a second dimension might be necessary. Some stars showed narrower absorption lines than other stars of the same class even though the pattern of spectral lines matched well. Between 1914 and 1935, Mount Wilson Observatory ordered spectra of the same class by the strength of certain spectral features. In time, it was realized that narrower lines resulted from a lower density in the atmosphere of these stars. Because the temperatures are the same, their atmospheres are much larger than those of normal main sequence stars. Therefore, these narrow-line stars are brighter than their main sequence counterparts. (See Equation J.23.)

These spectral "anomalies" are in fact luminosity indicators. W. W. Morgan, P. C. Keenan, and E. Kellman[8] developed a second dimension to the spectral classification now in general use. This M-K system introduces luminosity classes as follows:

I: Supergiants
II: Bright giants
III: Giants
IV: Subgiants
V: Main sequence (dwarfs)
VI: Subdwarfs

The location of these groups in the Hertzsprung-Russell (H-R) diagram is shown in Figure 2.4. Classes I to V may be subdivided by using the suffix a (brightest), or ab, or b (dimmest). The luminosity criteria are based on line strengths, ratios of line strengths, and widths of hydrogen lines. The low density in the atmospheres of the larger stars alters the percentage of atoms that are ionized. This in turn alters the line strengths and makes the spectrum appear to belong to a hotter star, and therefore to an earlier spectral type. Note in Figure 2.4 that the spectral classes of the giants and supergiants occur to the right of the same classes for stars on the main sequence. However, the color index is

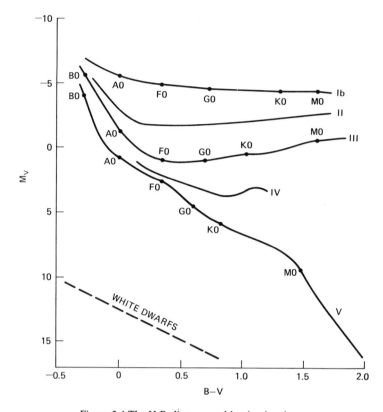

Figure 2.4 The H-R diagram and luminosity classes.

largely unaffected by the changes in the lines and therefore indicates a different temperature.

2.4 THE M-K SYSTEM AND *UBV* PHOTOMETRY

One of the most important aspects of the *UBV* system is its close ties to the M-K spectral classification system. As stated earlier, the zero points for the color indices were defined by stars classified as A0 V on the M-K system. This allows the colors of the *UBV* system to be related directly to an M-K spectral type and temperature. Figures 2.5a and b and Table 2.1 show these relationships for main sequence stars. These apply to stars that are not viewed through significant quantities of interstellar dust. This dust selectively absorbs more blue light than red light making a star appear redder than it actually is.

TABLE 2.1. Color Indices and Temperatures for Main Sequence Stars

Spectral Type	$(B - V)$	$(U - B)$	Effective Temperature (°K)
O5	−0.32	−1.15	54,000
B0	−0.30	−1.08	29,200
B5	−0.16	−0.56	15,200
A0	0.00	0.00	9600
A5	+0.14	+0.11	8310
F0	0.31	0.06	7350
F5	0.43	0.00	6700
G0	0.59	0.11	6050
G5	0.66	0.20	5660
K0	0.82	0.47	5240
K5	1.15	1.03	4400
M0	1.41	1.26	3750
M5	1.61	1.19	3200

SOURCE: Novotny, E. 1973. *Introduction to Stellar Atmospheres and Interiors.* New York: Oxford University Press, p. 10.

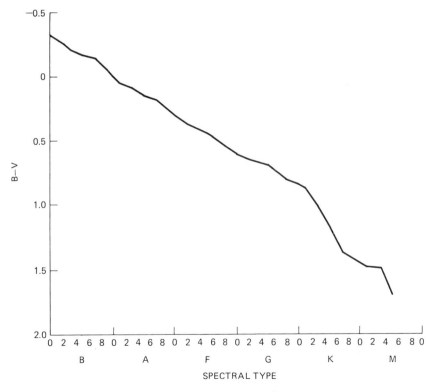

Figure 2.5a B-V versus spectral type.

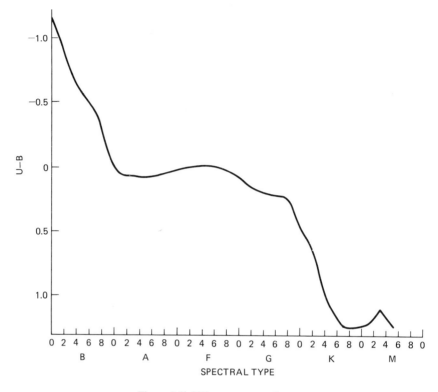

Figure 2.5b U-B versus spectral type.

Johnson and Morgan established the relationship between color indices and absolute magnitude by a two-step process. The first step was to measure the color indices of nearby stars with accurately known parallaxes. From the parallax and apparent magnitude, the absolute magnitude can be calculated directly. Unfortunately, there are very few early-type stars with accurately measured parallaxes because these stars are relatively rare. Even the A, F, and G stars are not as common as one would like for calibration purposes. The second step was to fill these gaps in spectral types using stars in nearby galactic clusters. A color versus *apparent* magnitude plot can be made for these clusters after correcting the magnitudes for interstellar absorption. Because all the stars within the cluster are nearly the same distance away, the apparent magnitude for each star differs from the absolute magnitude by some additive constant. If it is assumed that there is no difference between the main sequence of nearby field stars and that of a cluster, then the plot for the cluster can be slid vertically (in magnitude) on top of the plot for the field stars until the main sequences match. Then, the absolute magnitude of the cluster stars and the distance to the cluster is defined. The clusters used for this process were NGC 2362, the Pleiades, and the Praesepe. The completed diagram appears in Figure 2.6, which is an H-R diagram using a color index instead of the M-K type. Because of the effect discussed at the end of Section 2.3, the relation between color index and spectral class depends slightly on the luminosity class.

Figure 2.7 shows a plot of $(U - B)$ versus $(B - V)$, for (unreddened) main-sequence stars. This is a so-called *color-color diagram*. Note that the $(U - B)$ color gets smaller as you move upwards in the plot (the star is becoming brighter in U than in B). Blackbodies of various temperatures follow a nearly linear relation (upper curve). However, stars (lower curve) deviate significantly from a blackbody. These two curves differ because of the absorption lines in stellar spectra. From type O to A0, the hydrogen absorption lines increase in strength and so does the Balmer discontinuity. The flux seen in the U filter decreases causing $(U - B)$ to become larger. (Remember, magnitudes are larger if a star is fainter.) After A0, the Balmer lines (and the discontinuity) grow weaker and $(U - B)$ begins to decrease. After class F5, however, the metal lines and molecular bands become strong. Many of these

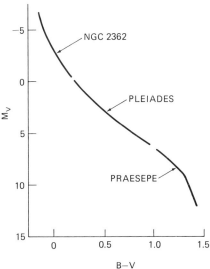

Figure 2.6 Main sequence matching.

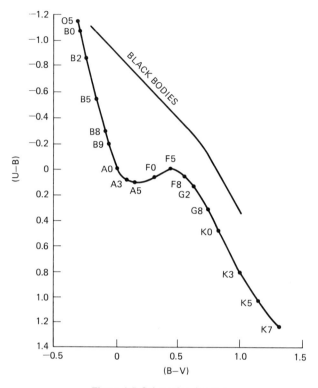

Figure 2.7 Color-color diagram.

absorption features are in the ultraviolet and cause the $(U - B)$ color to become large again. There is a complication for stars near the bump in the color-color plot at F5. Because the value of $(U - B)$ depends on the amount of absorption by metal lines, an abnormal metal abundance can have a significant effect on this color. A low metal abundance causes the star to plot higher than a normal star.

Figure 2.8 shows the color-color plot again, but this time to illustrate the effect of interstellar reddening. Reddening causes a star to move to the right nearly parallel to the reddening line in the figure. As an example, if an observed star plots at point A in the diagram, it can be assumed that extrapolating to the left, parallel to the reddening line, yields its intrinsic colors. The amount of color change produced by the dust is called the *color excess* and is denoted by $E(B - V)$ and $E(U - B)$, as labeled in Figure 2.8. The slope of the reddening line is given by

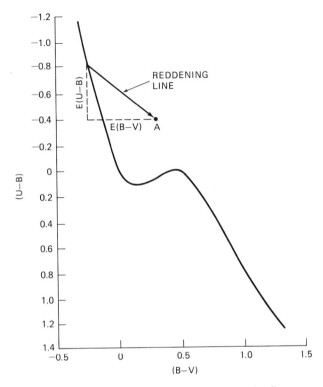

Figure 2.8 Reddening displacement on the color–color diagram.

$$\frac{E(U-B)}{E(B-V)} = 0.72 - 0.05(B-V) \qquad (2.12)$$

For early type stars $(B-V)$ is nearly zero, so the second term is very small and

$$\frac{E(U-B)}{E(B-V)} \simeq 0.72 \qquad (2.13)$$

For stars that are later than type A0, it is not possible to use a color-color plot to determine the intrinsic color unambiguously. This is true because the color-color curve turns upward at A0. Extrapolating to the left along the reddening line will result in two intersections of the color-color curve. For stars later than A0, the color excesses must be obtained by comparing the spectral class implied by the observed colors to that

48 ASTRONOMICAL PHOTOMETRY

obtained by spectroscopy. The latter is not affected by the reddening because it is based on the pattern of spectral lines.

For stars from type B0 to A0, there is yet another way to deal with interstellar reddening. The quantity Q is defined as

$$Q = (U - B) - 0.72(B - V) \qquad (2.14)$$

where $(U - B)$ and $(B - V)$ are the observed colors. Q is independent of reddening. To see this, note that

$$E(B - V) = (B - V) - (B - V)_i \qquad (2.15)$$

and

$$E(U - B) = (U - B) - (U - B)_i \qquad (2.16)$$

where $(B - V)_i$ and $(U - B)_i$ are the intrinsic colors of the star. Now we solve these two equations for $(B - V)$ and $(U - B)$, respectively, and substitute into Equation 2.14. Then,

$$Q = E(U - B) + (U - B)_i - 0.72\,[E(B - V) + (B - V)_i] \qquad (2.17)$$

$$Q = (U - B)_i - 0.72(B - V)_i + E(U - B) - 0.72(B - V). \qquad (2.18)$$

Now substitute Equation 2.13, which results in

$$Q = (U - B)_i - 0.72(B - V)_i, \qquad (2.19)$$

independent of reddening. Equation 2.19 is then used to produce Figure 2.9. The observed colors of a reddened star can be used to calculate Q by Equation 2.14. Figure 2.9 then yields the intrinsic spectral type.

The total absorption in the visual magnitude can be estimated in the following way. Define a quantity R as

$$R = \frac{A_V}{A_B - A_V} \qquad (2.20)$$

where A_V and A_B are the absorption, in magnitudes, in V and B, respec-

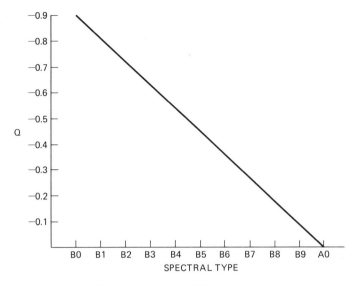

Figure 2.9 Q versus MK spectral type.

tively. The observed magnitudes are related to the intrinsic magnitudes by

$$B = B_i + A_B \tag{2.21}$$
$$V = V_i + A_V. \tag{2.22}$$

Substituting these two expressions into Equation 2.15 gives

$$E(B - V) = [(B_i + A_B) - (V_i + A_V)] - (B - V)_i \tag{2.23}$$
$$E(B - V) = A_B - A_V. \tag{2.24}$$

Thus, Equation 2.20 becomes

$$R = \frac{A_V}{E(B - V)} \tag{2.25}$$

or

$$A_V = RE(B - V). \tag{2.26}$$

The value of R has been found to be about 3.0 for most directions in our galaxy. However, there appear to be some regions where the nature

50 ASTRONOMICAL PHOTOMETRY

of the interstellar dust is different and R may reach values as high as 12. If $E(B - V)$ is determined by one of the above methods and R is assumed to be 3.0, it is then possible to calculate A_V and correct the apparent visual magnitude by

$$V_i = V - A_V \qquad (2.27)$$

to obtain the intrinsic visual magnitude.

*2.5 ABSOLUTE CALIBRATION

It is sometimes helpful or necessary to convert a measured magnitude into an actual flux measurement. The process of absolute calibration is both difficult and tedious. The process has recently been discussed by Lockwood et al.[9] No attempt to explain the process is made here, but a means of approximately transforming UBV measurements into flux is shown. Recall Equation 1.7:

$$m_\lambda = q_\lambda - 2.5 \log F_\lambda.$$

Assume that the atmospheric extinction correction has been made. Denote this by an added subscript on m, that is,

$$m_{\lambda 0} = q_\lambda - 2.5 \log F_\lambda, \qquad (2.28)$$

or more explicitly,

$$V_0 = q_v - 2.5 \log F_v \qquad (2.29)$$
$$B_0 = q_b - 2.5 \log F_b \qquad (2.30)$$
$$U_0 = q_u - 2.5 \log F_u. \qquad (2.31)$$

Johnson[10] has determined the q's as they appear in Table 2.2. Thus, the flux can be determined by

$$F_V = 10^{-0.04(V_0 - q_v)} \qquad (2.32)$$
$$F_B = 10^{-0.4(B_0 - q_b)} \qquad (2.33)$$
$$F_U = 10^{-0.4(U_0 - q_u)} \qquad (2.34)$$

TABLE 2.2. Absolute Zero-Point Constants

Filter	Approximate Effective Wavelength (Angstroms)	q_λ
U	3600	−38.40
B	4400	−37.86
V	5500	−38.52
R	7000	−39.39
I	9000	−40.2
J	12,500	−41.2
K	22,000	−43.5
L	34,000	−45.2
M	50,000	−46.6
N	102,000	−49.8

NOTE: Filters *I* through *N* will be explained in Section 2.7a.

The constants that appear in Table 2.2 are simply 2.5 times the logarithm of the flux of a zero magnitude star, in watts per square centimeter per Angstrom. Because of the difficulties of the calibration process, these constants may contain errors between 10 and 20 percent.

Example: What is the flux reaching the earth from a star which has $V_0 = 3.0$? From Table 2.2, $q_v = -38.52$.

$$F_V = 10^{-0.4(3.0+38.52)}$$
$$F_V = 10^{-16.61}$$
$$F_V = 2.47 \times 10^{-17} \frac{\text{watts}}{\text{cm}^2 \text{ Angstrom}}$$

This is the flux at the effective wavelength of observation. (See Equation J.53.) The total flux measured in the V filter can be found, approximately, by multiplying this number by the width of the filter's bandpass (1000 Å). Thus,

$$F_V \approx 2.5 \times 10^{-14} \text{ watts/cm}^2.$$

The total power collected in the V filter by the telescope is obtained by multiplying by the collecting area, that is

$$P_V \approx 2.5 \times 10^{-14} (\pi R_t^2) \text{ watts},$$

52 ASTRONOMICAL PHOTOMETRY

where R_t is the radius in centimeters of the primary mirror or lens of the telescope.

2.6 DIFFERENTIAL PHOTOMETRY

The concept of differential photometry is outlined in Section 1.6. We now proceed to a more detailed discussion. The actual observations consist of a series of measurements, which are given in counts per second (pulse counting) or percent of full-scale deflection (DC) through each filter of both the variable and the comparison star. We represent the measurements through the V, B, and U filters by d_v, d_b, and d_u, respectively. We add a second subscript to indicate the variable (x) or the comparison star (c). The magnitude difference between the variable and the comparison star in each filter is given by

$$\Delta v = -2.5 \log \frac{d_{vx}}{d_{vc}} \qquad (2.35)$$

$$\Delta b = -2.5 \log \frac{d_{bx}}{d_{bc}} \qquad (2.36)$$

$$\Delta u = -2.5 \log \frac{d_{ux}}{d_{uc}} \qquad (2.37)$$

if pulse-counting electronics are used. If DC electronics are used, it is possible that the two stars may require a different amplifier gain. The above equations are then modified to read,

$$\Delta v = -2.5 \log \frac{d_{vx}}{d_{vc}} + G_{vx} - G_{vc} \qquad (2.38)$$

$$\Delta b = -2.5 \log \frac{d_{bx}}{d_{bc}} + G_{bx} - G_{bc} \qquad (2.39)$$

$$\Delta u = -2.5 \log \frac{d_{ux}}{d_{uc}} + G_{ux} - G_{uc} \qquad (2.40)$$

The additional terms give the difference in amplifier gain (in magnitudes) between the variable and the comparison star. In Chapter 8, the procedure of gain calibration is discussed.

It is also possible to use these same measurements to form differences in color indices between the variable and the comparison star. To see this, note that

$$\begin{aligned}\Delta(b - v) &= (b_x - v_x) - (b_c - v_c) \\ &= (b_x - b_c) - (v_x - v_c) \\ &= \Delta b - \Delta v.\end{aligned} \quad (2.41)$$

Likewise,

$$\Delta(u - b) = \Delta u - \Delta b. \quad (2.42)$$

The beginner need not carry the data reduction beyond this point. There are many worthwhile observing projects that can be done with differential photometry, some of which are discussed in Chapter 10.

In rare circumstances, it might be necessary to apply a small extinction correction to differential photometry. This should seldom be necessary, because the variable and comparison star are close together in the sky and have been viewed through essentially the same air mass. However, in practice, this sometimes does not occur. It may be that a suitable comparison star was not found within 1° of the variable or that the comparison star was not observed frequently. In the latter case, the earth's diurnal motion causes the two stars to be viewed through significantly different air masses. Equations 2.35 through 2.37 (or 2.38 through 2.40) must be corrected to give the magnitude difference above the earth's atmosphere by making use of Equation 2.6. Thus,

$$(\Delta v)_0 = \Delta v - k'_v(X_x - X_c) \quad (2.43)$$
$$(\Delta b)_0 = \Delta b - k'_b(X_x - X_c) \quad (2.44)$$
$$(\Delta u)_0 = \Delta u - k'_u(X_x - X_c), \quad (2.45)$$

where X_x and X_c are the air masses of the variable and comparison star, respectively, at the time of observation. The color index differences can be corrected using Equations 2.7 and 2.8. That is,

$$\Delta(b - v)_0 = \Delta(b - v) - k'_{bv}(X_x - X_c) - k''_{bv}\Delta(b - v)\overline{X} \quad (2.46)$$

and

54 ASTRONOMICAL PHOTOMETRY

$$\Delta(u - b)_0 = \Delta(u - b) - k'_{ub}(X_x - X_c), \quad (2.47)$$

where $\overline{X}$ is the average air mass of the variable and comparison star.

It must be stressed that all of the above magnitude and color differences are on the instrumental system of the photometer in use. It is possible to do differential photometry on the standard UBV system. The procedure is to observe your comparison star along with some UBV standards. You can then determine V, $(B - V)$, and $(U - B)$ of the comparison star. This need only be done on one night if it is done well. On all other nights, you only need observe your variable and the comparison star. The magnitude and color differences between your variable and comparison star on the standard system can be found by rewriting Equations 2.9 through 2.11 to obtain

$$\Delta V = (\Delta v)_0 + \epsilon \Delta(B - V) \quad (2.48)$$
$$\Delta(B - V) = \mu \Delta(b - v)_0 \quad (2.49)$$
$$\Delta(U - B) = \psi \Delta(u - b)_0 \quad (2.50)$$

Note that if the two stars have nearly the same color, the second term on the right of Equation 2.48 is nearly zero. Furthermore, μ and ψ are approximately equal to one for most photometers using the 1P21 photomultiplier tube and standard UBV filters. This justifies the earlier statement that an uncalibrated photometer gives nearly the same magnitude and color difference as a calibrated one. The real advantage of calibrating the comparison star is that you can use your observations to compute the actual standardized magnitude and colors of the variable star by

$$V_x = V_c + \Delta V \quad (2.51)$$
$$(B - V)_x = (B - V)_c + \Delta(B - V) \quad (2.52)$$
$$(U - B)_x = (U - B)_c + \Delta(U - B). \quad (2.53)$$

2.7 OTHER PHOTOMETRIC SYSTEMS

By no means is the UBV system the only photometric system. There are many other valuable systems. A complete discussion of all these systems is beyond the scope of this text. However, we describe briefly an intermediate-band and a narrow-band system following the discussion of an extension to the *UBV* system.

2.7a The Infrared Extension of the *UBV* System

In order to expand the usefulness of the *UBV* system to the classification of cool stars, the system has been extended with bandpasses in the infrared. Table 2.2 lists the letter designation of each filter and its approximate effective wavelength. Photometry with filters *U*, *B*, *V*, *R*, *I* can be accomplished using an S-1, an extended-red S-20 photomultiplier,[11] or a photodiode as a detector. Wavelengths in the range of J through N require specialized detectors, such as those using lead sulfide, and cooling to liquid-helium temperatures. These techniques are beyond the scope of this text. Interested readers are referred to a review by Low and Rieke.[12]

2.7b The Strömgren Four-Color System

The Strömgren system[13] is an intermediate-band width system that overcomes many of the shortcomings of the *UBV* system and provides astrophysically important information. Table 2.3 contains the filter designations, central wavelengths, and bandwidths of the four filters.

Unlike the *UBV* system, the Strömgren system is almost totally filter-defined. The *y* (yellow) filter matches the visual magnitude and corresponds well with *V* magnitudes. This filter transmits no strong spectral features in early-type stars. The red limit is set by the filter and not by the detector as in the case of the *UBV* system. The *b* (blue) filter is centered about 300 Å to the red of the *B* filter of the UBV system to reduce the effects of "line blanketing." For stars of spectral types later

TABLE 2.3. Filters Used in the Strömgren System

Filter	Central Wavelength (Angstroms)	Full Width at Half Transmission (Angstroms)
y	5500	200
b	4700	100
v	4100	200
u	3500	400

than A0, absorption lines of metals become strong. A filter that is centered in a wavelength region where such lines are common transmits less flux than it would if the lines were absent. This blanketing effect is a temperature indicator in that it becomes strong in later spectral types. To get a clear measure of its strength, it is necessary to measure a star's flux in a region relatively free of blanketing and compare it to a region where blanketing is strong. For early type stars, the b and y filters are free from blanketing. In later type stars, the two filters are affected almost equally. The violet (v) filter is centered in a region of strong blanketing but longwards of the region where the hydrogen lines begin crowding together near the Balmer limit. The u (ultraviolet) filter measures both blanketing and the Balmer discontinuity. Unlike the U filter in the UBV system, this filter is completely to the short wavelength side of the Balmer discontinuity. Yet, it is centered far enough from the atmospheric limit near 3000 Å so that the observing site plays no role in defining the wavelength region observed. Hence, the system is nearly filter-defined and insensitive to the detector used. There are essentially no effects due to the filter's bandwidth. That is, there are no second-order color terms in the extinction corrections or the transformation equations. This is a simplification compared with the UBV system.

Figure 2.10 schematically illustrates a stellar spectrum and the placement of the four filters. The color indices in the Strömgren system are very useful quantities. Because both the b and y filters are relatively free from blanketing, the index $(b - y)$ is a good indicator of color and

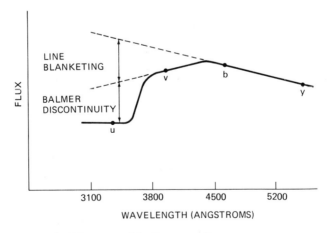

Figure 2.10 Placement of the Strömgren filters.
Note: For the sake of clarity, stellar absorption lines are not shown.

effective temperature. A color index is essentially the slope of the continuum. In the absence of blanketing, the continuum slope would be roughly constant and $(b - y)$ approximately equals $(v - b)$. Because $(v - b)$ is affected by blanketing, the difference between these two indices indicates the strength of blanketing. Hence a metal index, m_1, can be defined as

$$m_1 = (v - b) - (b - y). \qquad (2.54)$$

To determine how the continuum slope has been affected by the Balmer discontinuity, the index c_1 is defined as

$$c_1 = (u - v) - (v - b). \qquad (2.55)$$

This index measures the Balmer discontinuity, nearly free from the affects of line blanketing. To see this, note that the u measurement contains the effects of both blanketing and the Balmer discontinuity. The v filter contains only the effect of blanketing which is roughly one-half as strong as in the u filter. Further note that c_1 has been defined so that Equation 2.55 can be rewritten as

$$c_1 = (u - 2v + b). \qquad (2.56)$$

Subtraction of the $2v$ term essentially cancels the blanketing, leaving the effects of the Balmer discontinuity.

In summary, the Strömgren system provides a visual magnitude, a measure of the effective temperature, a measure of the strength of metal lines, and a measure of the Balmer discontinuity. Furthermore, it is filter-defined, independent of any one detector and requires no second-order terms in extinction or transformation equations. The only major drawback of the system is that the smaller bandpasses make faint stars more difficult to measure. *The Astronomical Almanac*[14] has a list of standard stars for the Strömgren system.

2.7c Narrow-Band Hβ Photometry

As our example of narrow-band photometry, we discuss briefly a frequently used extension of the four-color system, Hβ photometry. In this system, a narrow interference filter that is centered on the Hβ line is

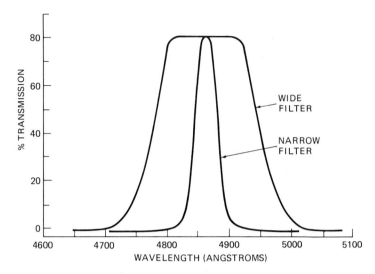

Figure 2.11 Filter responses, Hβ system.

used. In early type stars, this is a strong absorption line. The amount of light flux passed by the filter is heavily dependent on the line strength. The strength of Hβ is a luminosity indicator in stars of spectral type O to A and a temperature indicator in types A to G.

This system actually requires two filters since a small amount of detected flux could mean a strong Hβ absorption line or, simply, a faint star. Thus, a second, broader filter that measures much of the adjacent continuum is used. The ratio of the measurements through the two filters indicates the strength of Hβ with respect to the continuum. Figure 2.11 shows the response of the filters.

Obviously, effective narrow-band photometry requires a large telescope. Furthermore, interference filters are expensive and require thermostating at the telescope. It is for these reasons we do not go into further depth on this topic. The interested reader is referred to Chapter 5 in Golay.[1]

REFERENCES

1. Golay, M. 1974. *Introduction to Astronomical Photometry*. Boston: D. Reidel.
2. Johnson, H. L., and Morgan, W. W. 1951. *Ap. J.* **114**, 522.
3. Johnson, H. L., and Morgan, W. W. 1953. *Ap. J.* **117**, 313.
4. Abell, G. 1978. *Exploration of the Universe*. New York: Holt, Rinehart and Winston.

5. Swihart, T. L. 1968. *Astrophysics and Stellar Astronomy.* New York: John Wiley and Sons.
6. Smith, E., and Jacobs, K. 1973. *Introductory Astronomy and Astrophysics.* Philadelphia: W. B. Saunders Co.
7. Keenan, P. C. 1963. In *Basic Astronomical Data.* Edited by K. Aa Strand. Chicago: Univ. of Chicago Press, chapter 8.
8. Morgan, W. W., Keenan, P. C., and Kellman, E. 1943. *An Atlas of Stellar Spectra.* Chicago: Univ. of Chicago Press.
9. Lockwood, G. W., White, N. M., and Tüg, H. 1978. *Sky and Tel.* **56**, 286.
10. Johnson, H. L. 1965. *Comm. Lunar and Planetary Lab.* No. 53.
11. Fernie, J. D. 1974. *Pub. A.S.P.* **86**, 837.
12. Low, F. J., and Rieke, G. H. 1974. In *Methods of Experimental Physics.* New York: Academic Press **12**, chapter 9.
13. Strömgren, B., 1966. *Ann. Rev. Astr. Ap.* **4**, 433. Palo Alto: Annual Review Inc.
14. *The Astronomical Almanac.* Washington, D.C.: Government Printing Office. Issued annually.

CHAPTER 3
STATISTICS

If we measure 23,944 counts from a source in 10 seconds, will we measure the same number of counts in the next 10-second interval? How many counts are needed to achieve 1 percent accuracy for the measurement? How do I analyze my data for errors? These are but a few of the questions that need to be answered before any data reduction is complete.

Experimental observations always have inaccuracies. The role of the experimenter is to know the extent of these inaccuracies and to account for them in the best manner. You must know how to combine observations and errors to compute a result. If your observations are to be compared to theoretical predictions, it is necessary to know something about the accuracies of both calculations if you want to make an intelligent comparison of their agreement.

This chapter attempts to answer some questions about errors and the field of statistics in general. The derivations and advanced concepts can be found in Appendix K. The majority of the material presented in this chapter comes from texts by Young and Bevington (see Section 3.8), both of which are available in paperback and are highly recommended.

3.1 KINDS OF ERRORS

Errors come in different types. Most errors occur in three major categories: illegitimate, systematic, and random. These are discussed in turn.

Illegitimate errors are not directly concerned with the data itself. Instead, these include mistakes in recording numbers, setting up the equipment incorrectly, and blunders in arithmetic. They cannot be represented by any theoretical model and must be eliminated by the observer through careful work.

Systematic errors are errors associated with the equipment itself or with the technique of using the equipment. For instance, if an analog amplifier has an offset voltage, the resulting chart recorder deflection will be in error. Another example is not removing the U filter's red leak from your data. Red stars will then appear brighter in the U filter than they really are.

Very often in experimental work, systematic errors are more important than random errors. However, they are also much more difficult to deal with. Always compare your results with the standard system values and other observers whenever possible to calibrate your equipment and observing procedures.

Random or chance errors are produced by a large number of unpredictable and unknown variations in the experimental situation. They can result from small errors in judgment on the part of the observer, such as in reading a chart recorder record. Other causes are unpredictable fluctuations in conditions, such as nearly invisible cirrus clouds or variations in a photomultiplier tube's high-voltage power supply. It is found empirically that such random errors are frequently distributed according to a simple law. This makes it possible to use statistical methods to treat random errors.

Because random errors can be modeled, they form the basis for much of the remaining material in this chapter. Illegitimate and systematic errors must be eliminated by the experimentalist wherever possible.

3.2 MEAN AND MEDIAN

The actual value of what you are trying to measure is unknown. No one knows the exact magnitude of a star, just as the speed of light, although well measured, is not known exactly. If you determine the magnitude of a star on five separate occasions, you are likely to get five different values. Intuitively, you would suspect that the most reliable result for the star's magnitude would be obtained by using all five measurements rather than only one of them. You can approximate the true value by taking these measurements (a sample) and determining the *average* or *sample mean* by summing all of the measurements and dividing by the number of them. In a more general mathematical notation,

$$\bar{x} = \frac{1}{N} \sum_{i=1}^{N} x_i, \qquad (3.1)$$

where x_i are the values of the individual measurements and N is the total number of measurements taken. The summation sign in Equation 3.1 is just a shorthand way of saying "the sum of x values from $i = 1$ to N." All such summations in the rest of the chapter have similar limits, and we may drop the $i = 1$ and N from the summation sign at times.

Example: On five occasions, you measured the visual magnitude of Mizar to be 4.50, 4.65, 4.55, 4.45, and 4.60. What is Mizar's mean visual magnitude from this data?

$$\bar{x} = \frac{1}{5}(4.50 + 4.65 + 4.55 + 4.45 + 4.60)$$
$$\bar{x} = \frac{1}{5}(22.75)$$
$$\bar{x} = 4.55$$

Sometimes we want to compute the average of a set of values in which some of the numbers are more important than others. For instance, measurements taken on a cloudy night or at low altitudes are probably less important than those taken on a crystal-clear night near the zenith.

A procedure that suggests itself is to assume that the clear zenith observation was made more than once. Suppose we have a cloudy observation, a low observation, and the clear zenith value. We include the clear zenith value twice to account for its supposed better accuracy. Then, of course, we must divide by the total number of observations, which is now four. More generally, if we have several observations with different degrees of reliability, we can multiply each by an appropriate "weighting factor," and then divide the sum of these products by the sum of all of the weighting factors. This is the concept of the *weighted mean,* and can be represented mathematically by

$$\bar{x} = \frac{\sum_{i=1}^{N} w_i x_i}{\sum_{i=1}^{N} w_i}. \tag{3.2}$$

Note that if all of the weights are unity (or more generally, if they are all equal), the weighted mean reduces to the mean as previously defined by Equation 3.1. The problem with weighted means is determining the weights in a rigorous manner without any observer bias. That is, is an observation on a clear night two, three, or only one and a half times as good as a cloudy night observation? Unless you can decide on a consistent scheme, it is probably better to just take a straight mean and use as many observations as possible.

The *median* of a sample (or set of observations) is defined as that value for which half of the observations will be less than the median and half greater. For our five-observation example, we first order the observations in increasing order: 4.45, 4.50, 4.55, 4.60, and 4.65. The median is then the mid-value or 4.55. If we had six observations, the median would fall between the third and fourth values. To compute the median, we would average these two values.

Example: We now make a sixth observation of Mizar and obtain a visual magnitude of 4.90. What is the mean and median of our sample?

$$\text{mean} = \frac{1}{6}(4.50 + 4.65 + 4.55 + 4.45 + 4.60 + 4.90)$$
$$\text{mean} = \frac{1}{6}(27.65)$$
mean = 4.61
ordered values: 4.45, 4.50, 4.55, 4.60, 4.65, 4.90
median = (4.55 + 4.60)/2
median = 4.57

Note that the mean and median do not have to agree, as they are independent estimates of the best value for the sample. Usually the mean is used but there are cases in which the median is a better indicator of the sample.

3.3 DISPERSION AND STANDARD DEVIATION

Now that we have a method of determining the best value from our sample of observations, we need some indication of how much faith we have in that value.

The *deviation* (d_i) or *residual* of any measurement x_i from the mean $\bar{x}$ is defined as the difference between x_i and $\bar{x}$. Mathematically,

$$d_i = x_i - \bar{x} \tag{3.3}$$

the deviation is a measure of the quality of the observations, so intuitively you would think that taking their sum and dividing by the number of values would give an average deviation. The problem is that some of the deviations are positive and others negative, and because of the way we defined the mean and the deviations, their sum is exactly equal to zero. One method to get around this problem is to use the absolute value of each deviation in the sum (i.e., all negative deviations are now positive). This defines the *average* or *mean deviation*:

$$\bar{d} = \frac{1}{N-1} \sum_{i=1}^{N} |x_i - \bar{x}| \tag{3.4}$$

We divide by $N - 1$ rather than N because we use at least one measurement to determine the mean, $\bar{x}$, and therefore get an unrealistic deviation of zero for one measurement. The average deviation is a measure of the *dispersion* (or spread or scatter) of the observations around the mean. The presence of the absolute value sign makes the average deviation cumbersome to use in practice. It is not correct to call d_i the error in measurement x_i because $\bar{x}$ is only an approximation to the true value. However, this is a fine point that few observers obey.

A parameter that is easier to use analytically and is theoretically justified is the *standard deviation*, σ. It is obtained by first squaring each deviation, thereby removing any minus signs, and obtaining the mean squared, which is the *variance*, σ^2:

$$\sigma_x^2 = \frac{1}{N-1} \sum_{i=1}^{N} (x_i - \bar{x})^2. \qquad (3.5)$$

The standard deviation (σ or s.d.) is just the square root of the variance:

$$\sigma_x = \sqrt{\frac{1}{N-1} \sum_{i=1}^{N} (x_i - \bar{x})^2}. \qquad (3.6)$$

By rearranging, an easier computational form for σ_x can be obtained:

$$\sigma_x = \sqrt{\left[\frac{1}{N-1} \sum_{i=1}^{N} x_i^2\right] - \bar{x}^2}. \qquad (3.7)$$

Thus, the standard deviation is the *root mean square* of the deviations. Note that the standard deviation is always positive and has the same units as x_i.

The standard deviation defined in this manner tells us the amount of dispersion to be expected in any single measurement. Clearly, a single measurement in the sample has a larger deviation than the deviation of the mean of the sample. In other words, if you measure the magnitude of a star on five separate occasions, you have confidence that the mean of these five observations is accurate. If you make a *single* additional measurement, it may deviate significantly from the mean. However, the mean of five additional observations lies close to the first mean.

We can therefore define the standard deviation of the mean ($\sigma_{\bar{x}}$). It can be shown that σ_x is very closely related to σ_x, and is given by

$$\sigma_{\bar{x}} = \frac{1}{\sqrt{N}} \sigma_x \qquad (3.8)$$

or

$$\sigma_{\bar{x}} = \sqrt{\frac{\sum_{i=1}^{N} (x_i - \bar{x})^2}{N(N-1)}}. \qquad (3.9)$$

Authors often quote $\sigma_{\bar{x}}$ and thereby indicate greater accuracy in their results than is correct. In general, use σ_x unless you thoroughly understand the difference between these two values and know when to use $\sigma_{\bar{x}}$.

3.4 REJECTION OF DATA

To understand the significance of the standard deviation, we must first know the expected distribution of observations. The probability distribution is found by taking a large number of observations and seeing how probable it is to obtain any given value. This has been performed both experimentally and analytically and it has been found that most experiments have a common probability distribution, called the *normal* or *Gaussian distribution*. It is defined by the equation

$$f(x) = \frac{1}{\sigma\sqrt{2\pi}} \exp\left[\frac{-(x-\bar{x})^2}{2\sigma^2}\right] \qquad (3.10)$$

and is shown in Figure 3.1. We are not going to describe this distribution in detail, but you should notice that it is symmetrical about the mean and that the width of the peak depends on σ, the standard deviation. A smaller σ will yield a sharper peak. It can be shown that if the data can be represented by a Gaussian distribution, then 68 percent of the data will fall within 1σ of the mean, 95 percent within 2σ, and 99.7 percent within 3σ. This says that if a data point falls more than 3σ from the mean, there is a 99.7 percent probability that it is faulty.

One other term, called the *most probable error* (*p.e.*), is frequently used. If the data can be represented by a Gaussian distribution, 50 percent of the data falls within one probable error of the mean. Expressed in terms of the standard deviation,

$$p.e. = 0.675\sigma. \qquad (3.11)$$

Most scientific calculators are furnished with programs that calculate the mean and standard deviation of a set of numbers. *Learn always to quote an error when presenting data.* A number by itself is almost useless!

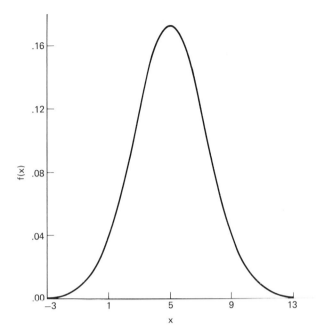

Figure 3.1 Gaussian distribution for $\bar{x} = 5.0$, $\sigma = 2.3$.

Example: Using our set of six measurements of Mizar, compute the standard deviation and present the mean.

$$\begin{aligned}
\sigma_x^2 &= \frac{1}{5} [(4.50 - 4.61)^2 + (4.65 - 4.61)^2 + (4.55 - 4.61)^2 \\
&\quad + (4.45 - 4.61)^2 + (4.60 - 4.61)^2 + (4.90 - 4.61)^2] \\
&= \frac{1}{5} [0.0121 + 0.0016 + 0.0036 + 0.0256 + 0.0001 \\
&\quad + 0.0841] \\
\sigma_x^2 &= 0.0254 \\
\sigma_x &= \sqrt{\sigma_x^2} = 0.16
\end{aligned}$$

visual magnitude of Mizar = 4.61 ± 0.16 (s.d.)

All of the discussion in this section has been leading up to the question of the rejection of data. Our sixth data point of 4.90 deviates by

2σ from the mean. It is tempting to regard this large deviation as a blunder or mistake rather than a random error. Should we remove it from the sample?

This is a controversial question and has no simple answer. It could be that Mizar is an unknown eclipsing binary and you caught it during eclipse. Throwing out that data value throws away the eclipse information! In any event, removing measurements constitutes tampering with or "fudging" the data.

Unless you can confidently state that a given measure is in error because it was a cloudy night or some similar obvious problem, you will have to make some sort of decision on data obviously out of bounds. There are two differing points of view that should be considered. It is your choice as to what method should be employed.

At one extreme, there is the point of view that there is never any justification for throwing out data, and to do so is dishonest. If you adopt this point of view, there is nothing more to say. You can remove much of the effect of one bad point by taking additional data, or you could mention any extraneous points when you report your results.

The other point of view is to reject a measure if its occurrence is so improbable that it would not be reasonably expected to occur. The usual criterion is that data should be more than 2σ or 3σ from the mean. The best way may be not to use the errant value in your calculations but report it so others may make their own choice. In any case, never iterate, that is, remove data, calculate a new mean and standard deviation, and remove data again.

Of course, there is a third possibility. It is possible that the data are *not* represented by a simple Gaussian, and that the wings of the distribution are larger than those of the Gaussian that fits the peak. Thus, the measure may be correct after all. You must decide this either by theoretical considerations or by taking enough measures to map out the wings of the distribution.

3.5 LINEAR LEAST SQUARES

The method of *least squares* or *regression analysis* is almost exclusively used in fitting lines to experimental data. In examining a plot of experimental data, the human mind will "eyeball" a line that roughly splits half of the data above the line and half below. In a crude fashion, the mind is approximating a least-squares line.

*3.5a Derivation of Linear Least Squares

If a straight line is to be fitted to data, then the line has the functional form

$$\hat{y} = a + bx \tag{3.12}$$

where $\hat{y}$ is the *calculated y* value for a given value of x. The fit is called *simple linear regression*, that is a linear function of only one variable, x. The deviations of the individual data points from this line can be defined as

$$\Delta y_i = (y_i - \hat{y}_i) \tag{3.13}$$

or

$$\Delta y_i = y_i - (a + bx_i) \tag{3.14}$$

Equation 3.14 is known as the *equation of condition*. Just as in the case of standard deviation, we square the deviations and try to minimize their sum. This yields the line with the least error, or the *least squared deviation*. If M is the sum of the squared deviations,

$$M = \sum_{i=1}^{N} (\Delta y_i)^2 \tag{3.15}$$

$$M = \Sigma y_i^2 + b^2 \Sigma x_i^2 + Na^2 + 2ab\Sigma x_i \tag{3.16}$$
$$- 2b\Sigma x_i y_i - 2a\Sigma y_i.$$

How do we minimize the sum? Mathematically, this is accomplished by taking the partial derivative of the function M with respect to each variable of concern, and then setting these derivatives equal to zero. The reader must take care to realize that the variables in this problem are a and b (not x and y). So set

$$\frac{\partial M}{\partial a} = \frac{\partial M}{\partial b} = 0.$$

The equations derived in this manner are called the *normal equations*. For our simple linear-regression example,

$$\frac{\partial M}{\partial a} = 2Na + 2b\Sigma x_i - 2\Sigma y_i = 0$$

$$\frac{\partial M}{\partial b} = 2b\Sigma x_i^2 + 2a\Sigma x_i - 2\Sigma x_i y_i = 0$$

or, rearranging,

$$(N)a + (\sum_{i=1}^{N} x_i)b = \sum_{i=1}^{N} y_i \qquad (3.17)$$

$$(\sum_{i=1}^{N} x_i)a + (\sum_{i=1}^{N} x_i^2)b = \sum_{i=1}^{N} x_i y_i. \qquad (3.18)$$

3.5b Equations for Linear Least Squares

Equations 3.17 and 3.18 can be solved simultaneously to yield after using some identities:

$$\text{Slope: } b = \frac{N\Sigma x_i y_i - \Sigma x_i \Sigma y_i}{N\Sigma x_i^2 - (\Sigma x_i)^2} = \frac{\Sigma(x_i - \bar{x})(y_i - \bar{y})}{\Sigma(x_i - \bar{x})^2} \qquad (3.19)$$

$$\text{Intercept: } a = \frac{\Sigma x_i \Sigma x_i y_i - \Sigma x_i^2 \Sigma y_i}{N\Sigma x_i^2 - (\Sigma x_i)^2} = \bar{y} - b\bar{x} \qquad (3.20)$$

Fitting a straight line to experimental data is the most common use of least squares. A FORTRAN routine to perform simple linear regression is given in Section I.4. This method can be generalized to any power of x or function M. For example,

$$\hat{y} = a_0 x^0 + a_1 x^1 + a_2 x^2 + \cdots$$

$$M = \Sigma[y_i - (a_0 x_i^0 + a_1 x_i^1 + a_2 x_i^2 + \cdots)]^2,$$

or

$$\hat{y} = a \cos(x)$$

$$M = \Sigma[y_i - a \cos(x_i)]^2.$$

Some comments are in order at this stage in the procedure. There are two basic reasons why the least-squares method is used rather than a freehand drawing. First, different people draw the freehand curve slightly differently because of observer bias. Second, the freehand method does not allow a quantitative measure of the goodness of the fit, an estimate of our confidence in the fitted line. The *standard error*, σ_e, of the least-squares estimate is given by

$$\sigma_e = \sqrt{\frac{1}{N-2} \sum_{i=1}^{N} (y_i - \hat{y})^2} \qquad (3.21)$$

This tells us the error expected at any point along the line. The *goodness of fit*, r, is given by

$$r = \frac{\frac{1}{N-1}\Sigma(x_1 - \bar{x})(y_i - \bar{y})}{\sqrt{\frac{1}{N-1}\Sigma(x_i - \bar{x})^2}\sqrt{\frac{1}{N-1}\Sigma(y_i - \bar{y})^2}} \qquad (3.22)$$

The four values determined in the least-squares analysis (slope, b; intercept, a; standard error, σ_e; goodness of fit, r) are used so commonly that several calculators are preprogrammed to perform the analysis. To be safe, always plot the data and draw the calculated line. Any significant deviations or a trend to the errors that may cause a lack of confidence in the fit then become obvious. A plot also serves as a check that all the data were entered correctly.

Example: We have just built a DC amplifier. Because it is transistorized, we suspect that its gain is temperature-sensitive. An experiment is devised, where a constant current is fed to the amplifier and the temperature of the amplifier is changed, with the chart-recorder reading recorded at various temperatures. The values obtained are listed below in the first two columns.

ASTRONOMICAL PHOTOMETRY

Temperature, °C (x)	Amplifier Gain (y)	$\hat{y}$
0	0.12	0.117
5	0.15	0.148
10	0.16	0.179
15	0.21	0.210
20	0.25	0.241
25	0.29	0.272
30	0.30	0.303
35	0.34	0.334
40	0.34	0.365
45	0.40	0.395
50	0.43	0.426

The data are shown in Figure 3.2. By visual inspection, it is obvious that the gain is increasing with increasing temperature. A linear least-squares line can be fit to the data and slope becomes the amplifier gain change per Celsius degree.

$$N = 11$$

$$\Sigma x_i = 275 \qquad \Sigma y_i = 2.99$$

$$\Sigma x_i y_i = 91.75 \qquad \Sigma x_i^2 = 9625$$

$$b = \frac{11 \times 91.75 - 275 \times 2.99}{11 \times 9625 - 275 \times 275} = 0.006182$$

$$a = \frac{1}{11}(2.99 - 0.006182 \times 275) = 0.1173$$

$$\Sigma(y_i - y)^2 = 0.001490$$

$$\sigma_e = \sqrt{\frac{1}{9}(0.001490)} = 0.013$$

The plotted straight line is the linear least-squares fit, whose numerical values are listed under $\hat{y}$ in the table above.

Multiple linear regression occurs when solving for more than one slope. That is the case where

$$y_i = a + b_1 x_i + b_2 w_i + b_3 z_i + \cdots$$

The solution to this problem is slightly more complicated because it involves putting the equation into matrix form and solving it by vector

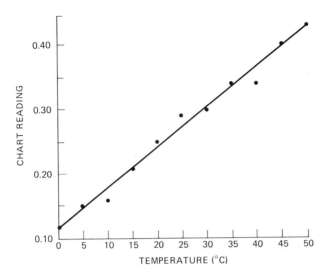

Figure 3.2 Amplifier temperature sensitivity.

differentiation and inversion. The solution is particularly useful in the least-squares solution of the standard *UBV* transformation equations and in the extinction calculation, but a discussion of this method is more involved than warranted for this chapter. Appendix K gives a complete discussion of the multivariate least-squares method. The average observer may find it more accurate to treat the transformation equation problem as a series of simple regression cases, thereby allowing better control over each step of the process.

3.6 INTERPOLATION AND EXTRAPOLATION

Often in photometry it is necessary to *interpolate,* that is, to find the value of some function between two base points. An example is when performing differential photometry in which several variable star observations are sandwiched between consecutive comparison star measures and you want the approximate comparison star reading at the time of each variable measure.

Consider the function in Figure 3.3. The solid line represents the true values of the function, with points identified at the base values x_0, x_1, x_2, and x_3. The function $y(x)$ might represent the star's intensity as it crosses the meridian, and x the time of the observation. We want to know its intensity between values x_1 and x_2. There are two usual approaches: exact and smoothed interpolation.

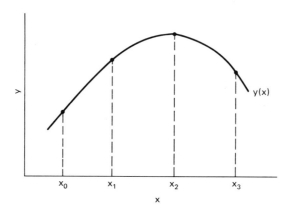

Figure 3.3 Interpolating between points.

3.6a Exact Interpolation

Exact interpolation makes use of the fact that there is one and only one polynomial of degree n or less which assumes the exact values $y(x_0)$, $y(x_1)$, ..., $y(x_n)$ at the $n + 1$ distinct base points $x_0, x_1, ..., x_n$. Therefore, to find $y(x)$ between x_1 and x_2, we could use a linear polynomial with points x_1 and x_2, a quadratic with x_0, x_1, and x_2, or a cubic with all four points. Exact only means that the polynomial fits the known data points exactly, not that it will interpolate exactly between those points. For example, consider linear interpolation for our desired value between x_1 and x_2. If we use a straight line between x_1 and x_2, we get a value reasonably close to the correct answer. If we use x_0 and x_3 instead as our end points and draw a straight line between them, the resultant answer lies considerably below the correct one.

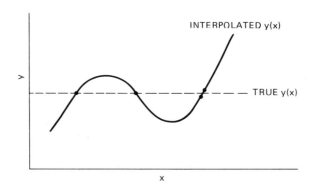

Figure 3.4 High-order interpolation.

Two problems arise in using high-order interpolating polynomials. First, increasing the order increases the number of "wiggles" between data values. This is shown in Figure 3.4, where we are using a cubic polynomial to fit data that essentially lie on a straight line. Second, interpolating polynomials using unequally spaced base points get very complicated with higher order. Therefore, unless you know that the answers should lie on a cubic or quartic line, use linear or at most quadratic interpolation.

To interpolate between points linearly, the function

$$y = a + bx \qquad (3.23)$$

is used, where b is the slope and a the intercept of the linear interpolating polynomial. To evaluate the slope and intercept:

$$b = \frac{y_2 - y_1}{x_2 - x_1} \qquad (3.24)$$

$$a = y_1 - bx_1 \qquad (3.25)$$

So,

$$y = a + bx$$
$$y = y_1 - bx_1 + bx$$
$$y = y_1 + b(x - x_1)$$
$$y = y_1 + \frac{y_2 - y_1}{x_2 - x_1}(x - x_1) \qquad (3.26)$$

Example: The count rate of Mizar was measured to be 200,000 counts per second at 03:00 UT and 300,000 counts per second at 04:00 UT. What is the best linear guess as to the count rate at 03:30 UT?

03:30 UT = 3.5 hours UT

$$y = 200{,}000 + \frac{(300{,}000 - 200{,}000)}{(4 - 3)}(3.5 - 3)$$

$$y = 200{,}000 + \frac{100{,}000}{1}(0.5)$$

$$y = 250{,}000 \text{ counts per second at 03:30 UT.}$$

76 ASTRONOMICAL PHOTOMETRY

Working in a similar manner, we can derive the interpolating formula for the second-order or quadratic polynomial between points x_0, x_1, and x_2 as

$$y = y_0 + \left(\frac{y_1 - y_0}{x_1 - x_0}\right)(x - x_0) + \left[\left(\frac{y_2 - y_1}{x_2 - x_1}\right) - \left(\frac{y_1 - y_0}{x_1 - x_0}\right)\right]\left[\frac{(x - x_0)(x - x_1)}{(x_2 - x_0)}\right] \quad (3.27)$$

You can see that even the quadratic interpolating polynomial is getting complicated. Usually, higher-order polynomials are evaluated by computer. Note that Equation 3.27 looks like the linear form with an added term. This extra term can be considered the error that exists if linear interpolation were used instead, and can be used to give an approximate error when presenting the interpolated value.

3.6b Smoothed Interpolation

So far, we have investigated polynomials that passed exactly through the base values. As seen from Figure 3.4, this can cause large errors if the base values have some inaccuracies built into them. The best method of interpolating under these circumstances is to use some sort of least-squares polynomial through the base points, and interpolate with this approximate function. For photometric data, better accuracy can be achieved with smoothed interpolation, but with increased complexity. An example is to use several comparison star observations and fit a linear least-squares line to the observed count rate. Then this line can be used along with the variable star measurements to derive the intensity differences for differential photometry. You usually achieve greater accuracy than if you use the comparison star observation closest to the variable star measure because you are using the information contained in earlier and later measurements.

In general, we suggest that you use exact linear interpolation whenever interpolation is necessary. If you have a large number of base values, smoothed linear linear interpolation could be used. Remember however, that interpolation by nature is not exact, and requires data on either side of the value to be calculated.

3.6c Extrapolation

We have neglected the case of *extrapolation,* that is, the determination of $y(x)$ where x lies beyond any of the observed values. This is because extrapolation is very, very risky and should be avoided at all costs!

An example of the errors that can arise from extrapolation is the measurement of the atmospheric extinction. If you follow an extinction star from the zenith to, say 45° above the horizon, determine extinction from it, and use your values for an observation on the horizon, your results may be good. But there also may be a cloud bank or smoke layer on the horizon, making extinction there much different than near the zenith.

The rule of thumb for extrapolation is that if the data point is close to the last base value, you can extrapolate, but should consider this extrapolated value as having very low weight.

3.7 SIGNAL-TO-NOISE-RATIO

It is intuitively obvious that the longer we continue to gather data on a star during a single observation, the more accurate our results become. We would like a quantitative measure of this accuracy. Experimental scientists commonly use a quantity known as the *signal-to-noise ratio,* or S/N, which tells us the relative size of the desired signal to the underlying noise or background light. The noise is defined as the standard deviation of a single measurement from the mean of all of the measurements made on a star.

Astronomers typically consider a good photoelectric measurement as one that has a signal-to-noise ratio of 100, or in other words, the noise is 1 percent of the signal. For photon arrivals, the statistical noise fluctuation is represented by the Poisson distribution, and for bright sources where the sky background is negligible,

$$\frac{S}{N} = \frac{\text{total received counts}}{\sqrt{\text{total received counts}}}$$

$$\frac{S}{N} = \sqrt{\text{total received counts}} \qquad (3.28)$$

Therefore, for a *S/N* of 100, we must acquire 10,000 counts. A *S/N* of 100 means that the noise causes the counts to fluctuate about the mean

78 ASTRONOMICAL PHOTOMETRY

by an amount equal to one hundredth of the mean value. To compute this error in magnitudes, we compare the mean number of counts, c, to the maximum or minimum values induced by the noise, that is

$$\Delta m = -2.5 \log \left(\frac{c \pm \frac{c}{100}}{c} \right)$$

$$\Delta m = -2.5 \log \left(1 \pm \frac{1}{100} \right)$$

$$= \pm 0.01 \text{ magnitude.}$$

In other words, a S/N of 100 implies an observational error of 0.01 magnitude. More detail on both the signal-to-noise ratio and the Poisson distribution can be found in Appendix K.

3.8 SOURCES ON STATISTICS

Listed below is a sample of statistics and numerical analysis texts that may be of interest to the reader. This list is by no means complete as we have not examined the dozens of available texts, but the sources listed do appear to present the material in a manner useful to the astronomer.

- Bevington, A. R. 1969. *Data Reduction and Error Analysis for The Physical Sciences.* New York: McGraw-Hill. Nice beginning college-level text with FORTRAN programs. Highly recommended.
- Bruning, J. L., and Kintz, B. L. 1977. *Computational Handbook of Statistics.* 2d ed. Glenview, Illinois: Scott, Foresman & Co. No least squares, but takes a computational approach. Includes FORTRAN programs.
- Carnahan, B., Luther, H. A., and Wilkes J. O. 1969. *Applied Numerical Methods.* New York: John Wiley and Sons. One of the best FORTRAN numerical analysis books. Explanation is at college level.
- Ehrenberg, A. S. C. 1975. *Data Reduction: Analyzing and Interpreting Statistical Data.* New York: John Wiley and Sons. Good beginning college text.
- Harnett, D. L. 1975. *Introduction to Statistical Methods.* 2d ed. Reading, Mass. Addison Wesley. Beginning college level.

- Meyer, S. L. 1975. *Data Analysis for Scientists and Engineers.* New York: John Wiley and Sons. Beginning college level.
- Young, H. D. 1962. *Statistical Treatment of Experimental Data.* New York: McGraw Hill. Very nice advanced high school level text with lots of explanations.

CHAPTER 4
DATA REDUCTION

There are three stages in the treatment of photoelectric data, which we call *acquisition, reduction,* and *analysis.* The techniques for data acquisition are presented in Chapter 9 and should be thoroughly studied before raw data are acquired. We treat data reduction now, rather than after Chapter 9, because intelligent data acquisition requires a knowledge of the types of observations necessary for the reduction process. The reduction of data from counts or meter deflections into magnitudes tied to the standard system can be a complicated process, but one that is required by many research projects. Careful reading of this material and the examples in the appendices will enable you to reduce any *UBV* observations and place them on the standard system. Most of the third stage, data analysis, is left up to the individual. Analysis involves the calculation of such quantities as periods, orbital elements, and in general all calculations beyond the determination of magnitudes and colors. The analysis depends greatly on the purpose of the investigation and should be obtained from other sources.

4.1 A DATA-REDUCTION OVERVIEW

You have some raw instrumental measurements of stars and sky background. What are the steps necessary to complete the reduction? There are many different ways that data reduction can proceed. A general outline that fits most situations follows:

1. If you are pulse counting, correct your values to one consistent set, that is, counts per second rather than counts per various arbitrary time intervals. The count rates should be corrected for dead time.

For DC photometry, the amplifier gain settings must be corrected to true gain using the gain table, which is discussed in Section 8.6.
2. Subtract the sky background from each stellar measurement. This must be done before the numbers are converted into logarithmic values (magnitudes).
3. Calculate the instrumental magnitude and colors. For differential photometry, calculate the magnitude differences between the variable and the comparison star.
4. Determine the extinction coefficients and apply the extinction correction. This step is often unnecessary for differential photometry. If you intend to leave your differential photometry on the instrumental system, skip to step 7.
5. Use the standard stars to determine the zero-point constants and, if necessary, the transformation coefficients.
6. Transform your instrumental measurements to the standard system.
7. Estimate the quality of the night by comparing the transformed standard-star magnitudes and colors with their accepted values. For differential photometry, check the reproducibility of the comparison star measurements after correcting them for extinction.
8. Perform any ancillary calculations such as time conversions that are necessary to make your observations useful and publishable.

Steps 1, 2, 3, 5, 6, and 7 are illustrated by a worked example in Appendix H. An example of step 4 is found in Appendix G. Step 8 is covered in detail in Chapter 5. In what follows, we review the concepts and difficulties of some of these steps and present a worked example of the data reduction associated with differential photometry.

4.2 DEAD-TIME CORRECTION

One of the major drawbacks of a pulse-counting system is its inability to count closely spaced pulses with accuracy. After the photomultiplier tube, preamp, or counter detects a pulse, there is a short time interval in which the device is unable to respond to an additional pulse. If two or more pulses arrive at any of the major components in an interval shorter than the so-called *dead time* of the component, these pulses will be detected as a single count. Incident photons from bright sources will on the average be more closely spaced in time than those from fainter sources. But these photons do not arrive in evenly spaced time intervals. From a bright source, four pulses may arrive in the first 10 nanoseconds, none in the next 10 nanoseconds, etc. The manufacturer's specifications on the photomultiplier tube, preamp, or counter dead times should not be relied upon, as those figures are based on evenly spaced, uniform pulses that are never found except in the testing laboratory.

The component with the longest dead time is the major contributor to the inaccuracy, so in general use a counter with at least a 100-MHz response and the fastest preamp possible. At Goethe Link Observatory, the Taylor preamp (see Chapter 7) is the slowest component. In general, the photomultiplier tube dead time is insignificant compared to that of the preamp or counter. The dead-time problem makes pulse counting nonlinear for bright sources.

The equation for the dead-time correction is simple in form, but difficult to solve. The equation can be written as

$$n = Ne^{-tN} \qquad (4.1)$$

where

n = observed count rate in counts per second
N = "true" count rate for a perfect system in counts per second
t = dead-time coefficient defined as $t = 1/N$ when observed count rate falls to $1/e$ of the true count rate.

Equation 4.1 can be rearranged to yield

$$\frac{n}{N} = e^{-tN}. \qquad (4.2)$$

Taking the natural logarithm of each side yields

$$\ln(n/N) = -tN$$

or

$$\ln(N/n) = tN. \qquad (4.3)$$

If we graph $\ln(N/n)$ versus N, then t is the slope of the best-fitted line.

Our problem, though, is that we do not know N and therefore cannot solve for t.

The technique for finding t takes advantage of the fact that for low count rates, the dead-time correction is negligible. Suppose we have some device that can attenuate the light reaching the photomultiplier tube by a known factor, which we will designate as b. Then, when the attenuator is in place, only $1/b$ of the light reaches the photomultiplier tube. (The nature of the attenuator is discussed later.) If we observe a light source or star with the device in place, the observed count rate, n_L, will be low and will very nearly equal the true count rate, N_L. If the attenuator is removed, the *true* count rate, N_H, will increase b times. That is,

$$N_H = bN_L \simeq bn_L. \tag{4.3a}$$

However, the *observed* rate increases by some smaller factor because of dead-time losses. A comparison of the observed rate, n_H, to bn_L gives a measure of the dead-time coefficient. Equation 4.3 can now be rewritten as

$$\ln\left(\frac{bn_L}{n_H}\right) = tbn_L. \tag{4.4}$$

If several light sources of different brightnesses are observed both with and without the attenuator, a plot of $\ln(bn_L/n_H)$ versus bn_L yields a line with slope t.

There are three methods of attenuation commonly used. Each method is considered in turn.

1. *Using aperture stops.* In this method, the telescope is pointed at a bright star and the front of the tube is covered by a piece of cardboard with a small circular opening. The count rate recorded through this opening is n_L. The count rate recorded through a larger opening in a second piece of cardboard is n_H. The factor b is just the ratio of the areas of the two apertures. There is a disadvantage with this method if a reflecting telescope is used. In this case, the apertures must be made small enough to be placed off the optical axis so that the secondary mirror support does not

block incoming light. If this is not done, then one must be careful to account for the area of the secondary support in the calculation of b. This may not be difficult to do if the mirror cell has a circular cross section and the support vanes are thin enough to ignore.

2. *Using the photometer diaphragms.* In this method, the light source cannot be a star. It must be an extended object with uniform surface brightness. For this purpose, a uniformly illuminated white card can be placed in front of the telescope. The diaphragms in the photometer head can then be used, a small one to measure n_L and a large one to measure n_H. The ratio of the diaphragm areas is b. The problem with this method is that the card must be uniformly illuminated, the diaphragm sizes must be known accurately, and the light source must be variable if more than one observation is to be made with the available diaphragms. The daytime sky can be used as the light source if extreme care is used to prevent current overload of the photomultiplier tube.

3. *Using a "neutral" density filter.* This method uses a neutral density filter placed in the filter slide. The star need only be measured once with and once without the filter in the light path. The density of the filter is then b. The problem with this method is that no filter is truly "neutral." This means that the amount of light transmitted by the filter depends, at least weakly, on the color of the star. This would seem to make this method very cumbersome. However, once this color effect is calibrated, the neutral density filter offers a convenient way to find the dead-time coefficient. To calibrate the color dependence of the filter, several stars of widely different colors are observed both with and without the filter. The stars selected (see Appendix C) should be relatively faint so that the dead-time correction is negligible. For each star, a magnitude difference $(v_1 - v_0)$ is calculated by

$$v_1 - v_0 = -2.5 \log (n_1/n_0), \qquad (4.5)$$

where n_1 is the count rate with the filter in place and n_0 is the rate without the filter. Then $(v_1 - v_0)$ is plotted versus $(B - V)$ for each star. The resultant graph should be a nearly horizontal line. In the case of a filter used at Indiana University, a least-squares fit resulted in

$$v_1 - v_0 = -0.008 (B - V) + 3.934.$$

It can be seen that this relation depends very weakly on color. The factor, b, is the ratio of light intensity without the filter to that transmitted. That is,

$$v_1 - v_0 = -2.5 \log (I_1/I_0) \qquad (4.6)$$

or

$$b = I_0/I_1 = 10^{0.4(v_1 - v_0)}. \qquad (4.7)$$

The dead-time coefficient can then be found using the observations of bright stars.

An example of a dead-time coefficient determination can be found in Appendix F. Even when t is known for a given count rate, an iterative technique is required to solve Equation 4.1. First substitute the observed count rate, n_0, for N and calculate a corrected count rate, n_1. This new value is then substituted for N and the process is repeated until n_k approaches n_{k+1} to the accuracy required. A FORTRAN subroutine to iterate this equation is given in Section I.1, in addition to an example shown in Appendix H.

4.3 CALCULATION OF INSTRUMENTAL MAGNITUDES AND COLORS

Section 1.7 derives the equations necessary to formulate instrumental magnitudes and colors, along with a physical understanding of the constants involved. Writing Equations 1.10 and 1.12 explicitly for the *UBV* system, we have

$$v = c_v - 2.5 \log d_v \qquad (4.8)$$
$$b - v = c_{bv} - 2.5 \log (d_b/d_v) \qquad (4.9)$$
$$u - b = c_{ub} - 2.5 \log (d_u/d_b), \qquad (4.10)$$

which relate the observed deflections or counts, d, to the instrumental magnitudes and colors. Because these instrumental values are used in Section 4.5 to evaluate the zero-point shifts in the transformation equations, the constants, the c's, are arbitrary. For DC work, the usual forms of Equations 4.8 through 4.10 are:

$$v = -2.5 \log (d_v) + G_v \qquad (4.11)$$
$$b - v = -2.5 \log (d_b/d_v) + G_b - G_v \qquad (4.12)$$
$$u - b = -2.5 \log (d_u/d_b) + G_u - G_b \qquad (4.13)$$

where G is the relative gain for each filter and d is the chart-recorder deflection at that gain setting. For pulse counting, the equations become:

$$v = -2.5 \log (\dot{C}_v) \qquad (4.14)$$
$$b - v = -2.5 \log (\dot{C}_b/\dot{C}_v) \qquad (4.15)$$
$$u - b = -2.5 \log (\dot{C}_u/\dot{C}_b) \qquad (4.16)$$

where $\dot{C}$ is the count rate in counts per second through each filter. Worked examples of both forms of magnitude calculations can be found in Appendix H.

4.4 EXTINCTION CORRECTIONS

Remember that extinction represents the loss of starlight while traversing the earth's atmosphere. All published photometric results correct for this and essentially give the apparent magnitude of the star outside of the earth's atmosphere, called the *extra-atmospheric magnitude*. The equations discussed in Section 1.8 and derived in Appendix J are the basis for the treatment of extinction. Much of the material in this section makes use of the results obtained by Hardie.[1] Most extinction corrections account for first-order extinction, along with the associated air mass calculation. For greater accuracy, second-order extinction should be taken into account.

4.4a Air Mass Calculations

At altitudes more than 30° above the horizon, the simple plane-parallel approximation, derived in Appendix J, for the amount of atmosphere between an observer and a star is accurate to within 0.2 percent. When a star's altitude is greater than 30° or correspondingly, the zenith distance, z, is less than 60°, this approximation gives

$$X = \sec z \qquad (4.17)$$

where

$$\sec z = (\sin \phi \sin \delta + \cos \phi \cos \delta \cos H)^{-1} \quad (4.18)$$

where ϕ is the observer's latitude, δ the declination of the star, and H is its hour angle in degrees. The mass of the air traversed by the starlight is X. This quantity is at a minimum when a star is directly overhead, or

$$\sec z = \sec 0° = 1.$$

This amount of air is called one *air mass* for convenience, rather than trying to remember some large-numbered column density.

For zenith distances greater than 60°, the plane-parallel approximation breaks down. An equation that more closely approximates the effects of the spherical earth must be used. The most common polynomial approximation was made to data collected by Bemporad in 1904 and is given in Hardie[1] as:

$$X = \sec z - 0.0018167 (\sec z - 1) - 0.002875 (\sec z - 1)^2 \\ - 0.0008083 (\sec z - 1)^3 \quad (4.19)$$

where z is the *apparent*, not true, zenith distance. Equation 4.19 fits Bemporad's data to better than 0.1 percent at an air mass of 6.8. That is only 10° from the horizon and closer than you would want to observe. Because these data represent only average sky conditions at one location at the turn of the century, we cannot expect that the actual accuracy of the air mass calculation is 0.1 percent.

Other methods of determining air mass involve the use of tables or nomographs and are not presented here because they are less practical than solving the equations above with a scientific calculator.

Example: An observer located at 40° north latitude locates Sigma Leo (RA = $11^h20^m6^s$, Dec. = 6°8'21") at an apparent hour angle of 3^h, which is 45°. What is the air mass between the observer and Sigma Leo?

From Equation 4.18, we have:

$$\sec z = [\sin(40°)\sin(6.1392°) + \cos(40°)\cos(6.1392°)\cos(45°)]^{-1}$$
$$\sec z = 1.6466$$

From Equation 4.17, we have:

$$X = \sec z = 1.6466$$

For comparison, we can now use Equation 4.19 to improve our accuracy:

$$X = 1.6466 - 0.0018167(0.6466) - 0.002875(0.6466)^2 - 0.0008083(0.6466)^3$$
$$X = 1.6440$$

Notice that the two methods agree to about 0.1 percent.

4.4b First-order Extinction

Extinction is very difficult to model exactly because of the many variables that play important roles in the absorption of light in the earth's atmosphere. To do a first-order approximation is to account for the largest contributor, the air mass variation. In this approximation, the following equations hold:

$$v_0 = v - k'_v X \quad (4.20)$$
$$(b - v)_0 = (b - v) - k'_{bv} X \quad (4.21)$$
$$(u - b)_0 = (u - b) - k'_{ub} X, \quad (4.22)$$

where k' is the principal extinction coefficient in units of magnitudes per air mass and the subscript 0 is used to denote an extra-atmospheric value. Rearranging these three equations, we obtain

$$v = k'_v X + v_0 \quad (4.23)$$
$$(b - v) = k'_{bv} X + (b - v)_0 \quad (4.24)$$
$$(u - b) = k'_{ub} X + (u - b)_0. \quad (4.25)$$

The values of the extinction coefficients can then be found by following one star through changing air masses and plotting the color index or magnitude versus X. The slope of the line is the extinction coefficient and the intercept the extra-atmosphere magnitude or color index.

What has been presented is an ideal case. In reality, by the time the air mass in the direction of a star has changed appreciably, the sky may have undergone considerable change. Even if the atmosphere is static, the extinction in various parts of the sky is not constant. The change may be due to a local fluctuation, giving rise to scatter about the mean curve, or it may be on a large scale, such as an east-west variation. Moreover, the atmosphere varies daily depending on its moisture and dust content. As is explained in Section 4.4c another complication arises as a result of the extinction itself being color-dependent.

Do not be discouraged by all of the problems discussed above. The accurate measurement of extinction is a tough problem and if determined to high precision would leave little time for the actual observations! Therefore, knowing the value of the extinction coefficients to an accuracy of 2 or 3 percent is considered acceptable by most professional astronomers. Observational recommendations for extinction can be found in Chapter 9. Several methods of determining the principal extinction coefficients are available. The two most common are using comparison stars, and using a sample of A0 stars near your variables. These are discussed below.

Because of the spatial variation possible in extinction, the best possible choice for your extinction star is the comparison star for the observed variable. This choice has two advantages: the extinction measurements are never far in time or space from the variable star values, and if you are using comparison stars of the same color as your variables, second-order effects become negligible. This method is only suited to observing programs in which a single variable is observed most of the night. This is the only way to collect enough measurements of the comparison star, over a wide range of air mass, in order to determine the extinction coefficients.

Another method of determining the principal extinction coefficients is to use a sample of A0 stars covering the region of the sky in which you are observing. The extinction can then be determined by using least-squares analysis directly. This technique has the advantage of requiring a smaller amount of observing time. However, the analysis is more complicated and if a computer program is used, there is a temp-tation not to plot the measurements. In which case, one bad measurement because of haze moving in during your observations may not be apparent, yet it can affect your results drastically.

*4.4c Second-order Extinction

If we include the color dependence of the extinction coefficients, as explained in Section J.3, then we can modify the principal coefficients to

$$k'_v \rightarrow k'_v + k''_v(b - v) \qquad (4.26)$$
$$k'_{bv} \rightarrow k'_{bv} + k''_{bv}(b - v) \qquad (4.27)$$

Equations 4.20 and 4.21 become

$$v_0 = v - k'_v X - k''_v(b - v)X \qquad (4.28)$$
$$(b - v)_0 = (b - v) - k'_{bv}X - k''_{bv}(b - v)X. \qquad (4.29)$$

Equation 4.22 remains the same because of the definition of k''_{ub} as zero. We can solve Equations 4.28 and 4.29 for the second-order coefficients by using a close optical pair having very different colors. Because of their proximity, the air mass remains constant between them and we can obtain

$$v_{01} - v_{02} = (v_1 - k'_v X - k''_v(b - v)_1 X)$$
$$- (v_2 - k'_v X - k''_v(b - v)_2 X)$$

or

$$\Delta v_0 = \Delta v - k''_v \Delta(b - v)X \qquad (4.30)$$
$$\Delta v = k''_v \Delta(b - v)X + \Delta v_0.$$

Similarly,

$$\Delta(b - v) = k''_{bv} \Delta(b - v)X + \Delta(b - v)_0 \qquad (4.31)$$

where Δ indicates the difference in colors or magnitudes of the two stars at each air mass. The solution of Equations 4.30 and 4.31 is performed easily by plotting Δv or $\Delta(b - v)$ versus $\Delta(b - v)X$; the slope is then the second-order extinction coefficient. Each pair so measured can also be used to determine the principal coefficients once the second-order values are known.

From experience and theory, the second-order coefficient for v is essentially negligible. In addition, the values found through the use of Equation 4.31 have been found to be relatively constant and probably do not need to be determined any more often than are the color transformation coefficients.

An extinction example for both principal and second-order extinction can be found in Appendix G. Suitable pairs of stars can be found in Appendix B. Most of these pairs are from a list prepared by Kitt Peak National Observatory and are near the equator. Note that the red stars are quite often faint and are more difficult to measure. Bright pairs are just hard to find!

4.5 ZERO-POINT VALUES

From Chapter 2, we have the equations

$$V = \epsilon(B - V) + v_0 + \zeta_v \quad (4.32)$$
$$(B - V) = \mu(b - v)_0 + \zeta_{bv} \quad (4.33)$$
$$(U - B) = \psi(u - b)_0 + \zeta_{ub}. \quad (4.34)$$

These three equations are the working equations for this and the following sections. Rearranging,

$$\zeta_v = V - v_0 - \epsilon(B - V) \quad (4.35)$$
$$\zeta_{bv} = (B - V) - \mu(b - v)_0 \quad (4.36)$$
$$\zeta_{ub} = (U - B) - \psi(u - b)_0. \quad (4.37)$$

In other words, the zero point is equal to the standard value minus the transformed extra-atmospheric value. The zero points are calculated by solving Equations 4.35 to 4.37 for each standard and then taking means.

Figure 4.1 plots the three zero points determined with pulse-counting equipment at Goethe Link Observatory over an 18-month period. The break in ζ_v occurred when the mirrors were realuminized. It indicates an approximate 0.5 magnitude gain in sensitivity. The scatter is caused

92 ASTRONOMICAL PHOTOMETRY

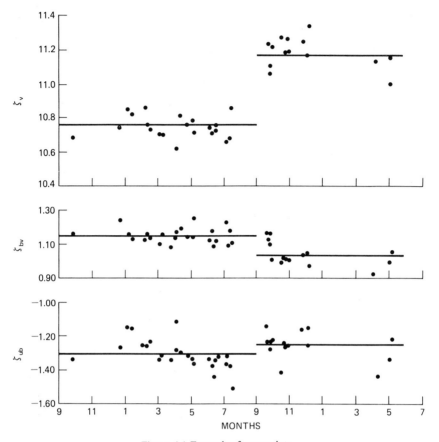

Figure 4.1 Example of zero points.

primarily by the small number of standard stars used in the zero-point calculation and by differing sky transparency. The zero-point values must be determined nightly.

4.6 STANDARD MAGNITUDES AND COLORS

Once the transformation coefficients have been determined, along with the nightly values of extinction and the zero-point shifts, Equations 4.28, 4.29, and 4.22 can be used to determine the extra-atmospheric instrumental magnitudes and colors. Substituting these values into Equations 4.32 through 4.34 yields the transformed standard magnitudes and colors. These values may not agree with accepted values for a constant star because of statistical scatter and the quality of the night, but means determined over several nights should yield good numbers.

4.7 TRANSFORMATION COEFFICIENTS

The transformation coefficients defined in Chapter 2 could be determined directly from Equations 4.32 to 4.34 by measuring several stars whose standard magnitudes and colors are known. In fact, that is the approach used for determining the V coefficient, ϵ:

$$V - v_0 = \epsilon(B - V) + \zeta_v. \tag{4.38}$$

The slope of the best-fitted line for a plot of $(V - v_0)$ versus $(B - V)$ will be the coefficient ϵ. Note that you are plotting the difference between the two magnitudes; this is more accurate than plotting V versus v_0 because it magnifies small variations in either number. For instance, the change in v from 8.79 to 8.80 is less than 1 percent. But if $V = 8.85$, then $V - v_0 = 0.06$ in one case and 0.05 in the other, a difference of 20 percent.

However, our two equations for the color indices are not in differential form. They can be converted to differential measurements by solving for the extra-atmospheric instrumental colors as shown for $(b - v)$:

$$(b - v)_0 = \frac{(B - V) - \zeta_{bv}}{\mu}. \tag{4.39}$$

Subtracting the extra-atmospheric instrumental value, $(b - v)_0$, from both sides of Equation 4.33;

$$\begin{aligned}(B - V) - (b - v)_0 &= \mu(b - v)_0 - (b - v)_0 + \zeta_{bv} \\ &= (\mu - 1)(b - v)_0 + \zeta_{bv}\end{aligned}$$

and then substituting Equation 4.39 into the right-hand side, yielding

$$(B - V) - (b - v)_0 = \left(1 - \frac{1}{\mu}\right)(B - V) + \frac{\zeta_{bv}}{\mu}, \tag{4.40}$$

94 ASTRONOMICAL PHOTOMETRY

and similarly for $(U - B)$,

$$(U - B) - (u - b)_0 = \left(1 - \frac{1}{\psi}\right)(U - B) + \frac{\zeta_{ub}}{\psi}. \quad (4.41)$$

Equations 4.40 and 4.41 along with Equation 4.38 are our working equations for determining the transformation coefficients. Plots of the left-hand sides of these equations versus either $(B - V)$ for Equations 4.38 and 4.40, or $(U - B)$ for Equation 4.41, yield slopes that are related to the transformation coefficients.

In practice, there are two methods commonly used to solve for the transformation coefficients:

1. Choose several standards from the Johnson standard list in Appendix C, measure them, and determine the transformation coefficients. With this method, you are able to select bright stars of widely differing colors. It has the disadvantage of requiring accurate knowledge of the extinction coefficients. Select at least 10 and preferably 20 or more standard stars with a wide range of $(B - V)$ and $(U - B)$ colors and try to observe them when they are near the zenith. Note that the Johnson list has internal errors of $\pm 0\overset{m}{.}02$, so do not expect to achieve results that are significantly more accurate.
2. Use one of the standard clusters from Appendix D for your standards. This eliminates the extinction problem, but adds the problem of faintness. Not only are most clusters fourth to sixth magnitude or fainter, but also the red stars are fainter than the hot blue stars of the cluster and add more error to the determination. Another problem is systematic errors in the cluster standard values that can make the coefficient determination from one cluster different from another. This is a minor effect, but it should be kept in mind.

We suggest that you use method 1 with a telescope aperture of 25 centimeters (10 inches) or less. For larger telescopes, use method 2. The normal procedure is to determine carefully the transformation coefficients at the beginning and end of your observing season, as well as once or twice in between. Resolve to spend half of a good night for each of these determinations. The coefficients change very slowly with time, and mean values are generally sufficient. For method 2, the different determinations should be made using different clusters, if possible. Transformation examples using method 1 for DC photometry and method 2 for pulse counting are given in Appendix H.

4.8 DIFFERENTIAL PHOTOMETRY

The reduction of differential photometry data is treated somewhat differently from the descriptions found in previous sections. Because differential photometry is the starting point for most newcomers to photometry, we explain the reduction process in detail. We assume that the observations were made in accordance with the recommendations of Section 9.3c. Variable star observations are bracketed by those of the comparison stars.

Table 4.1 contains a partial list of observations of the eclipsing binary UZ Leonis. In this example, a DC photometer was used. Columns 1, 2, and 3 contain the object's name, universal time of observation, and the filter designation. Column 4 contains the amplifier gain in magnitudes. Columns 5 and 6 contain the chart recorder pen deflection of the star and sky through each filter.

1. The first step is to subtract the sky background from each stellar measurement. The results appear on column 7. If a pulse-counting photometer had been used, and if the stars were fairly bright, a dead-time correction would be applied (Section 4.2) before subtracting the sky background.
2. The observations in column 7 follow a pattern such as

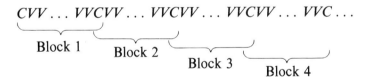

where C and V represent a measurement through each filter of the comparison and variable star, respectively. The data have a block structure where each string of several variable star measurements is sandwiched between two comparison star measurements. Each block is reduced separately by averaging the comparison star measurement at the beginning and end of the block. Equations 2.38 through 2.40 or 2.35 through 2.37 are used to pro-

TABLE 4.1. Differential Photometry Data

Object	UT	Filter	Gain	Star	Sky	Net	Δv	Δb	Δu	ΔV	$\Delta(B-V)$	$\Delta(U-B)$
Comp.		V	10.505	40.2	12.0	28.2						
		B	10.505	50.0	12.0	38.0						
		U	11.938	44.3	24.0	20.3						
UZ Leo	2:40	V	11.034	41.2	15.1	26.1	0.628			0.628		
		B	11.034	48.4	15.1	33.3		0.689			0.057	
		U	12.432	50.7	33.6	17.1			0.717			0.033
	2:42	V	.	41.1		26.0	0.633			0.633		
		B	.	48.4		33.3		0.689			0.052	
		U	.	50.3		16.7			0.743			0.064
	2:45	V		41.5		26.4	0.616			0.616		
		B		48.8		33.7		0.676			0.056	
		U		51.0		17.4			0.698			0.026
Comp.		V	10.505	41.0	12.0	29.0						
		B	10.505	51.2	12.0	39.2						
		U	11.938	46.2	24.5	21.7						
UZ Leo	2:59	V	11.034	41.8	15.3	26.5	0.634			0.634		
		B	11.034	49.2	15.3	33.9		0.687			0.049	
		U	12.432	51.4	34.6	16.8			0.762			0.075
	3:01	V	.	42.0		26.7	0.626			0.626		
		B	.	49.5		34.2		0.677			0.047	
		U	.	52.5		17.9			0.693			0.016
	3:04	V		42.1		26.8	0.622			0.622		
		B		50.0		34.7		0.661			0.036	
		U		52.5		17.9			0.693			0.038
Comp.		V	10.505	41.2	11.8	29.4						
		B	10.505	51.0	11.8	39.2						
		U	11.938	45.3	24.0	21.3						

Average: $\Delta V = 0.626 \pm 0.007$ (s.d.)
$\Delta(B-V) = 0.049 \pm 0.007$
$\Delta(U-B) = 0.042 \pm 0.024$

Block 1 spans Comp. (first) through UZ Leo 2:45.
Block 2 spans Comp. (second) through Comp. (last).

duce a value of Δv, Δb, Δu for each variable star measurement within the block. The process is then repeated until all the blocks have been reduced. Note that each comparison star measurement does "double duty" because the last comparison star measurement in one block is also the first one in the next block. Columns 8, 9, and 10 contain the calculated magnitude differences. Check a few of these entries with your calculator.

In many cases, reduction stops at this point because extinction corrections are often ignored in differential photometry. Conversion to the standard photometric system is unnecessary for many types of research projects. However, if the variable and comparison are separated by more than a degree it may be wise to apply an extinction correction. A worked example of this correction to differential photometry can be found in Appendix G. The data in Table 4.1 do not require this correction.

If the transformation coefficients are known, it is possible to convert the magnitude differences from the instrumental to the standard system.

3. In this example, $\epsilon = -0.004$, $\mu = 0.927$, and $\psi = 1.178$. Equations 2.41 and 2.42 were used to compute $\Delta(b - v)$ and $\Delta(u - b)$ for each stellar measurement. Equation 2.49 and 2.50 can then be used to compute $\Delta(B - V)$ and $\Delta(U - B)$. Finally, Equation 2.48 can be used to compute ΔV. Check a few of the entries in columns 11 through 13. Note that the difference between the magnitude and colors in the instrumental and standard systems is practically negligible. This means that the detector and filters used match the standard system well.

If the comparison star has been standardized, Equations 2.51 through 2.53 can be used to calculate the actual magnitude and color of the variable. In this particular example, the comparison star was standardized on a previous night with the following results:

$$V = 8.950 \pm 0.038 \text{ (s.d.)}$$
$$(B - V) = 0.287 \pm 0.012$$
$$(U - B) = 0.039 \pm 0.058.$$

If we average the values of ΔV, $\Delta(B - V)$, and $\Delta(U - B)$ in columns 11, 12, and 13 we can compute the magnitude and colors of UZ Leonis to be

$$V = 8.95 + 0.63 = 9.58$$
$$(B - V) = 0.29 + 0.05 = 0.34$$
$$(U - B) = 0.04 + 0.04 = 0.08.$$

The probable errors for these values are found by adding the standard deviations of the comparison and variable star in quadrature, that is

$$\text{p.e.} = 0.675 \sqrt{\sigma_{comp}^2 + \sigma_{var}^2}.$$

The final quoted results are then

$$V = 9.58 \pm 0.03 \text{ (p.e.)}$$
$$(B - V) = 0.34 \pm 0.01$$
$$(U - B) = 0.08 \pm 0.04.$$

The larger error in $(U - B)$ reflects the fact that both the comparison star and UZ Leonis are very faint in the U filter.

*4.9 THE $(U - B)$ PROBLEM

In Chapter 2, the second-order extinction coefficient for $(U - B)$ was arbitrarily defined as zero. However, k''_{ub} can be a larger correction than k''_{bv}, because the u extinction depends on:

1. the Balmer discontinuity
2. the second-order color term
3. systematic nonlinear deviations resulting from the assumption that k''_{ub} was constant in the the original UBV data.

Because of these problems, the $(U - B)$ color term is inaccurate and poorly defined. Unfortunately, use of the existing system is so traditional that it would be extremely difficult to redefine the UBV system. The best solution is to calculate your $(u - b)_0$ values correctly, accounting for all first- and second-order effects, and then transform your ubv data to the existing, but nonideal, standard UBV system.

The remainder of this section presents one such method of transformation as derived by Moffat and Vogt.[2] This kind of correction is complicated and should only be attempted by those who are experienced in photometry.

Moffat and Vogt found that, for any given star, the residuals in $(U - B)$ vary linearly with air mass, X. That is,

$$\Delta[(U - B) - (u - b)] = \gamma_1 + \gamma_2 X. \tag{4.42}$$

A plot of this equation indicates that γ_1 and γ_2 are linearly related. That is,

$$\gamma_2 = \beta \gamma_1$$

where

$$\beta \simeq -0.27.$$

We can therefore rewrite Equation 4.42 as

$$\Delta[(U - B) - (u - b)] = \gamma_1(1 + \beta X) \tag{4.43}$$

where β is a constant and γ_1 is a function of spectral type or color. The problem of correcting for $(U - B)$ differences is then reduced to one of determining γ_1. However, γ_1 is a nonlinear function of $(U - B)$ and, in addition, is not a unique function of $(U - B)$. But it is a linear and unique function of another parameter, q, that is similar to the reddening-free parameter of Johnson and Morgan:[3]

$$\gamma_1 = \rho q \tag{4.44}$$

where

$$q = (U - B) - 1.05(B - V). \tag{4.45}$$

The procedure to follow in correcting $(U - B)$ is:

1. For each standard star, determine q from Equation 4.45.

2. For each standard, determine γ_1 from

$$\gamma_1 = \frac{(U - B) - (u - b)}{1 + \beta X} \qquad (4.46)$$

3. Plot γ_1 versus q and determine ρ from Equation 4.44.
4. For all future stars, correct $(U - B)$ by

$$(U - B) = \psi(u - b)_0 + \zeta_{ub} + \rho q (1 + \beta X) \qquad (4.47)$$

This procedure will reduce the mean external error in $(U - B)$ to about $0^m.02$. Without it, the mean error would be approximately three times higher. Include k''_{ub} in all equations in a similar manner as k''_{bv} is included in $(B - V)$ equations.

REFERENCES

1. Hardie, R. H. 1962. In *Astronomical Techniques*. Edited by W. A. Hiltner. Chicago: Univ. of Chicago Press, chapter 8.
2. Moffat, A. F. J., and Vogt, N. 1977. *Pub. A.S.P.* **89**, 323.
3. Johnson, H. L., and Morgan, W. W. 1953. *Ap. J.* **117**, 313.

CHAPTER 5
OBSERVATIONAL CALCULATIONS

There are a number of calculations that are useful for obtaining and reducing observational data. These include determining when an object appears above the horizon on a given night, precession of coordinates, and the calculation of date and time quantities. These and other calculations are discussed in this chapter. The subjects are recommended reading even if photometry is not attempted, as they are also involved in most visual observations.

5.1 CALCULATORS AND COMPUTERS

Observational calculations and data reduction are very tedious without the use of a calculator or computer. The scientific calculator has now been in existence for a decade. Since the introduction of the Hewlett-Packard HP-35, calculators have made great strides in capability with a reduction in cost. Programmable varieties are excellent and card-programmables are the ultimate, as programs and data can be stored for later recall. The mode of operation (whether reverse Polish notation (RPN), algebraic, or even a high-level language such as BASIC) is unimportant as long as you are comfortable with your choice. We suggest the use of a programmable calculator with seven to 10-digit accuracy and the capability of converting degrees-minutes-seconds into decimal degrees, a real blessing to astronomers! The well-known brands, such as Hewlett-Packard and Texas Instruments, should be your first choice. A review of calculators is given in *Sky and Telescope*.[1]

A calculator is all that is required to perform the data reduction. However, some people may prefer something more advanced. The next step up from a programmable scientific calculator is the microcomputer. Large-scale integration has advanced sufficiently that the purchase of a microcomputer should be considered if much data reduction or instrumentation control is anticipated. The eight-bit microcomputer is the industry standard at this time. The advantage of the eight-bit machines based on the Intel 8080 or the Motorola 6800 is that large amounts of peripheral computing power is available. Most peripherals seem to be designed for either the 8080/S-100 or 6502/Apple® busses, and use of one of these busses would probably be the best approach to take for the next few years.

The disadvantages of an eight-bit microcomputer are numerous. The heart of it was designed for instrument control in industry, not for number manipulation. It is slow and does not have hard-wired arithmetic (except for integer addition and subtraction). The ability to accept or transmit only eight bits of data at a time limits high-speed resolution to only 256 steps. For instance, if a photometer acquires 10,433 counts per second, and dumps data to the computer once per second, these data must be brought into the computer in eight-bit chunks rather than all at once. Also, performing floating-point arithmetic with software is slow and occupies valuable storage space.

The new 16-bit microprocessors, such as the Intel 8086, the Zilog Z8000, and the Motorola 68000, are designed to be computer-processing units and have much more capability in instrument control and data reduction as well. These new chips have the potential of revolutionizing the microcomputer market. Software and accessories may not appear on the market for some time, though, and proper use of these microprocessors may require more knowledge of computer software and hardware than you wish to acquire.

Arithmetic operations used in data reduction can be performed in software at the expense of speed and space. Since most eight-bit microprocessors do not have even hardware integer multiply and divide, extensive data reduction using software emulation can be very slow. A better approach is to use a hardware arithmetic chip to perform the operations for you. The American Micro Devices AM-9511 and AM-9512 arithmetic chips are approximately 100 times faster than 8080 emulation and should be definitely considered if you intend to spend a large amount of computer time reducing your data. An even faster chip is available from Intel as a co-processor for the 16-bit 8086. Called the 8087, it performs double precision floating point additions in 14 microseconds and overall has approximately 1000 times the throughput of 8080 software emulation.

If you wish to purchase a microcomputer, we suggest the following:

1. Get a mainframe using the S-100 bus from a reputable manufacturer. This bus is currently used for most scientific and commercial applications because of the wide variety of processors and peripherals available on it. In addition, it will probably be used by most of the 16-bit processors, so upgrading in the future will be relatively simple. A less radical approach for the beginner is to purchase an Apple computer. Its bus structure will allow the addition of peripheral units at a later date. The Apple® and other computers such as from Radio Shack, Commodore, and Heathkit are attractive because of their large number of owners and service centers, and should be seriously considered if future expansion is not foreseen.
2. Use a high-level language to do your programming, such as a good FORTRAN, PASCAL, BASIC, or FORTH, preferably compiled rather than interpreted and with relocatable object code. Your computer should have the capability of being programmed in machine language for dedicated tasks or input/output commands.
3. Have some type of mass storage device. Cassette recorders can be used. Though much cheaper, they are slower and less reliable than disk drives or computer tape decks. A magnetic disk drive is the best method of mass storage. Because the drive is mostly mechanical, you pay for mechanical reliability.
4. A hard-copy device like a line printer or a teletype is extremely handy. Again, this is mechanical and there will be less chance of mechanical failure with devices with fewer moving parts. Simple units use the Diablo or Cume print wheels and are quite desirable for word processing as well. Dot matrix printers are useful in that most incorporate dot-addressable graphics. Converted IBM Selectrics give a nice copy, but have hundreds of moving parts and require many adjustments.

At present, a very good microcomputing system can be purchased for around $2500. Granted, this is expensive, but the cost is still decreasing and some of this cost is defrayed by the fact that the computer can

104 ASTRONOMICAL PHOTOMETRY

replace considerable equipment, such as the counter and magnetic tape unit of the photon-counting system. Find a computer store and ask the dealer to help you in your computer selection.

Another approach to data reduction involves the use of a computing center found at many universities. Here all of the hardware is maintained for you, and high-speed, reliable, and efficient programming languages and equipment are available. In addition, computing centers may have programming support and plotting or graphics capabilities on hand. Some centers allow outsiders to use the system at minimal cost if it is being used for research. Approach the center directly, or talk to someone in the astronomy, physics, or computer science departments about your needs and desires.

5.2 ATMOSPHERIC REFRACTION AND DISPERSION

When we observe the sun and stars near the horizon, the atmosphere bends the rays and makes the object appear higher in the sky than it really is, as shown in Figure 5.1. This effect reaches a maximum on the horizon, where an object appears to be 35 arc minutes above its actual location. This means that when the sun appears to touch the horizon, in reality it has already set! Atmospheric refraction affects images in three ways: it changes the measured zenith angle and therefore the air mass, it disperses the image so that each star looks like a miniature spectrum, and it changes the apparent right ascension and declination of a star. These effects must be accounted for accurately when viewing objects near the horizon.

5.2a Calculating Refraction

The simplest method of calculating refraction is to assume a plane-parallel atmosphere made of layers, each with a differing index of refrac-

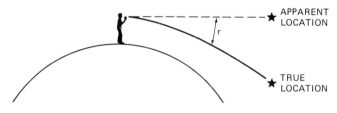

Figure 5.1 Refraction.

tion decreasing uniformly outward. Using Snell's law at each boundary, we find the angle of refraction, r, is approximately equal to

$$r = 60\rlap{.}{''}4 \tan z_{tr}, \tag{5.1}$$

where z_{tr} is the true zenith distance.

This equation has an error of about 1 arc second at $z = 60°$. An empirical improvement was derived by Cassini and Bessel in the seventeenth century and is of the form

$$r = 60\rlap{.}{''}4 \tan z_{tr} - 0\rlap{.}{''}06688 \tan^3 z_{tr}. \tag{5.2}$$

Equation 5.2 is accurate to better than 1 arc second at $z = 75°$, or $15°$ above the horizon. A more accurate equation which accounts for atmospheric pressure, temperature, and the observer's elevation is given by Doggett et al.[2] A short listing of the refraction correction is presented in Table 5.1. Remember that this correction is subtracted from the true zenith distance to get the apparent zenith distance, z_{ap}.

TABLE 5.1. Atmospheric Refraction (760 mm Hg, 10°C, 5500 Å)

$z_{ap}°$	r''	$z_{ap}°$	r''	$z_{ap}°$	r''
5	5	58	93	75	214
10	10	59	97	76	229
15	16	60	101	77	247
20	21	61	105	78	267
25	27	62	109	79	291
30	34	63	114	80	319
35	41	64	119	81	353
40	49	65	124	82	393
45	59	66	130	83	444
46	60	67	136	84	508
48	65	68	143	85	592
50	69	69	151	86	704
52	74	70	159	87	865
54	80	71	168	88	1105
55	83	72	177	89	1494
56	86	73	188	89.5	1790
57	89	74	200	90	2189

Example: The true zenith distance of BX And was determined to be 69°. What is its apparent location?

1. From Equation 5.1 we have:

$$r = 60\rlap{.}''4 \tan(69°)$$
$$= 157\rlap{.}''3$$
$$r = 2'37''$$
$$z_{ap} = z_{tr} - r \qquad (5.2a)$$
$$= 69° - 2'37''$$
$$z_{ap} = 68°57'23''$$

2. From Equation 5.2:

$$r = 60\rlap{.}''4 \tan(69°) - 0\rlap{.}''06688 \tan^3(69°)$$
$$r = 156\rlap{.}''2$$
$$z_{ap} = 68°57'24''$$

3. From Table 5.1:

$$r = 143''$$
$$z_{ap} = 68°57'37''$$

Because z_{ap} approximately equals z_{tr}, you can use either value in Equation 5.1 or 5.2 with minimal error.

5.2b Effect of Refraction on Air Mass

If you use the apparent zenith distance, the air mass calculated from either Equation 4.18 or 4.19 will be correct. However, you must assume that the hour angle setting circle is correct, or you must include the refraction numerically in the hour angle calculation. Table 5.2 shows the error involved in neglecting refraction at several zenith distances. For z greater than 60°, the error is significant enough that it cannot be ignored. Generally, you can neglect refraction above an altitude of 30°, but be sure to include the correction at lower altitudes.

TABLE 5.2. Refraction Air Mass Errors

z_{ap} °	X_{ap}	X_{tr}	%
0	1.000	1.000	0.00
30	1.154	1.154	0.00
60	1.994	1.996	0.10
65	2.356	2.359	0.13
70	2.904	2.910	0.21
75	3.816	3.830	0.37
80	5.598	5.645	0.83
85	10.211	10.468	2.46

5.2c Differential Refraction

The index of refraction for glass, air, or any material is not constant with wavelength. This spreads the light from a star into a miniature spectrum, as if the earth's atmosphere were a prism. This dispersion is obvious when looking at stars near the horizon, as they appear blue on the top and red below. It also gives rise to the "green flash" of the setting sun. Table 5.3 gives the separation angle of the red and blue images at various zenith distances. For z greater than 75°, the images are far enough apart that they are no longer centered in a small diaphragm. They cause erroneous measurements unless a correction is made. In other words, do not observe within 15° of the horizon unless it is absolutely necessary! A secondary effect of differential refraction is that the red and blue rays that make up the observed stellar image are separated by the earth's atmosphere and thus give rise to a color-dependent scintillation, manifested in red and blue flashes. More detail on various

TABLE 5.3. Red and Blue Image Separation

z	r''
0	0.00
30	0.35
45	0.60
60	1.04
75	2.24
90	29.00

refractive effects can be found in Tricker[3] or Humphreys.[4] Both make quite interesting reading.

5.3 TIME

Time is the most accurate piece of data the astronomer has and should be treated accordingly. A 1 percent error in the determination of the magnitude of a star is considered excellent, yet a digital watch has an accuracy of 1 second in a day (0.001 percent), or 1000 times more accurate.

Because astronomers are located worldwide and have been observing for centuries, certain conventions are observed in order that measurements made by two different observers can be correlated with minimal effort.

This section assumes some knowledge of the various time systems involved: solar time, sidereal time, and so on. More detail can be found in any good introductory astronomy text such as Abell.[5]

5.3a Solar Time

An *apparent solar day* is defined as the length of time between two successive transits of the sun, from astronomical noon until astronomical noon the following day. The length of time is dependent on three factors: the rotation of the earth on its axis, the obliquity of the ecliptic, and the speed of the earth in its revolution around the sun. Because the latter two items have effects which are variable throughout the year, the length of the apparent day is not constant. *Mean solar time,* the length of an average day, one year divided by 365¼ days, eliminates this problem. Apparent solar time is measured by sundials; mean solar time is the time measured by clocks, with which everyone is familiar.

Another problem exists because noon does not occur simultaneously at all places on earth. Time zones were created to eliminate this problem, with each zone being about 15° wide or approximately ¹⁄₂₄ of the earth's daily rotation. Greenwich, England is defined as the arbitrary zero point; an observer located at 15° west longitude is one hour behind Greenwich, an observer at 30° west longitude is two hours behind Greenwich, and so forth. This is convenient for daily living but not for the astronomer. If an eclipse was observed at 3 p.m. local time in

Hawaii, what time was it where you live? *All astronomical observations must be recorded in Universal Time (UT), the local mean time in Greenwich, England.* For an observer in San Francisco, eight hours are added to Pacific Standard Time (PST) before recording data. UT is kept on a 24-hour clock to eliminate a.m.-p.m. ambiguities.

Example: A meteor is seen at 10:30 pm EST in New York on January 1, 1981. What UT should be recorded?

1. Convert EST to 24-hour time.
$$10:30 \text{ pm} = 22:30 \text{ EST}$$
2. Add 5 hours for time-zone difference.
$$22:30 \text{ EST} + 05:00 = 27:30 \text{ UT}$$
3. Subtract 24 hours (it is the next day in Greenwich).
$$27:30 - 24:00 = 03:30 \text{ UT}$$
4. Add 1 day to the date because of step 3.
$$\text{January } 1 + 1 = \text{January } 2$$

The observation was made on January 2, 1981 at 03:30 UT.

5.3b Universal Time

Observations should be recorded to within the nearest minute, or more precisely for rapidly varying objects. This accuracy cannot be reliably obtained from your AM radio or local bank sign. The best method is to use a shortwave receiver and listen to one of the national time signals broadcast by WWV in the United States, CHU in Canada, and similar services in other countries. A complete list is published by the British Astronomical Association.[6] WWV broadcasts at 2.5, 5, 10, and 15 MHz, and CHU broadcasts at 3.330, 7.335, and 14.670 MHz.

To receive these signals, buy a new portable multiband radio or a single-frequency radio such as the Radio Shack TIMEKUBE® or buy a used general-coverage receiver. Check local ads, amateur radio dealers and clubs, or the classified ads of a magazine such as *QST* or *Ham Radio*. If the radio you obtain works on batteries only, consider buying an AC adapter as batteries are adversely affected by cold weather and may become inoperative. A word of warning: if you intend to use a microcomputer with your system, the harmonics generated by its internal clock will interfere with the WWV frequencies, but generally not those of CHU.

A digital 24-hour clock is extremely convenient for the observatory, as no conversion is necessary to record the time. Several commercial clocks are available on the market at reasonable cost. Instructions for building your own can be found in back issues of many electronics magazines. Look for those using direct drive to each digit to eliminate RF multiplexing noise. Clocks with internal calendars, such as the CT7001 by Cal-Tex, or with BCD output, such as the MM5313 by National, may be used in microcomputer applications.

5.3c Sidereal Time

Sidereal time (ST) is the time kept by the stars, or more precisely, the right ascension of a star currently on the meridian. The length of a *sidereal day* is defined as the time between successive transits of the vernal equinox. A sidereal day is about 4 minutes shorter than a solar day. Knowledge of this time is essential to be able to point your telescope to the right region of the sky.

For many telescopes, the normal setting circles measure declination and *hour angle,* the distance from the celestial meridian. Hour angle can be expressed as

$$\text{Hour angle (HA)} = \text{Local sidereal time (LST)} - \text{Right ascension (RA)}. \quad (5.3)$$

We use the terms *local sidereal time* and *sidereal time* interchangeably in this chapter. The RA of a star in this equation is for the *present* epoch, that is not 1900, 1950, and so forth. Given the ST of an object and its coordinates, you can then point your telescope to the correct direction in space. Correspondingly, given the ST and RA for an object, the hour angle can be calculated for use in computing air mass.

Calculation of sidereal time can be accomplished in four basic ways: (1) from the HA and RA of some easily found object such as a bright star; (2) from tables given in *The Astronomical Almanac;*[7] (3) through use of a programmable calculator if the UT and date are known; or (4) from a sidereal rate clock. These four methods are explained below.

OBSERVATIONAL CALCULATIONS   111

1. *From HA and RA.* Pick some bright star and set your telescope on it. By reading the HA from the setting circles, set a solar rate clock to ST from the equation

$$ST = RA + HA \qquad (5.4)$$

This method is as accurate as your setting circles. The solar rate clock keeps nearly sidereal time for about a night's observations.

2. *From tables.* The Astronomical Almanac[7] gives the sidereal time at 0^h UT for every day of the year. Abell[5] includes a coarser table. Converting this time to the sidereal time at another location is a fairly complicated procedure. Examples are given in the *Almanac*. You need to know the day, UT, and your longitude, which can be found from Goode's World Atlas, or from a road atlas, a topographic map, or a plot survey of your observatory.

Example: What is the ST at 03:00 UT on July 7, 1973, for an observer at longitude 86°23′.7 west?

LST at Greenwich from the *Almanac*	$18^h\ 59^m\ 16^s$
− your longitude (86°23′.7 at 15° per hour	$-5^{\ h}\ 45^m\ 35^s$
+ UT difference from 0^h	$+3^{\ h}\ 00^m\ 00^s$
+ ST/UT difference over 3^h period at 10^s per hour	$+\ \ \ \ \ \ \ \ \ \ \ 30^s$
Sidereal Time at 03:00 UT:	$16^{\ h}\ 14^m\ 11^s$

3. *With a calculator.* Given the longitude, UT, and Julian date (see Section 5.3d), there is a simple equation to obtain sidereal time:

$$ST = 6.6460556 + 2400.0512617\ (JD - 2415020)/36525$$
$$+ 1.0027379\ (UT) - \text{longitude (hours)}. \qquad (5.5)$$

This equation takes into account the fact that there is approximately one extra sidereal day in a year, or 2400 extra hours in a century. We want $0 \leq ST \leq 24^h$, so after solving the above equation, we must subtract off that multiple of 24 hours (extra sidereal days) that leaves a remainder in this range. A FORTRAN subroutine that calculates sidereal time can be found in Section I.6.

Example: Calculate ST for 03:00 UT on July 7, 1973, from longitude 86° 23′.7 west ($5^h45^m35^s$).
The Julian date is 2441870.5 at 0^h UT from the *Almanac*. Then

$$ST = 6.6460556 + 2400.0512617(2441870.5$$
$$- 2415020)/36525 + 1.0027379(3) - 5.75972$$
$$= 1768.23611 \text{ hours } (73 \times 24 = 1752 \text{ hours})$$
$$= 1768.23611 - 1752.0$$
$$ST = 16^h14^m10^s$$

4. *From a sidereal rate clock.* There are two varieties of this specialized clock: an electric clock, with a sidereal rate motor that is very expensive, and an electronic digital clock. Both can be purchased commercially. The electronic version can also be built from plans published in *Sky and Telescope*.[8] It is basically the same as a pulse counter, counting one pulse per sidereal second. This can be achieved either by using a crystal oscillator or by adding extra pulses to a 60-Hz clock. Setting the sidereal rate clock should be performed using methods 2 or 3 above and should be checked often in case of power failures or oscillator drift.

5.3d Julian Date

Just as time zones cause problems between widely spaced observers, differing dates of observations can be a real headache when using the standard calendar. How many days have passed since you were born? The simplest approach is to use a running count of the number of elapsed days. This count was proposed by J. J. Scaliger in 1582 and is called the *Julian date* (JD) of an observation or event. The zero point was set far enough in the past that all recorded astronomical events have a positive JD. Scaliger suggested the use of Julian date $0 = 12^h$ UT on January 1, 4713 B.C. because several calendars were in phase on that day. The JD begins at noon because most active observers in the sixteenth century were in Europe and no date change would occur during a night's observations for them.

The Julian date for any given day can be found in the *Almanac* or through use of the following equation:

$$JD\ (0^h\ UT) = 2415020 + 365\ (\text{year} - 1900)$$
$$+ (\text{days from start of year}) + (\text{no. of leap years since 1900}) - 0.5. \quad (5.6)$$

A FORTRAN subroutine for this calculation can be found in Section I.2. Note that most observations are recorded in JD units including fractions of a day, instead of separate JD and UT.

Example: An observation was made on June 15, 1973, at 11:40 UT. What Julian date should be recorded?

$$JD\ (0^h\ UT) = 2415020 + 365\ (1973 - 1900) + 166 + 18 - 0.5$$

$$JD\ (0^h\ UT) = 2441848.5$$

$$11:40\ UT = 11.6667\ \text{hours}\ UT/24 = 0.4861\ \text{day}\ UT$$

$$JD = 2441848.5 + 0.4861$$

$$JD = 2441848.9861.$$

*5.3e Heliocentric Julian Date (HJD)

Any time recorded in the process of observing is *geocentric,* that is, made from a site on the earth. Because the earth revolves around the sun, an observer is closer to or further from a particular star at different times of the year. Six months after the earth's closest approach to the star, it is two astronomical units farther away (or less if the star is not on the ecliptic). Because of the finite speed of light, up to an additional 16 minutes are required for its light to traverse this extra distance. This light-travel-time effect causes scatter around the mean light curve of a variable, as compared with observations made from a relatively stationary object like the sun. Astronomers therefore prefer to record all observations as though made from the sun by adding or subtracting the light's travel time, depending on whether the earth is farther or closer, respectively, to the object than the sun is. The date derived in this manner is called the *heliocentric Julian date* (HJD).

Before the advent of the small computer, the heliocentric correction was made through the use of laboriously precomputed tables. Examples of these are Prager,[9] which is difficult to find, Landolt and Blondeau,[10] and Bateson,[11] which is coarser and less accurate than the others. It is now simpler to compute the correction than to use tables.

The most thorough description of the geometry involved is presented by Binnendijk,[12] and the reader is encouraged to glance at his figures and derivations. Basically, the problem is one of projection. There are two planes involved: the earth's equatorial plane, where right ascension, declination, and solar X, Y, Z Cartesian coordinates are defined; and the earth-sun-object plane, where we wish to know the projection of the earth's distance from the sun on the sun-object line. If the projections are carried through properly, one arrives at

$$\text{HJD} = \text{JD} + \Delta t \tag{5.7}$$

where

$$\Delta t \text{ (days)} = -0.0057755[(\cos \delta \cos \alpha) X + (\tan \epsilon \sin \delta + \cos \delta \sin \alpha) Y] \tag{5.8}$$

and X, Y are the rectangular coordinates of the sun for the date in question; α, δ are the right ascension and declination of the star for that date, respectively; and ϵ is the *obliquity of the ecliptic,* $23°27'$. Equation 5.8 holds as long as epochs are consistent. The values of X and Y can be obtained from the *Almanac*[7] or by the trigonometric series given by Doggett et al.[2]

The method of Doggett et al.[2] is shown below because it is relatively easy to program on a small computer or programmable calculator. See Doggett et al. for definitions of the various terms used in this method.

1. Determine the relative Julian century by

$$T = (\text{JD} - 2415020)/36525 \tag{5.9}$$

2. Obtain the mean solar longitude from

$$L = 279°\!.696678 + 36000.76892\,T + 0.000303\,T^2 - p \quad (5.10)$$

where

$$p = [1.396041 + 0.000308(T + 0.5)][T - 0.499998] \quad (5.11)$$

The value p is the precession from 1950 to date and therefore is subtracted from the present epoch longitude in Equation 5.10 to obtain 1950.0 longitude.

3. Obtain the *mean solar anomaly* by

$$G = 358°\!.475833 + 35999.04975\,T - 0.00015\,T^2 \quad (5.12)$$

4. Finally, obtain X and Y for 1950.0 through the expansions

$$\begin{aligned}
X = &\ 0.99986 \cos L - 0.025127 \cos(G - L) \\
&+ 0.008374 \cos(G + L) \\
&+ 0.000105 \cos(2G + L) + 0.000063\,T \cos(G - L) \\
&+ 0.000035 \cos(2G - L) \quad (5.13) \\
Y = &\ 0.917308 \sin L + 0.023053 \sin(G - L) \\
&+ 0.007683 \sin(G + L) \\
&+ 0.000097 \sin(2G + L) - 0.000057\,T \sin(G - L) \\
&- 0.000032 \sin(2G - L). \quad (5.14)
\end{aligned}$$

Only the first few terms of the X and Y expansions need to be kept for the accuracy required in this application. A slightly more accurate FORTRAN routine that calculates X, Y, and Z coordinates is presented in Section I.7.

Example: On June 15, 1973, at 11:40 UT, an observation was made of V402 Cygni. What is the appropriate heliocentric correction?

1. $T = (2441848.9861 - 2415020)/36525$
   $T = 0.7345376$ century

2. $L = 279°.696678 + 36000.76892(0.7345376)$
   $+ 0.000303(0.7345376)^2 - [1.396041$
   $+ 0.000308(0.7345376 + 0.5)](0.7345376 - 0.499998)$
   $L = 26723°.2879$ (subtract multiples of 360° to get 83°.2879)
3. $G = 358°.475833 + 35999.04975(0.7345376)$
   $- 0.00015(0.7345376)^2$
   $G = 26801°.1315$ (subtract multiples of 360° to get 161°.1315)
4. Compute X and Y from Equations 5.13 and 5.14

$$X = 0.108021$$
$$Y = 0.926683$$

5. Calculate coordinates

$$\delta = 37°0'.33 = 37°.00556 \ (1950.0)$$
$$\alpha = 20^h07^m15^s = 20^h.12083 = 301°.8125 \ (1950.0)$$

6. Finally, calculate Δt and the heliocentric date

$\Delta t = -0.0057755\{[\cos (37°.00556) \cos (301°.8125)]0.108021$
$+ [\tan (23°27') \sin (37°.00556) + \cos (37°.00556)$
$\sin (301°.8125)] \times 0.926683\}$
$\Delta t = -0.0057755(0.045473 - 0.386915)$
$\Delta t = 0^d.00197$
$\quad\quad\quad HJD = 2441848.9861 + 0.00197$
$\quad\quad\quad HJD = 2441848.9881$

This calculation is much easier when programmed on a calculator or computer!

5.4 PRECESSION OF COORDINATES

The coordinates of a star or other celestial object do not remain constant with time. Listed below are some of the major contributing factors, along with their *maximum* values. This should give you an idea how difficult it is to set a telescope accurately.

1. *Precession:* 50 arc seconds per year.
2. *Nutation:* 9 arc seconds over 19 years.

3. *Aberration:* 20 arc seconds over 1 year.
4. *Heliocentric parallax:* 0.75 arc second over 1 year.
5. *Refraction:* 60.4 tan z arc seconds daily.
6. *Variation in latitude:* 0.3 arc second.
7. *Proper motion:* 10 arc seconds per year.
8. *Barycentric parallax of solar system:* a few arc seconds.
9. *Geocentric parallax:* at most a few arc seconds even for solar system objects.

Most of these are short-term, cyclic aberrations, except for precession and proper motion. Proper motion is not discussed because it has a significant effect only on the closest stars.

Precession is caused by the gravitational pull on the earth's equatorial bulge by the sun and planets. The coordinates go through a complete cycle in 26,000 years, so the change per year is small, but the effect is cumulative. Catalogs and atlases give the mean coordinates of stars, usually at the beginning of some year, such as Equinox 1950.0. However, to identify a star on the Bonner Durchmusterung (BD) atlas (see Section 9.1a), which is Equinox 1855.0, with 1950 coordinates can be difficult unless one precesses the coordinates to the equinox of the atlas. More importantly, a telescope measures the *present equinox coordinates* and not those of 1900.0, 1950.0, or any other equinox given in standard catalogs.

Because the precession of the earth's axis is well known, equations to precess the coordinates are available. Their basic form is

$$\Delta\alpha = (m + n^s \tan \delta_m \sin \alpha_m)(t_f - t_0) \quad (5.15)$$
$$\Delta\delta = (n'' \cos \alpha_m)(t_f - t_0) \quad (5.16)$$

where $\Delta\alpha$ is in seconds, $\Delta\delta$ is in arc seconds and

t_o = equinox of the known coordinates (years)
t_f = equinox to precess to (years)
$$m = 3^s.07234 + 0^s.00001863 t_m \quad (5.17)$$
$$n^s = 1^s.336457 - 0^s.00000569 t_m \quad (5.18)$$
$$n'' = 20''.04685 - 0''.0000853 t_m \quad (5.19)$$

118 ASTRONOMICAL PHOTOMETRY

t_m = mean time with respect to 1900:

$$t_m = (t_f + t_0)/2 - 1900 \qquad (5.20)$$

This value is negative for equinoxes before 1900. The subscript m for α and δ indicate that the coordinates should be for the *midpoint* of the precession. That is, if you are precessing from 1950 to 1980, the coordinates on the right-hand side of the equations should be for 1965. Also, because you are taking the sine and cosine of right ascension, α should be converted to degrees.

For most calculations, the above equations will give reasonably good results, even if α_m and δ_m are replaced by α_0 and δ_0 (the original equinox).

Example: Precess γ Aql from 1953.0 to 1975.0.

γ Aql 1953.0: $\alpha = 19^h 44^m 01\overset{s}{.}458 = 19^h.733738 = 296\overset{\circ}{.}00607$
$\phantom{\gamma \text{ Aql 1953.0: }}\delta = 10°29'50\overset{''}{.}83 = 10\overset{\circ}{.}497453$

γ Aql 1975.0:

$t_f - t_0 = 1975 - 1953 = 22$ years

$t_m = (1975 + 1953)/2 - 1900 = 64$ years

$m = 3.07234 + 0.00001863(64) = 3\overset{s}{.}073532$

$n^s = 1.336457 - 0.00000569(64) = 1\overset{s}{.}3360928$

$n'' = 20.04685 - 0.0000853(64) = 20\overset{''}{.}041391$

$\Delta\alpha = [3.073532 + 1.3360928 \tan(10.497453) \sin(296.00607)](22)$
$\Delta\alpha = 62\overset{s}{.}723$

$\Delta\delta = [20.041391 \cos(296.00607)](22)$
$\Delta\delta = 193\overset{''}{.}32$

$$\alpha(1975) = \alpha(1953) + \Delta\alpha$$
$$= 19^h44^m01^s.458 + 62^s.723$$
$$\alpha(1975) = 19^h45^m04^s.181$$
$$\delta(1975) = \delta(1953) + \Delta\delta$$
$$= 10°29'50''.83 + 193''.325$$
$$\delta(1975) = 10°33'4''.15$$

For comparison, here are the values from the 1975 *Almanac:*

$$\gamma \text{ Aql } (1975.0): \alpha = 19^h45^m4^s.2$$
$$\delta = 10°33'5''$$

Note that within the accuracy of the *Almanac*, our answers are correct even though the mean coordinates were not used.

When high accuracy is needed, an analytical method of obtaining the coordinates by integrating the precession equations is used rather than using the mean coordinates with iteration. This more rigorous method is presented in the explanation in the *Almanac*,[7] and a FORTRAN subroutine using it is given in Section I.3.

5.5 ALTITUDE AND AZIMUTH

One of the most common calculations made by astronomers is determining where an object is above the horizon and where along the horizon it lies. These two coordinates are called *altitude* and *azimuth*, and constitute the most natural coordinate system.

The equations for determining altitude and azimuth can be found in Smart,[13] who gives diagrams, and Doggett et al.,[2] who give the equations used in the *Almanac*. Neither source gives a complete explanation of how the equations are derived. Though these equations can be used by themselves, it is useful to know their derivation.

*5.5a Derivation of Equations

Figure 5.2 shows the situation where the altitude and azimuth of a star is unknown. This is a problem in spherical trigonometry because we are

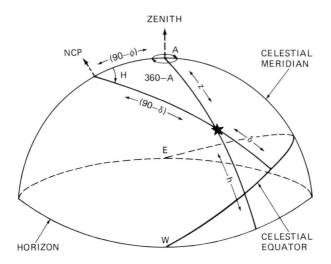

Figure 5.2 Altitude-azimuth sphere.

measuring angles on the celestial sphere. We are looking for h, the altitude above the horizon, and A, the azimuth measured east from due north. Three great circle lines are shown, one representing the celestial equator and two passing through the star from the zenith to the horizon and from the north celestial pole (NCP) to the celestial equator. The latter two lines and the celestial meridian define the spherical triangle needed to obtain h and A.

A spherical arc is defined by

$$D = R\theta \tag{5.21}$$

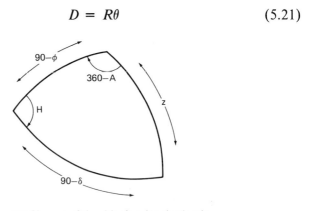

Figure 5.3 Close-up of the altitude-azimuth triangle.

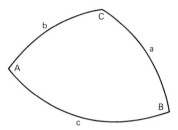

Figure 5.4 A general spherical triangle.

where θ is the angle subtended by the arc and R is the spherical radius. In the case of a *unit sphere* ($R = 1$), the sides of the spherical triangle are equal to the angle subtended by the arc. Therefore, the three sides are: (1) from the NCP to z, an angle of $90° - \phi$, ϕ = latitude; (2) from the zenith to the star, the zenith distance or $90° - h$; and (3) from the NCP to the star, an angle of $90° - \delta$, δ = declination. The enclosed angles of concern are the hour angle, H, measured from the celestial meridian to the star, and the coazimuth, $360° - A$. See Figure 5.3.

The solution of the triangle makes use of the law of cosines, and can be found in books such as the CRC tables.[14] Referring to Figure 5.4,

$$\cos a = \cos b \cos c + \sin b \sin c \cos A. \quad (5.22)$$

This works with any appropriate permutation. For altitude, we note:

$$\cos(90° - h) = \cos(90° - \phi)\cos(90° - \delta) \\ + \sin(90° - \phi)\sin(90° - \delta)\cos H$$

or

$$\sin h = \sin\phi \sin\delta + \cos\phi \cos\delta \cos H \quad (5.23)$$

where H is given in degrees by

$$H = 15(\text{LST} - \alpha). \quad (5.24)$$

For azimuth, we use the other known enclosed angle:

$$\cos(90° - \delta) = \cos(90° - \phi)\cos(90° - h)$$
$$+ \sin(90° - \phi)\sin(90° - h)\cos(360° - A)$$

or

$$\sin \delta = \sin \phi \sin h + \cos \phi \cos h \cos A. \qquad (5.25)$$

Solving for the azimuth,

$$\cos A = (\sin \delta - \sin \phi \sin h)/(\cos \phi \cos h). \qquad (5.26)$$

5.5b General Considerations

Equations 5.23 and 5.26 are our working relations for altitude and azimuth. One problem exists in solving for A. All calculators and computers return values of cosines only between 0 and 180°, yet azimuth extends over the entire range of 0 to 360°. We can remove the ambiguity by noting that when H is greater than 0°, A is greater than 180°. Therefore, for $-180° \leq H \leq 180°$,

$$A = A \quad \text{when } H \leq 0, \text{ and}$$
$$A = 360° - A \quad \text{when } H > 0.$$

Another solution, by Doggett et al.[2] solves for tan A. While this gives some computational simplicity in that the altitude is not needed, it adds the complexity that tan A is undefined at $+90°$ and $-90°$. It is suggested that Equation 5.26 be used, unless there is good reason to use another method.

Example: What is altitude and azimuth for AS Cas on February 5, 1979, at 01:00 UT from Goethe Link Observatory?
For that date and time, LST = 4^h12^m = $4^h.2000$.

$\alpha(1979) = 0^h24^m23^s = 0^h.4064$
$\delta(1979) = 64°6'.5 = 64°.1083$
$\phi = 39°18' = 39°.30$
$H = 15\ (4.2000 - 0.4064) = 56°.9042$

$\sin h = \sin (39°3) \sin (64°1083)$
$\quad + \cos (39°3) \cos (64°1083) \cos (56°9042)$
$\sin h = 0.75432$

$h = 48°97$

$\cos A = [\sin (64°1083) - \sin (39°3) \sin (48°97)] /$
$\quad [\cos (39°3) \cos (48°97)] = 0.83037$

$A = 33°86.$

But, since $H = 56°$ ($H > 0$),

$A = 360° - 33°86$

$A = 326°14.$

REFERENCES

1. Staff, 1979. *Sky and Tel.* **58**, 25.
2. Doggett, L. E., Kaplan, G. H., and Seidelmann, P. K. 1978. *Almanac for Computers for the Year 1978.* Washington, D.C.: Nautical Almanac Office.
3. Tricker, R. A. R. 1970. *Introduction to Meteorological Optics.* New York: American Elsevier.
4. Humphreys, W. J. 1940. *Physics of the Air.* New York: McGraw-Hill.
5. Abell, G. O. 1975. *Exploration of the Universe.* New York: Holt, Rinehart and Winston.
6. *The Handbook of the British Astronomical Association.* England: Sumfield and Day, Ltd. Published yearly.
7. *The Astronomical Almanac.* Washington, D.C.: Government Printing Office. Issued annually.
8. Reid, F., and Honeycutt, R. K. 1976. *Sky and Tel.* **52**, 59.
9. Prager, R. 1932. *Klein. Veroff. Univ. Sternw. Berlin-Babelsberg,* no. 12.
10. Landolt, A. U., and Blondeau, K. L. 1972. *Pub. A. S. P.* **84**, 784.
11. Bateson, F. M. 1963. In *Photoelectric Astronomy for Amateurs.* Edited by F. B. Wood. New York: Macmillan, p. 97.
12. Binnendijk, K. L. 1960. *Properties of Double Stars.* Philadelphia: Univ of Pennsylvania Press, pp. 228–232.
13. Smart, W. M. 1962. *Text-Book on Spherical Astronomy.* Cambridge: Cambridge Univ. Press.
14. S. M. Selby, ed. *CRC Standard Mathematical Tables.* Cleveland: The Chemical Rubber Co. Published yearly.

CHAPTER 6
CONSTRUCTING THE PHOTOMETER HEAD

Careful design and construction of the photometer head is very important and it requires substantial comment. The goal of this chapter is to supply you with enough background information to allow you to design and construct a photometer head intelligently. We have not included detailed construction plans because the requirements of individual observatories and researchers varies greatly. Amateur and professional astronomers approach the construction of photometers somewhat differently. The professional intends to mount the completed photometer head on a rather large telescope. Hence the components can be made from heavy metal stock and the tube can be totally enclosed in a cold box. The total weight of the finished photometer can be as much as 45 kilograms (100 pounds), which exceeds the weight of many an amateur's telescope! Obviously, the amateur must keep weight and size as primary restrictions on the design. Our emphasis in this chapter is on the needs of this small telescope user. We first make some comments on design and construction. We then describe a simple, lightweight design in Section 6.5.

We discuss briefly designs utilizing a photodiode as a detector because such designs are difficult to find in the literature. We refer interested readers to papers by Persha[1] and De Lara et al.[2] Finally, a photodiode photometer is available commercially from Optec, Inc.[3]

6.1 THE OPTICAL LAYOUT

The first step in designing a photometer head is to position the optical elements on a drawing showing the correct relative sizes and spacing.

This layout depends on the F-ratio of the telescope to be used, because this determines the rate at which the light cone formed by the telescope's objective diverges from the focal point. For instance, if you are using an F/8 telescope, the light cone will have a 1-centimeter diameter at a distance of 8 centimeters from the focus, a 2-centimeter diameter at a distance of 16 centimeters, and so forth. A small F-ratio telescope causes special problems because the cone diverges very rapidly, forcing the photometer components to be placed within an uncomfortably small distance from the focal point. If you plan to use a small F-ratio telescope, say F/5 or less, consider the design described by Burke and Pippin.[4]

The optical layout procedure is best accomplished by using a large sheet of graph paper so that the drawing can be made full size. Pick a spot near the left side of the paper to be the focal point. Draw a horizontal line through the focal point across the page to represent the optical axis of the photometer. Now draw the expanding light cone from the focal point according to your F-ratio as described above. The diaphragm is of course located at the focal point. The filters are positioned at a distance from the focal point where the light cone has a diameter of 5 to 10 millimeters. This is a compromise between the desire to use small filters and the need to illuminate an area of the filter large enough so that dust or small defects in the filter do not affect the light cone significantly. The amount of available space between the filter and diaphragm is now fixed. For an F/8 light cone, this distance is 80 millimeters for a 10-millimeter spot size on the filter. Within this space, the photometer builder must find room for the filter assembly and the flip mirror with its associated lenses. In Figure 6.1, we have positioned the diaphragm and filter. Consult this figure as you read the remainder of this section.

The next element to be positioned is the Fabry or field lens. This lens has a very important function. It focuses an image of the telescope's objective, illuminated by the light of the star on the photocathode. This spot of light remains at the same place on the photocathode no matter where the star drifts within the diaphragm. This is important because no photocathode can be made with uniform sensitivity. Without the Fabry lens, the photomultiplier output could change considerably depending on the position of the star in the diaphragm. The desired spot size depends on the size of the photocathode of the photomultiplier. If the spot is too large, light will be thrown away because some of it will

126 ASTRONOMICAL PHOTOMETRY

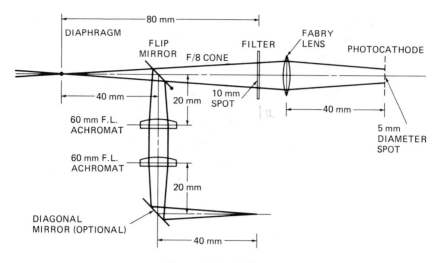

Figure 6.1 Optical layout.

miss the photocathode. If the spot is too small, local defects in the photocathode may have a degrading effect on the tube's output. For the 1P21 photomultiplier, a spot size of about 5 millimeters is about right. To calculate the Fabry lens's focal length, recall from elementary optics that for a thin lens the ratio of the object size to the object distance equals the ratio of the image size to the image distance. The object size is the diameter of the telescope's objective, D, and the object distance is essentially the telescope's focal length, f_t, so the relation becomes

$$\frac{D}{f_t} = \frac{a}{\text{image distance}}$$

where a is the image size. Because the focal length of the telescope is very large compared to that of the Fabry lens, the image distance is very nearly equal to the focal length, f, of the Fabry lens. That is,

$$\frac{D}{f_t} = \frac{a}{f}$$

or

$$f = a \cdot F, \tag{6.1}$$

where F is the focal ratio of the telescope, f_t divided by D. For the F/8 telescope used in our example, the Fabry lens must have a focal length of 40 millimeters for a 5-millimeter spot size. The spot size can be adjusted slightly to give a focal length of a common lens that can be found easily in optics catalogs. The spacing between the Fabry lens and the photocathode is now fixed as the focal length of the Fabry lens. The diameter of the lens depends upon its distance from the diaphragm. This distance is not critical. We have arbitrarily placed it 12 millimeters behind the filter in Figure 6.1. Once the lens is positioned on your drawing, the size of the light cone at this point gives the size of the lens. The lens should actually be made a little larger to allow for small misalignment after construction and the fact that the lens mount covers the edges. The lens itself is just a simple double convex lens. For proper ultraviolet transmission, this lens should be made of crown glass or, preferably, quartz.

The flip mirror can now be added to the drawing. It is placed between the diaphragm and the filter at a 45° angle to the optical axis. This mirror directs light into the first lens of the diaphragm-viewing optics. This lens is an achromat of sufficient diameter to accept the total light cone. It is positioned at a distance from the telescope's focal point equal to its own focal length. This results in a collimated beam of light after passing through the lens. If you cannot afford the luxury of custom-made optics, it is best to consult an optics catalog to find which focal lengths are available before positioning this lens in your drawing. A 60-millimeter focal length lens is available from Edmund Scientific[5] or A. Jaegers.[6] We have positioned this lens at a distance of 60 millimeters from the diaphragm. The next lens is an identical achromat that can be positioned at any comfortable distance from the first lens. The second lens reconverges this light for the viewing eyepiece. This optical arrangement gives a focused image of both the diaphragm and the star within it. Figure 6.1 shows the optical elements with the dimensions indicated.

A look at Figure 1.8 should convince you that the head for a photo-diode photometer is somewhat simpler. Because the detector is at the telescope's focus, the flip mirror can direct the light cone directly to a viewing eyepiece without the need of the two lenses discussed above. There is also no Fabry lens or diaphragm. The size of the active area of the photodiode defines the "diaphragm size." Unfortunately, this can not be adjusted. Without a diaphragm, it is necessary to have illuminated cross hairs in the viewing eyepiece. They are aligned such that if a star is centered, when the flip mirror is removed the star's light will fall directly on the photodiode. The mechanical design of the head must allow for the components to be adjusted to achieve this alignment. It is also necessary for the eyepiece focus to be adjusted and locked so that when a star appears focused in the eyepiece, the photodiode will be at the telescope's focus. These design problems are no more difficult to solve than those encountered in a photomultiplier-type photometer.

There is quite a large jump between laying out the optical components and performing the actual mechanical construction. We now offer some specific comments to make that jump seem a little smaller.

6.2 THE PHOTOMULTIPLIER TUBE AND ITS HOUSING

There are many manufacturers of photomultiplier tubes, each offering a wide array of devices. Suppliers of tubes frequently used by astronomers are EMI Gencom,[7] ITT,[8] and RCA.[9] The tube you select depends upon the spectral response required by your research. For example, Figure 1.5 shows that you should not try to use a tube with an S-4 response to measure stars at 8000 Å. An S-20 or S-1 tube response should be used instead. If you wish to obtain measurements on the *UBV* system, the choice of tube response is very crucial. As discussed in Chapter 2, the *UBV* system is defined by both the tube response and the filter transmission. For example, the *U* filter, and to a lesser extent the *B* and *V* filters, transmit light beyond 7000 Å. This means that these filters do not isolate the single spectral region they were meant to measure. However, the 1P21 has very little sensitivity beyond 7000 Å and the filter's red leaks, except for a small amount from the *U* filter, can be ignored. If the 1P21 is replaced with a tube with an S-1 or S-20 response, the filter set must be changed to "plug" the red leaks. Fernie[10] has used a single red-extended S-20, EMI 9658R tube to make *UBVRI* measurements. However, the *UBV* filters had to be altered from the type usually used with the 1P21. The point to be made here is that care must be taken to match detectors and filters to insure proper transformation to the standard system.

There are many tubes newer than the 1P21 that have been successfully used to make *UBV* measurements. The EMI 6256 has an S-11 response that is only slightly more red-sensitive than an S-4 response. This tube has a peak quantum efficiency of 21 percent compared to 13 percent for the 1P21. Tubes such as the ITT FW-118 (S-1), and FW-130 (S-20) and the RCA 7102 (S-1) have also been used successfully for *UBVRI* observations, with the appropriate filter modifications. Another tube of interest is the EMI 9789, which has a bialkali (cesium-potassium antimonide) surface with a spectral response very similar to S-4. However, it has a peak quantum efficiency of 20 percent and a smaller photocathode than the 1P21. A smaller photocathode has less area for thermionic emission, in this case resulting in about one-fifth of the dark current of the 1P21 at room temperature. The tube appears to transform well to the *UBV* system with the standard filters.

There are many different kinds of photomultipliers in successful use today. Despite all the advances in photomultiplier tube technology since the introduction of the 1P21, there are some good reasons to use this tube for UBV observations. The first is the fact that this tube was used to define the *UBV* system and requires little experimentation with filters. While many astronomers have used other tube and filter combinations successfully to make *UBV* observations, it may not always be apparent to the novice when there is a problem with the transformation. Another important reason for using the 1P21 is cost. The EMI 9789 and 6256 cost over $300 and $600, respectively. The 1P21 costs less than $100. The 1P21 has a companion tube called the RCA 931A, identical in all respects except that the 931A has a lower sensitivity and lower cost. This tube sells for less than $20 and is ideal for testing a new photometer and learning observational techniques. If the tube is somehow damaged, it will not cost hundreds of dollars to replace. Once confidence has been gained in both the photometer and the observer's abilities, the 931A can be replaced with a 1P21 without any modifications to the photometer. The 931A represents a good investment for the newcomer to astronomical photometry. If you are constructing a photometer of small size, the Hamamatsu Corporation[11] has introduced a line of miniature photomultipliers. Their R869 tube is equivalent to the RCA 1P21 but at one-half the size and about the same cost. In the following discussion of photometer construction, we assume that the detector is a 1P21 or a 931A. Modifications for other tubes can be made by the reader, based on the manufacturer's specification sheets.

Figure 6.2 shows a photograph of the 1P21. The photocathode is just behind the grid of wires seen near the front of the tube. Below the glass envelope is a base with 11 electrical pin contacts and a center post for positioning the tube in the socket. This center post has a key that must

130 ASTRONOMICAL PHOTOMETRY

Figure 6.2 A 1P21 photomultiplier tube.

face the incident light for proper tube orientation. Figure 6.3 shows the physical dimensions of the tube taken from the RCA specification sheets. The photocathode is about 5 millimeters in front of the central axis of the tube. You should account for this displacement when you position the Fabry lens.

The 1P21 has nine dynodes, a photocathode, and an anode. The photocathode is operated at a potential of about -1000 V with respect to ground. The first dynode is at a potential of about -900 V, or about 100 V more positive than the photocathode. This potential difference provides for the acceleration of the electrons released at the photocathode to the first dynode. Each succeeding dynode is 100 V less negative than its predecessor. Finally, the last dynode is -100 V with respect to

CONSTRUCTING THE PHOTOMETER HEAD 131

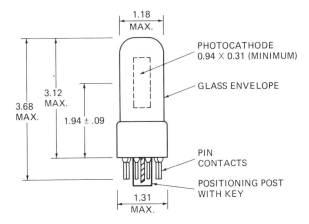

Figure 6.3 Physical dimensions of the 1P21 (all dimensions are in inches).

the anode that collects the cascade of secondary electrons. The pins on the tube base are provided so that the proper voltage can be applied to these tube elements. The voltage differences between dynodes are achieved by a simple voltage divider circuit that is wired directly to the tube socket. The tube socket is an Amphenol 78511T or equivalent. Figure 6.4 shows the tube socket with its voltage divider resistor string as viewed from the bottom. If you are pulse counting, it may be necessary to put a capacitor (0.01 mf, 1000 V, ceramic) between each pin (1 through 9) and ground. This reduces the instrumental sensitivity to external noise pulses.

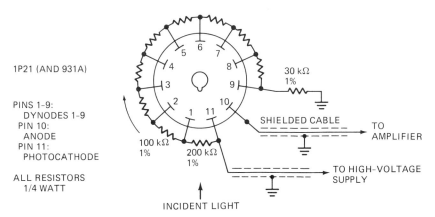

Figure 6.4 Wiring of tube socket (bottom view).

Note that the high voltage is applied to the photocathode on pin 11, which is next to pin 10, the anode pin from which the signal is collected. In order to prevent high voltage from leaking across to pin 10 and entering your amplifier, the socket must be made from a very high-resistance material such as mica or Teflon. Under no circumstances should a plastic socket be used.

After assembly, the socket should be cleaned with isopropyl alcohol to remove any fingerprints that may lead to leakage current. Because moisture at the tube socket can lead to leakage current, many astronomers seal the entire voltage divider string with silicon rubber. Prewired, sealed-tube sockets can be purchased from EMI and Hamamatsu. A hand-wired tube socket that is kept clean and dry is very adequate for a simple photometer.

The photomultiplier is a delicate device that should be handled with care. To avoid leakage-current problems, do not touch the tube at the base of the connector pins or the glass envelope. The photocathode is extremely sensitive and can be permanently damaged by exposure to bright light. The rule of thumb is never to allow light brighter than starlight to strike the tube when the high voltage is on. However, bright stars, especially on larger telescopes, can temporarily damage the tube, causing a condition known as *fatigue*. When a tube reaches the fatigue level, the tube seems to lose sensitivity and the output current drops. For the 1P21, fatigue occurs at an output current of about 1 μA. The sensitivity returns after the tube has been allowed to "rest" in total darkness. Fatigue can be avoided by lowering the high voltage applied to the tube, allowing brighter stars to be measured. Experience has shown that for most 1P21 tubes the transformation coefficients are not affected by a voltage change. There is no guarantee that this is true for all 1P21 tubes and it is certainly not true for many other tube designs. Even exposure to room light with the high voltage off can temporarily increase the dark current. Therefore, it is a good idea to handle the tube in subdued light during its installation in the tube housing. From then on, the dark slide should be kept closed except while observing. Another important precaution is to make sure that your high-voltage supply is connected properly so that negative high voltage is applied to the tube. The tube cannot be damaged by cold, but exposure to high temperatures for prolonged periods of time can lead to degraded performance. The best storage conditions for a photomultiplier tube are cool, dry, and dark.

The photomultiplier tube housing can be as simple as a pair of brass cylinders or as complex as a hermetically sealed thermoelectrically cooled chamber. The choice of housing depends on your commitment to serious observing and your budget. For the user of a small telescope who plans to observe brighter objects, we strongly recommend the simple approach. The function of the housing is to keep the tube in a dry, light-tight environment with the photocathode aligned to the optical axis of the photometer. The housing contains a dark slide which can be opened to allow the starlight to reach the photocathode or closed to keep the tube in total darkness when it is not being used. In a simple housing, one brass cylinder holds the tube socket, voltage divider resistors, and the cable connectors. This cylinder then snugly slides over the other cylinder, making a light-tight housing except for the small hole that allows light to reach the photocathode. The dark slide mechanism is mounted in front of this hole. A simple housing of this type will be shown in Section 6.5. The next step in complexity is to add a magnetic shield. This is a cylinder of mu-metal that fits into the housing and surrounds the photomultiplier. It shields the tube from external magnetic fields that might interfere with electrostatic focusing of the secondary electrons emitted from each dynode. External fields that would affect the tube significantly are rarely encountered. However, such a shield is a rather inexpensive precaution. Magnetic shields for the 1P21 can be supplied by the Hamamatsu Corporation[11] or Perfection Mica Company.[12]

The ultimate in photomultiplier tube housings incorporates a cooling system. Cooling the photomultiplier to dry-ice temperature can eliminate most of the tube's dark current. This has significant advantages for measuring faint stars that may produce a tube current comparable only to the dark current. Among professional astronomers, cooled tubes are the rule, not the exception. Cooling the 1P21 reduces the dark current from about 200 counts per second, for pulse counting, to less than one count per second. Essentially any output from the tube then results from starlight. A cooled tube has less sensitivity to red light, reducing the amount of the red leak through the U filter. This also means that the transformation coefficients of a tube change when it is cooled. If you determine these coefficients when the tube is uncooled, they will not be valid for observations made when the tube is cooled.

The usual means of cooling the tube is to replace the simple housing discussed previously with a cold box. The layout of a cold box is nicely illustrated in an article by Johnson.[13] A cold box consists of three boxes. The photomultiplier tube is sealed in an airtight container that has a small window allowing light to reach the photocathode. This inner container is surrounded by a larger box which holds one-half to one kilogram of dry ice. This box is, in turn, surrounded by the outermost box that holds styrofoam or polyurethane foam to insulate the dry-ice container. The light entrance to the cold box is a window and/or the Fabry lens. It is sometimes necessary to mount a small heating element near these glass components to prevent them from developing a coating of frost. Cold boxes can add considerable weight to the photometer. For this reason, special lightweight designs are required for telescopes less than 40 centimeters (16 inches) in aperture. Recently, thermoelectric cooling systems have begun to be used in astronomy. They can reduce the tube temperature typically by 20 to 40 Celsius degrees below the outdoor temperature. These cooling systems are large, heavy, and expensive. They probably work as well as a dry-ice cold box but may require a water supply or a fluid-circulation system. It is possible to purchase simple uncooled housings and dry-ice cold boxes. Suppliers are EMI Gencom,[7] Hamamatsu,[11] Pacific Precision Instruments,[14] EG & G Princeton Applied Research,[15] and Products for Research.[16] There is a word of caution to note before you place an order for a cooled photomultiplier tube housing. Most of the housings built by these companies are intended for laboratory use. Consequently, not every model can support its own weight properly if it is held to the photometer by a simple mounting flange around the light input port. Before placing an order, it is advisable to call the company and make certain that the model of your choice can be mounted to a telescope, that it can work in any position, and that it can function in the sometimes hostile environment of your observatory.

6.3 FILTERS

Most of the wide-band filters used in optical astronomy are made by Corning Glass Works[17] or Schott Optical Glass.[18] Table 6.1 lists a recommended filter set for *UBV* photometry using the 1P21. Note that it is important to order the specified thickness. Filters of different thicknesses have slightly different transmission curves. The *B* filter is actually a sandwich of two filters. The GG13 filter looks like clear glass

TABLE 6.1. Recommended *UBV* Filters

Bandpass	Filter and Thickness (Schott Filters)
U	UG2[a] (2 mm)
B	GG13 (2 mm) + BG12 (1 mm)
V	GG14 (2 mm)
Red leak	UG2[a] (2 mm) + GG14 (2 mm)

[a]These two filters are from the same melt.

but it is designed to prevent transmission of light shortward of the Balmar discontinuity. Some observers have reported difficulty in making the *U* filter transformations unless the filter matches the one used to define the *UBV* system. This filter is a Corning 7-54 made from Corning 9863 glass. Unfortunately, this filter has a larger red leak than the Schott UG2. The red-leak filter listed in Table 6.1 is just a sandwich made from a second *U* filter and a *V* filter. The *V* filter does not transmit ultraviolet light normally passed by the *U* filter. However, it does transmit any red light which the *U* filter transmits. The combination of the two filters transmits only the red light "leaked" by the *U* filter. After a star has been measured with the ordinary *U* filter, it is measured again with the red-leak filter. The red-leak measurement can then be subtracted from the *U* measurement to obtain a corrected *U* measurement. When ordering these *U* filters, you should request that they both come from the same melt or order a single piece from which you can cut two filters. This helps to insure that the red-leak properties of the two filters are as nearly identical as possible.

If you are using a photometer that places the filters inside the telescope's focal point, be sure to add a Schott cover glass to make all filters the same total thickness. The passage of the telescope's light cone through glass alters the focal point slightly. If the filters are of different thicknesses, they will each cause the light to focus at a different point. This is not of concern for any photometer design discussed in this book, with the exception of the photodiode photometer.

Because of its different spectral response, a photodiode must use a different set of filters to match the *UBV* system. De Lara et al.[2] used an EG & G Electro-Optics Division[19] SGD-040L PIN photodiode with the filters listed in Table 6.2. The thicknesses of the Corning filters are not specified because they are adjusted by the manufacturer to achieve the

TABLE 6.2. Filters Used with a Photodiode by De Lara et al.

Bandpass	Filters
B	Corning 5030 + Corning 9782 + Schott GG13 (2 mm)
V	Corning 9780 + Corning 3384
R	Corning 3480 + Corning 4600
I	Corning 2600 + Corning 3850

required bandpasses. It should be noted that De Lara et al. were not satisfied with this filter set. It is presented here as a starting point for those who wish to experiment with photodiode and filter combinations.

The *UBV* filters can be ordered in a 2.5 × 2.5 centimeter (1 × 1 inch) size at a cost of a few dollars per filter. They should be handled with care because they are very thin and made from soft glass that can be scratched easily. The surface of the *U* filter may appear to develop a water-spotlike pattern on its surface with age. This can be removed by polishing the surface with rouge or barnesite.

Normally, filters are mounted in a photometer in one of two ways. One technique uses a filter slide. The filters are held side by side in a long rectangular holder. The holder slides lengthwise so that the filters can be positioned one at a time in the light beam. This design presents a minor problem for a small photometer. The slide must be long enough to accommodate each filter and the width of the slide walls. For a four-filter photometer, the slide is at least 11.4 centimeters (4.5 inches) long for 2.5 centimeter (1 inch) wide filters. With either end filter in the light path, the slide sticks out about 10 centimeters (4 inches) to either side of the optical axis. That requires a total linear span of more than 19 centimeters (7.5 inches). This makes it very difficult to contain the filter slide within the main chassis of a small photometer head, which would simplify the design and make it easier to keep light-tight. One is forced to adopt a design that is frequently used by professional astronomers. Figure 6.5 illustrates the idea schematically. The filter slide fits within a rectangular housing which is light-tight except for holes on the top and bottom plates that allow the telescope's light cone to enter, pass through the filter, and exit. The outer housing is inserted into the photometer chassis and allows the slide to move its entire length, while eliminating the need to enlarge the photometer chassis. It is important to incorporate some sort of detent device so that each filter "clicks" into

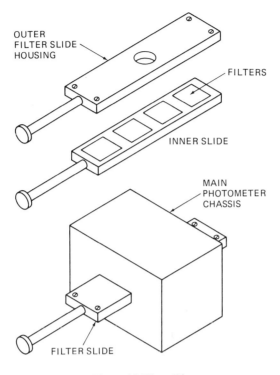

Figure 6.5 Filter slide.

place and is held in the light path during a measurement. This is usually accomplished by notching the positioning rod and mounting a spring-loaded ball bearing in the rod's bearing at the end of the outer housing.

The second approach to filter mounting is the filter wheel. A flat circular disk has filters mounted around its periphery. There is a small hole under each filter to allow light to pass through the disk. The filter wheel is aligned perpendicular to the optical axis of the photometer with its rotation axis offset from the optical axis so that each filter passes into the light beam as the wheel rotates. Looking ahead to Figure 6.16, you find a sketch of a filter wheel. Once again, a detent mechanism is necessary to insure proper positioning of the filters. Of course, there are other methods of mounting filters. In a design by Dick et al.[20] the filters are mounted on a four-sided carousel that rotates around the photomultiplier. This very simple design has the advantages of compactness and ease of construction.

6.4 DIAPHRAGMS

Photometer diaphragms are usually made by drilling small holes in a metal plate. The first step is to decide on the desired sizes. We speak commonly of the size of a diaphragm in terms of the angular size of the field of view it permits us to see in the telescope's focal plane. That is, a diaphragm which is said to be 20 arc seconds exposes a portion of the sky 20 arc seconds in diameter to our detector. To translate an angular size to the physical diameter of the hole we need to drill requires a knowledge of the plate scale of the telescope. If we took a photographic exposure of the full moon (0.5° across) at the focus of our telescope and it appeared 1 centimeter wide on the developed film, the plate scale would be 0.5° per centimeter. In other words, this is simply a statement of how angular sizes in the sky project to the telescope's focal plane. The plate scale depends upon focal length. Short focal length telescopes have large plate scales; that is, they are wide field instruments. Long focal length telescopes have very small plate scales, hence narrow fields and high magnification. The plate scale can be computed in units of arc seconds per millimeter by

$$\text{plate scale} = \frac{K}{f_t}$$

where K is 20626 (8120) if the focal length of the telescope is expressed in centimeters (inches).

Let us suppose that the F/8 telescope discussed in Section 6.1 has a 20.3-centimeter (8-inch) diameter. Its focal length is then 162.4 centimeters (64 inches) and by the above equation it has a plate scale of 127 arc seconds per millimeter. Table 6.3 contains a list of twist drill numbers in the U.S. system and their corresponding diameters. We can use this table together with the plate scale to predict the angular size of a diaphragm made with any of these drills. For instance, a number 75 drill has a diameter of 0.533 millimeters, which gives a diaphragm of 0.533 × 127 or 67.7 arc seconds for this telescope. It is desirable to make the diaphragms rather small to minimize the amount of sky background light seen. Typically, professional astronomers use a diaphragm which is 20 arc seconds or less. This is practically impossible with small telescopes because their focal lengths are shorter, making the necessary drill sizes impossibly small. Even with a number 80 drill, which is only

TABLE 6.3. Twist Drill Diameters

Drill No.	Inches	Millimeters	Drill No.	Inches	Millimeters
40	0.0980	2.489	60	0.0400	1.016
41	0.0960	2.438	61	0.0390	0.991
42	0.0935	2.375	62	0.0380	0.965
43	0.0890	2.261	63	0.0370	0.940
44	0.0860	2.184	64	0.0360	0.914
45	0.0820	2.083	65	0.0350	0.889
46	0.0810	2.057	66	0.0330	0.838
47	0.0785	1.994	67	0.0320	0.813
48	0.0760	1.930	68	0.0310	0.787
49	0.0730	1.854	69	0.0293	0.743
50	0.0700	1.778	70	0.0280	0.711
51	0.0670	1.702	71	0.0260	0.660
52	0.0635	1.613	72	0.0250	0.635
53	0.0595	1.511	73	0.0240	0.610
54	0.0550	1.397	74	0.0225	0.572
55	0.0520	1.321	75	0.0210	0.533
56	0.0465	1.181	76	0.0200	0.508
57	0.0430	1.092	77	0.0180	0.457
58	0.0420	1.067	78	0.0160	0.406
59	0.0410	1.041	79	0.0145	0.368
			80	0.0135	0.343

a few times the width of a human hair, the diaphragm is 44 arc seconds for our 20.3-centimeter (8-inch) telescope. The same diaphragm used with a 76.2-centimeter (30-inch) diameter F/8 telescope is 12 arc seconds. This points out a basic disadvantage that the small telescope user must face. The larger plate scale of the small telescope means that photometry must be done with diaphragms that admit a fairly large amount of sky background light. This leads to a poorer signal-to-noise ratio, with the result that faint stars cannot be measured as well.

Even a simple photometer should contain at least three different diaphragm sizes, one of which is fairly large. This allows the observer to use a larger diaphragm on nights of poor seeing conditions, or a small one when the moon is bright. An intermediate size probably is used the most and should be of such a size that your clock drive can keep a star within this diaphragm for at least 5 to 10 minutes. Additional observational considerations for diaphragm selection are discussed in Section 9.4.

Section 6.5 contains a description of a photometer designed for a 20.3-centimeter (8-inch) F/8 telescope discussed in this section. The diaphragm selection in this design is based on the above considerations and on the availability of small drills. The largest diaphragm was made with a number 47 drill, which for a plate scale of 127 arc seconds per millimeter yields 253 arc seconds, or 4.2 arc minutes. The remaining two diaphragms were made with number 61 and 76 drills, yielding sizes of 126 and 64.5 arc seconds, respectively.

The diaphragm holes can be drilled in a brass or steel plate. A drill press must be used to insure that the holes are drilled straight and to minimize drill breakage. This latter point becomes important because these drills are very thin and break easily. Drills of this size usually must be purchased at large hardware stores or dealers in machinists' supplies. Purchase at least two of the smaller drills as you are bound to break at least one. The chucks on many drill presses do not close down far enough to hold drills this small. It may be necessary to find a local machinist who can drill these holes for you. It is important to counterbore the holes to avoid a tunneling effect when looking through the diaphragm viewing eyepiece. The drill for the counterbore should be several times the size of the diaphragm drill. The counterbore should be made as deep as possible without enlarging the diaphragm hole. It is possible to make holes about half the size of a number 80 drill by pricking aluminum foil with a sharp needle. The foil is placed on a flat metal surface, and the hole is made by turning the needle between your fingers. Make several holes and inspect them with a magnifying glass. The one which appears most round can be used as a diaphragm by gluing the foil to a piece of metal that has a larger hole in it.

Like filters, the diaphragm position usually is selected by a diaphragm slide or wheel. A diaphragm slide is usually used because, unlike the case of filters, it can be very short and completely contained within the main photometer chassis. The diaphragm holes are drilled in a flat rectangular piece of metal. It is also necessary to provide some sort of detent system so each diaphragm "clicks" into position on the optical axis. A diaphragm slide assembly is described in Section 6.5.

There are two final points to be made. First, when you view a star in the diaphragm through the viewing eyepiece it may be very difficult to tell if it is centered. This is because the sky background light does not outline the diaphragm sufficiently. As a result, everything except the star looks black. An exception to this occurs if you are observing near an urban area. In this case, the sky background is such that the diaphragm appears as a gray circle in the eyepiece. This is about the only benefit that light pollution provides for astronomy! Hopefully, you will have the opportunity to observe from a site that requires an internal source of diaphragm illumination. The simplest method is to mount a very tiny light bulb or a light-emitting diode (LED) just above the diaphragm slide. The electric current to this lamp is controlled by a contact switch attached to the flip mirror. When the flip mirror is moved out of the light path, the lamp is automatically shut off. It is a good idea to have a potentiometer in the circuit in order to adjust the lamp brightness.

Second, make sure that the focus of the diaphragm viewing eyepiece is at least initially adjustable. A quick look at Figure 6.1 should convince you that the spacing and size of the optical components depends on the assumption that the diaphragm is placed at the telescope's focal plane. This is a condition that must be established at the beginning of each observing session. This is accomplished first by focusing the diaphragm eyepiece until the diaphragm appears with sharp definition. Then center a star in the diaphragm. If this star appears out of focus, the diaphragm plane and focal plane do not coincide. This is remedied by adjusting the *telescope's* focus until the star's image appears sharp. Many photometers are designed so that the eyepiece focus is adjusted once during construction and locked in place. When observing, it is then only necessary to adjust the telescope's focus to make the stellar image clear. The disadvantage of this procedure is that eyeglass wearers must use their glasses when looking through the eyepiece. If they focus the star without their glasses on, the diaphragm will not appear focused because this was adjusted in the workshop by someone who, presumably, had normal vision. For the eyeglass wearer who does not use his or her glasses at the eyepiece, it is best to leave the eyepiece focus adjustable.

6.5 A SIMPLE PHOTOMETER HEAD DESIGN

In this section, we describe a simple photometer head design suitable for use by amateur astronomers or any user of a small telescope. Detailed construction plans are not presented. Instead, the sketches and discussion presented are intended as a guide and a source of ideas for the photometer builder. We recommend that this person look at the designs of the AAVSO photometry manual,[21] Allen,[22] Burke and Pippin,[4] Code,[23] Dick et al.[20], Grauer et al.[24] and Nye.[45]

142 ASTRONOMICAL PHOTOMETRY

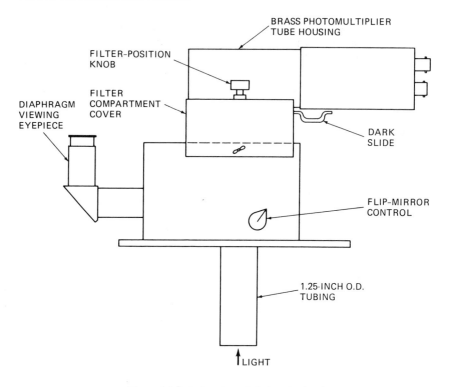

Figure 6.6 Exterior view of photometer head.

The design presented below was built originally for a 20.3-centimeter (8-inch) F/8 Newtonian telescope. Because it is difficult to counterbalance objects placed at a Newtonian focus, this photometer head was made as lightweight as possible, 1.4 kilograms (3.1 pounds). It has been used successfully for years and has produced a lot of observational data with the 20.3-centimeter (8-inch) and larger telescopes. The basic design is patterned after one by D. Engelkmeir.[21] Figure 6.6 shows a sketch of the exterior of the photometer head. The central box is made from a standard 12.7 × 10.2 × 7.6-centimeter (5 × 4 × 3-inch) aluminum electrical chassis. This choice was made to avoid machining and to provide a lightweight box at a cost of a few dollars. The interior was spray painted with flat black paint to reduce scattered light. The bottom of the box is mounted on a 19.0 × 19.0-centimeter (7.5 × 7.5-inch) base that is oversized to help mount the photometer head to the telescope securely. This is illustrated later. The starlight enters a 8.18-centimeter (1.25-inch) O.D. brass tube that fits the standard focusing drawtube of U.S.-made amateur telescopes. The diaphragm viewing eyepiece is attached to a diagonal mirror for ease of viewing. This is especially helpful when using Newtonian telescopes. The filters are contained in a separate compartment above the main box for easy access. The photomultiplier tube, a 1P21, is mounted in a simple, uncooled housing made from brass tubing. There are four external controls: the dark slide, the filter position, the diaphragm slide, and the flip mirror. Figure 6.7 shows a photograph of this head, seen in a perspective slightly different from Figure 6.6. Obviously, a photometer head should be made light-tight except for incident starlight.

Figure 6.8 shows a cut-away view of the interior of the photometer head. Wherever possible, plastic and aluminum have been used to reduce weight. The flip mirror is a quarter-wave optical flat. This mirror is mounted to a flat metal plate with double-sided adhesive tape. The metal plate is soldered to a rod that attaches to a knob on the outside of the box. The mirror directs light to the collimating and imaging lenses used for viewing the diaphragm. These lenses are identical achromats mounted in a single tube that helps to insure their mutual align-

Figure 6.7 The photomultiplier head.

144 ASTRONOMICAL PHOTOMETRY

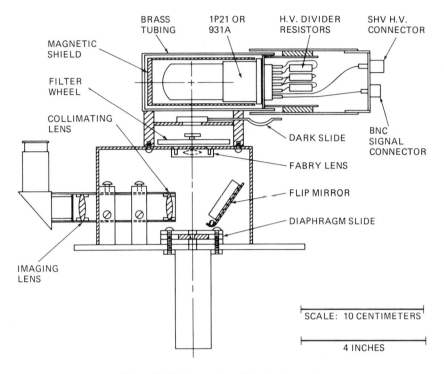

Figure 6.8 Cutaway view of photometer head.

ment. This tube is held in place by two plastic blocks. The tube passes through a hole in each block and is held in place by three adjustment screws similar to the way finder telescopes are mounted. In the workshop, the screws can be adjusted until the image of the diaphragm appears centered in the eyepiece. The ability to make this adjustment is valuable especially if you are unable to build components to machine-shop precision.

The top plate of the box is removable. On its interior side, the Fabry lens is held by a small plastic block. The exterior side holds a filter compartment and the tube housing. The left half of the housing, as shown in Figure 6.8, contains a magnetic shield and the photomultiplier tube. The right half of the housing is made from slightly larger tubing so that it slides over the left half. It holds the tube socket, voltage divider resistors, and cable connectors. The tube socket is held in place by a short length of tubing, which in turn is held snugly in place by a spacer between it and the interior housing wall. It is important to establish an electrical connection between the walls of the tube housing and electrical ground. This is necessary to provide good electrical shielding for the tube. The recommended BNC and SHV connectors should be installed so that the outside of the connector makes electrical contact with both the housing walls and the cable shield.

Just below the housing is the dark slide. This is simply a narrow compartment in which a small metal plate can be moved by a rod over the light opening. The floor of this compartment is lined with felt to make the slide move smoothly and to enhance the light seal.

Below the dark slide is the filter compartment. In this design, the filters were placed behind the Fabry lens rather than in front. This was purely a matter of convenience. It is preferable to place filters in a light beam that is nearly parallel so they do not deviate the beam. Both the telescope and Fabry lens have about the same F ratio, so it makes little difference where the filters are placed. Figure 6.9 shows some details of the filter assembly. The filters are mounted on a circular disk 6 centimeters (2.4 inches) in diameter. Because of this small size, the filters had to be cut smaller than their normal one-inch size. Access to the filter wheel is gained by removing a cover plate that is held in place by a single wing nut. The filter wheel is turned by two identical gears. One is mounted on the central shaft of the filter wheel, high enough to clear the filters. The second gear is mounted on a shaft that comes down from the top of the filter cover. This gear system was necessary since the filter wheel shaft is covered partially by the tube housing, leaving no room for a positioning knob. The two gears move this knob about 1.5 centimeters (0.6 inch) to the right in Figure 6.9. The detent positioning is accomplished by a scheme suggested by Burke and Pippin[4] and Stokes[25]. The shaft that holds the gear on the filter cover is actually the shaft from an electronic rotary switch. Such switches are inexpensive and contain an accurate detent system. All that is necessary is to remove the wafers containing the switch contacts. Usually these switches have a stop that prevents 360° rotation. This stop is often just a metal tab that can be bent out of the way. The switch positions should be spaced evenly around the knob. The number of switch positions should be a whole number times the number of filters. In this case, a 12-position switch was used with four filters.

Figure 6.9 also shows the diaphragm assembly. This slide consists of five pieces made from flat steel stock. The top and bottom plates are both 6.4 × 5.1 × 0.3 centimeters (2.5 × 2.0 × ⅛ inches) with a 0.64-

146 ASTRONOMICAL PHOTOMETRY

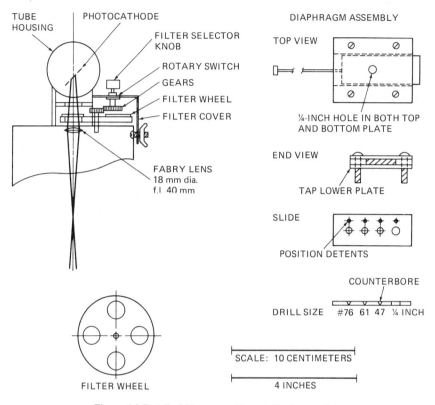

Figure 6.9 Detail of filter assembly and diaphragm slide.

centimeter (0.25-inch) central hole. There are two 6.4 × 1.3 × 0.3-centimeter (2.5 × 0.5 × ⅛-inch) metal strips separating the top and bottom plates and forming the 2.54-centimeter (1-inch) cavity for the diaphragm slide. The fifth piece is 6.4 × 2.5 × 0.3 centimeters (2.5 × 1.0 × ⅛ inches) and contains the diaphragm holes. There arc three diaphragms, as discussed in Section 6.4 and a 0.6-centimeter (0.25-inch) hole for a wide field of view. Detent positioning is accomplished by a spring-loaded ball bearing mounted in the bottom plate. This ball pushes against the bottom of the slide. As seen in Figure 6.9, there is a series of holes drilled just above each diaphragm position. When a diaphragm is on the optical axis, the spring pushes the ball into the detent hole and the slide locks into position. Because the detent holes are smaller than the ball bearing, pressurc on the diaphragm rod pushes the ball bearing down and the slide moves until the next detent hole is aligned.

While the photometer described here has worked well for many years, we recommend two design changes. First, the filter wheel and filter compartment should be made larger in order to accept standard 1-inch filters. The second change concerns the flip mirror. A look at Figure 6.8 reveals the problem. When the mirror is swung back to make a measurement, it reflects any room light entering the eyepiece into the Fabry lens. It has been found necessary to place a cap over the eyepiece when a measurement is being made. The solution is to hinge the mirror about a point in the upper left of Figure 6.8 so that when the mirror is swung out of the light cone it covers the front of the collimating lens. A foam-rubber ring placed around the front of the lens housing would cushion the mirror.

It is very important to mount the photometer head solidly to the telescope. Any flexure can cause misalignment and move the star out of the diaphragm. Supporting the photometer head solely with the drawtube of the typical small telescope is insufficient unless the telescope's focuser has been redesigned. Figure 6.10 shows the photometer head just described mounted at the Cassegrain focus of a 30.5-centimeter (12-inch) telescope. The large base plate of the head is attached to a mounting rail on the telescope. A similar mounting rail could be used with Newtonian telescopes. The Cassegrain telescopes used by professional astronomers usually focus by moving the secondary mirror rather than the rack-and-pinion devices used by amateurs. This allows the photometer to be bolted firmly to the tail piece of the telescope, eliminating most flexure. This focusing arrangement should be considered more seriously by amateur astronomers.

6.6 ELECTRONIC CONSTRUCTION

In this and succeeding chapters, we present designs for electronic circuitry. These designs represent in most cases prototype units that have been constructed and tested. However, we would like to make three points:

1. You should not attempt to construct these circuits unless you are familar with high-frequency, high-voltage, and high-gain construction techniques. Otherwise, we recommend purchasing commercial units.
2. While the prototype units work as reported, different layouts and components may require minor modifications to perform properly.

Figure 6.10 The photometer head mounted on a telescope.

3. Because you are building high-performance circuits, you should use the best quality components that you can afford.

Finding sources of components can be complicated. Other than Radio Shack, you can generally find components at large electronic distributors. These companies usually have a $25 minimum order. Other excellent sources are hamfests, especially the larger ones in Dayton and Atlanta. These are advertised in *QST* and other amateur radio magazines. For most of us, the primary and cheapest source of electronic components is the mail-order house. *The Radio Amateur's Handbook*[26] lists many sources of parts in the chapter on construction techniques. We suggest reading this chapter as it contains much information that the amateur astronomer can use.

6.7 HIGH-VOLTAGE POWER SUPPLY

The high-voltage power supply for the photomultiplier tube is one of the most important components of the photometer. It should be adjustable, in order that the correct operating point for each tube can be found, and it should be well regulated. A 1 percent change in the output voltage can make a much larger change in the photomultiplier tube output because the gain of each stage is multiplicative. This kind of systematic error should be kept to a value smaller than anticipated from observational error (photon statistics); that is, the power supply should be regulated to better than 0.1 percent. Most variations are caused by voltage fluctuations from the power line, and may occur on a very short time scale.

The requirements listed above would be relatively simple with modern technology if not for the magnitude of the voltage required. A 1P21 tube uses approximately 1000 V (1 kV) and some other tubes, such as the FW-118, may require 2 kV. No integrated circuit (IC) voltage regulator and few transistors can handle these voltages.

Of course, one can always purchase a commercial, adjustable high-voltage supply. Appropriate used units can be obtained from many firms that deal in reconditioned test equipment, such as the Ted Dames Co.[27] New units are available from companies such as EMI,[7] Kepco,[28] Lambda,[29] Ortec,[30] or Princeton Applied Research.[15] Be advised, however, that some of these new units cost $200 to $1000.

You can construct your own power supply. There is a dearth of up-to-date circuits and we try to present those that we could find or derive. However, these circuits handle *lethal* voltages, so be extremely careful in working with them.

There are three general approaches used in building adjustable high-voltage supplies. The simplest is to use a bank of high-voltage batteries. The easiest electronically regulated supply is to obtain approximately 2 kV from a filtered supply and then use a voltage divider to pick off the appropriate voltage. The third method is also the most elegant: amplitude-modulate an RF oscillator and rectify the output voltage. Each of these approaches is discussed in more detail.

6.7a Batteries

The use of a battery bank has the advantage of excellent regulation and noise immunity. In addition, the bank is portable and the current drain

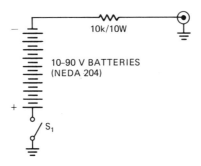

Figure 6.11 Battery supply.

is low enough that the batteries last their normal shelf life. The two disadvantages are that batteries are temperature-sensitive (keep them warm!) and expensive. Expect to pay about $1 per 10 V. Typical available batteries are the PX18 (45 V) and the NEDA 204 (90 V). Some high-voltage batteries used by the military can be obtained from surplus supply houses at a much lower cost; check electronics magazines for names and addresses. The U200 batteries (300 V) were available formerly and may still be found in local stores.

A typical battery-operated system is shown in Figure 6.11. The 10 K series resistor acts as a current limiter when an external short occurs. The case for this system should have insulation to prevent leakage paths to ground, moisture proofing to prevent leakage paths and damage to the batteries, and shielding on the outside for safety and noise immunity.

6.7b Filtered Supply

A schematic for a typical voltage divider supply is shown in Figure 6.12. This involves an AC transformer with a voltage doubler consisting of diodes D1 through D4. Each diode is bypassed with an RC combination to equalize voltage drops across the diodes and to guard against transients. Current-limiting resistors R1 and R4 protect the diodes against the initial turn-on surge and to allow some current to flow through the zener chain at all times. The output is filtered by capacitors C7 through C10, with equalizing resistors R7 through R10 doubling as bleeder resistors when the supply is shut off. The filtered output is fed to the zener bank. These diodes regulate the voltage and by switch or jumper

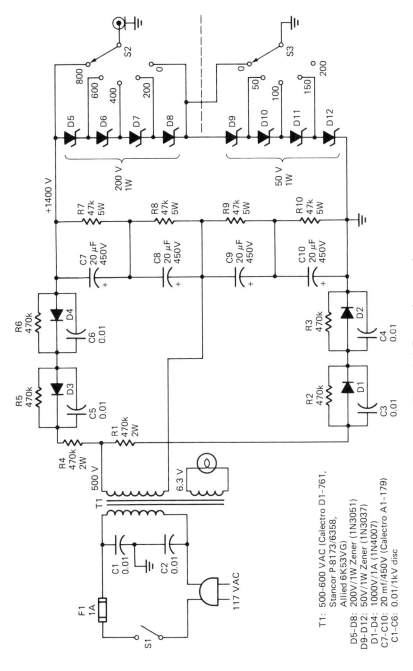

Figure 6.12 Zener voltage supply.

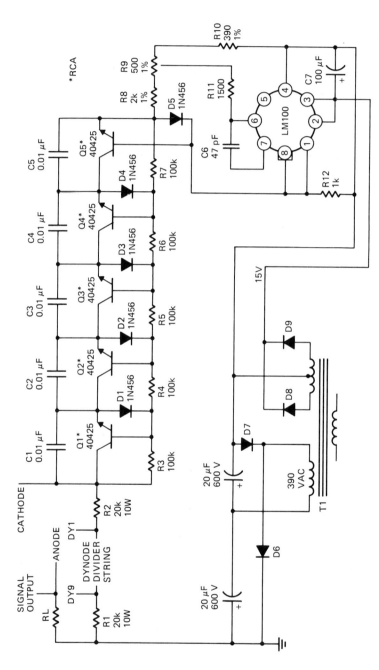

Figure 6.13 Current-regulated supply. (Copyright 1981 National Semiconductor Corporation)

selection can allow some voltage adjustment. A simplified chain of five 200 V zeners could be used if adjustment is not desired.

The regulation from this supply is not as good as might be expected. Because the zener knee is not infinitely sharp, there is some voltage change with current. In addition, zeners are very temperature-sensitive. The amount of regulation can be determined by comparing the current-limiting resistor to the dynamic resistance of the zeners, as this circuit is basically a voltage divider. If there is 20 percent ripple on the unregulated supply and the resistance ratio is 100 (typical values), then the regulation is approximately 0.2 percent. The advantage of this supply is that it is the least expensive to build of the electronically regulated supplies. If you do build it, try to thermostat the supply.

A novel variation of the filtered high-voltage supply was designed by Elkstrand[31] and is shown in Figure 6.13. Instead of trying to regulate the voltage, this circuit uses an LM100 IC regulator and regulates the current passing through the photomultiplier tube. A full-wave rectifier operating off one winding of the power transformer T1 provides a 15-V bias voltage for the LM100. The other winding is used in a voltage doubler as in Figure 6.12 with the output passing through the photomultiplier tube divider chain that develops the operating voltages for the cathode and dynodes. Five cascaded transistors, Q1 through Q5, are used as the pass transistors, each therefore passing one-fifth of the total voltage. This is the most economical solution to the problem of handling the required voltage levels. Base drive is provided for the cascade string by R3 through R7 in a manner that does not affect regulation. Capacitors C1 through C5 suppress and equalize transients across the pass transistors, and clamp diodes across the sensitive emitter-base junctions of the transistors prevent damage from voltage transients.

6.7c RF Oscillator

This approach was used by Code.[23] A schematic of this circuit is reproduced in Figure 6.14. Because of the high voltages involved, the only convenient oscillator involves vacuum tubes, and therefore the design is cumbersome. Basically, part of the rectified output voltage from an RF oscillator running at 100 kHz is used in a feedback loop to control the amplitude of the oscillator. By changing the amount of feedback voltage, the output voltage is adjustable. Because the current drain from a photomultiplier tube is negligible (less than 1 μA), the RF oscillator does not need much power capability.

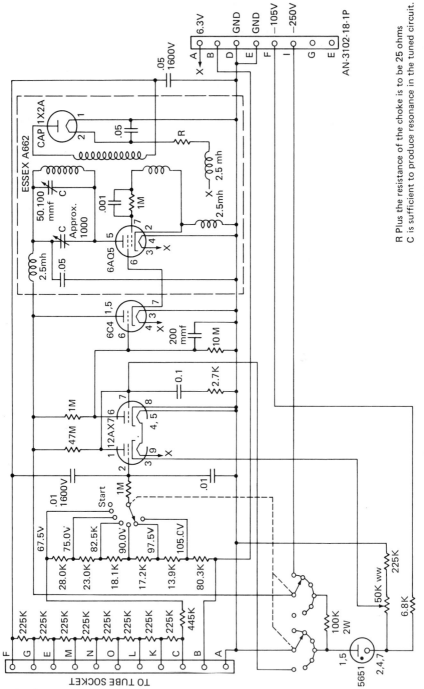

Figure 6.14 RF oscillator supply. (Reproduced from *Photoelectric Astronomy for Amateurs*, ed. F. B. Wood, Macmillan)

While obtaining the proper photomultiplier tube voltage in one step is impossible with transistors, there is another approach. Electronic flash units use an oscillator type of DC-DC supply that takes 4 to 6 V and transforms it to voltages in the 300 V range. These DC-DC supplies are available on the surplus market, and can be tied together in a series fashion to obtain the 1 kV voltage necessary for photomultiplier tubes. Regulation is then provided in the input voltages to the supplies, which can be obtained from standard IC regulators.

Commercial DC-DC supplies are available from companies such as RCA,[9] Venus,[32] Ortec,[30] EMI,[7] and Hamamatsu.[11] These supplies range from $100 to $500, and offer a small, convenient method of generating the high voltage for the photomultiplier tube. The input voltage is generally between 6 and 20 V to provide 900 to 1500 V output, with output regulation again controlled by the input voltage regulation. A DC-DC supply is an alternative to batteries for a small, portable supply.

6.7d Setup and Operation

To connect a high-voltage supply to the photomultiplier tube, coaxial cable such as RG58 can be used. For voltages above 1 kV, special high-voltage connectors of the SHV or MHV type should be used to gain immunity from possible dielectric breakdown. Always remember these precautions:

1. Never work on the high-voltage power supply with the power turned on.
2. Never disconnect the photomultiplier tube or the high-voltage cable to the photomultiplier tube with the high voltage turned on.
3. Bring the high voltage up and down slowly to avoid rapid changes.
4. Observe the correct polarity of the high-voltage power supply.

6.8 REFERENCE LIGHT SOURCES

As we mention in this and subsequent chapters, electronic equipment is generally temperature-sensitive. This means that if you use a set of A0 stars at the beginning of a night to determine the zero-point coefficients in the transformation equations, and then use these coefficients to reduce data taken many hours later, the results can be in error. There are other factors, such as telescope position dependence as a result of equipment flexure or magnetic fields, that also contribute errors to your results. To calibrate these irregularities out of the observations, astronomers use *standard light souces*. These come in three varieties: ground-mounted standard lamps, small radioactive sources, and stars.

Standard lamps are usually filament-operated devices with operating parameters kept constant. They cannot be mounted on the telescope because irregularities in the light output are common because of filament sag. In addition, they are difficult to use for astronomical purposes because of their relatively large output energy. To image the light source through the telescope requires that the energy collected by the telescope from the source must be equivalent in intensity to an average star. In general, such a lamp must be placed more than 100 meters away from the telescope. In addition, the filaments operate at a cool temperature in comparison to most stars and therefore temperature-sensitive errors tend to propagate in opposite directions.

Radioactive sources are usually weak beta emitters that are placed in contact with a phosphor. The fast electrons enter the phosphor crystals and produce light in one or more bands whose width is usually on the order of a hundred Angstroms. These sources can be made very small and are usually placed in the filter side or near the photomultiplier. This allows the observer to test the photometer for positional variations in addition to temperature and aging effects. The light output of these sources is usually blue-green, making calibrations a little easier than for the standard lamps. The phosphor in radioactive sources suffers from temperature effects equal to or greater than those of the photometer, and cannot be used for calibration unless thermostatted. In addition, the phosphor is highly light-sensitive, and there are long-term brightness variations that preclude calibrations over periods longer than a few months.

For the reasons listed above, few astronomers regularly use standard lamps or radioactive sources. They are used in initial calibration of the telescope and photometer and in cases where long-term accuracy (months for radioactive sources, longer for other types) is needed. An alternative to the ground-based calibration techniques is to use standard stars to monitor any equipment changes. This method has the advantage that it also measures atmospheric transparency changes. However, it suffers from increased complexity (extinction must be taken into account and often several stars are used), the problem of finding standard stars over the entire sky to measure flexure errors, and that calibrations take up observing time.

In general, we recommend using differential techniques in photometry. This eliminates almost all instrumental variations as they occur equally to both the comparison and the program star. If you intend to measure stars over the entire sky, to standardize your comparison stars for example, then use the North Polar Sequence stars to check transparency and temperature effects as their air mass (and therefore extinction) changes very little in the course of a night. In all cases, including extinction determinations, observe standards throughout the night, not just at the beginning or end.

6.9 SPECIALIZED PHOTOMETER DESIGNS

Earlier in this chapter we detailed the construction of a simple photometer that is certainly adequate for amateur and small telescopes. It is light, compact, and requires few specialized tools to construct. This section discusses other, more complicated designs that observatories have constructed. While the treatment is brief, it should be sufficient to inform you of other ideas and direct you to references where more detail can be found.

6.9a A Professional Single-beam Photometer

The photometer used at the Morgan Monroe Station of Goethe Link Observatory is a prime example of a professional instrument. Based on a design pioneered by William Hiltner, the photometer is shown in Figure 6.15. All external framework components are made of 0.635-centimeter (¼-inch) aluminum stock, milled with rabbet joints to form light-tight boxes. The assembled unit including the photomultiplier's cold box, weighs about 32 kilograms (70 pounds).

As seen from the top, where light enters the photometer from the telescope, we can identify the three major subdivisions of the instrument: *guide box, filter and diaphragm assembly,* and the *cold box.* The guide box has a military surplus eyepiece mounted on a movable X-Y platform for offset guiding and accurate star centering. This eyepiece and its associated mirror are placed in front of the diaphragm slide. The eyepiece has an illuminated reticle, with 30 arc second arms on the cross pattern. Once having centered the star in the diaphragm and adjusted the eyepiece position, the observer can repeatedly center stars in a 10 arc second diaphragm using the cross hairs without checking the diaphragm viewing eyepiece. Once a star is centered on the reticle the

158 ASTRONOMICAL PHOTOMETRY

Figure 6.15 The Goethe Link Observatory photometer.

viewing mirror is slid, parallel to its surface, until a hole in the mirror aligns with the optical axis and the starlight can enter the photometer. The eyepiece can be moved using its X-Y motion to find a guide star because the rest of the field is still available for viewing. The Erfle viewing eyepiece was chosen to provide sufficient eye relief so that an observer with eyeglasses can use the photometer easily.

The filter and diaphragm assembly begins with a detented filter slide that under normal circumstances contains an open hole and two neutral density filters, allowing the observer to perform photometry on stars as bright as zero magnitude. This slide can also be used to hold polarizing filters. The light beam then encounters a detented diaphragm slide with holes ranging from 10 to 120 arc seconds. Next the beam passes to a rotating filter wheel. Electronics control the motion of this wheel to move from one filter to another under programmed control. Manual motion is available with an externally mounted pushbutton, and the filter position number is shown on a seven-segment LED readout. The filter assembly is carefully machined and is removable with one screw, allowing convenient filter wheel and/or mechanism replacement.

The cold box has a self-contained dark slide and Fabry lens, allowing the observer to change photomultipliers during the night by switching cold-box assemblies. Through careful design, the Fabry lens does not require heating to remain frost-free. A single dry ice charge, about 1 kilogram (2.2 pounds), usually lasts an entire night. The pulse preamp and discriminator is mounted on the cold box, making good ground contact and matching the cold-box tube assembly with a preamp tailored to the photomultiplier tube's best operating conditions.

6.9b Chopping Photometers

The next level of complexity is to provide two light paths and to switch the detector back and forth between them. For instance, you could use two diaphragms in the focal plane and pick out the program star and a nearby comparison star. By directing the program or comparison star's light to the photomultiplier by the use of a mirror, you can "move" from one star to the other in a fraction of a second, nullifying any atmospheric changes. This setup allows differential photometry in one color during very poor sky conditions, such as uniform cirrus or broken cloud cover. An example of the dual-diaphragm arrangement is the photometer designed by Taylor[33] and shown in Figure 6.16. Here the photometer only chops between star and sky, which makes the mechanism simpler because the only moving element is the optical detent system; the filter and photometer assemblies remain the same. Because you are chopping between star and sky, you can only obtain accurate color indices. You must "intercompare" two stars to obtain differential magnitudes.

A similar system is used at Skibotn Valley in the Arctic Circle.[34] In this photometer the secondary mirror of the telescope is moved to pass the light of two separate stars, or the star and sky through a single diaphragm. This arrangement allowed the photometry of an eclipsing

160 ASTRONOMICAL PHOTOMETRY

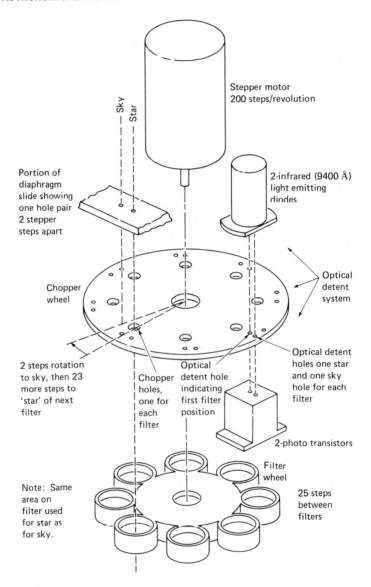

Figure 6.16 The Taylor dual-diaphragm photometer. (Courtesy of the *Publications* of the Astronomical Society of the Pacific)

binary star during an aurora that was as bright as a sixth-magnitude star and that varied by five magnitudes in 1 minute!

Chopping photometers work because clouds are relatively neutral in extinction; Serkowski[35] and Honeycutt[36] have shown that the effect of one magnitude of cloud extinction is only about 0.01 magnitude on the *UBV* colors. The biggest error is the variable transparency, easily accounted for by chopped photometers. Because only one detector is used, half of the time is usually spent observing the sky, decreasing the collecting efficiency.

An interesting variation of the chopping photometer was designed at Indiana University by De Veny.[37] This photometer uses two diaphragms of different sizes on the same star. By switching between the two diaphragms, you obtain a star and sky reading, and then a star and sky reading with a known additional amount of sky. You can then determine the sky background around the star mathematically and subtract it. The advantage of this system is that you are always measuring sky surrounding the star, while continuously observing the star, effectively multiplexing in the sky observations.

6.9c Dual-beam Photometers

A dual-beam photometer in this context means any system where two separate detectors are used. It can be built in one of two ways: either dividing the light from a single star into two components, or using the light from two stars separately.

The single-star photometer usually uses a dichroic beam-splitter to divide the beam into a blue and a red component. The response curve for such a beam splitter was calculated by Morbey and Fletcher[38] and is shown in Figure 6.17. Note that the division is not pure and sharp. This means that it is difficult to use a filter with the beam-splitter output and match the *UBV* colors exactly. Also, two separate detectors are used and their transformation coefficients must be known very accurately to obtain color indices. Half-aluminized mirrors could be used for the beam splitter, but then only half of the light in a given wavelength would be available, offering little gain over a chopping photometer. By using a dichroic splitter, you double throughput and prevent atmospheric variations from affecting the derived color index. An example of a single-star photometer as designed by Wood and Lockwood[39] is shown in Figure 6.18.

Basically, the dual-star photometer is two separate photometers mounted at the focal plane of the telescope. Usually one photometer is fixed on the optical axis and the other is movable in angle and radius to measure sky or a nearby comparison star. Because there is no beam splitter, no light is lost and no filter response change is involved. Usually

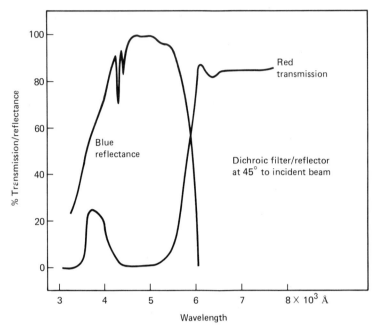

Figure 6.17 Response curves for dichroic filter/reflector. (Permission granted from the National Research Council of Canada.)

for accurate measurements, the role of the two photometers is interchanged to eliminate the instrumental response differences. An example of such a photometer as designed by Geyer and Hoffmann[40] is shown in Figure 6.19.

A novel approach to the dual-star photometer is the twin photometric reflector at Edinburgh.[41] Here two telescopes are contained on the same mount, with one continuously adjustable with respect to the other by several degrees. By pointing at one standard star with one telescope, the other telescope can determine a photoelectric sequence in a cluster in short order, or both telescopes can be used on a star for simultaneous two-color photometry.

PIN photodiodes would make an excellent dual-beam photometer. Small and lightweight, such a photometer would be feasible for moderate-sized amateur telescopes. This photometer would work well in areas with few photometric nights, such as the eastern U.S. The authors would like to hear about the design of this type of photometer for possible inclusion in future editions of this text.

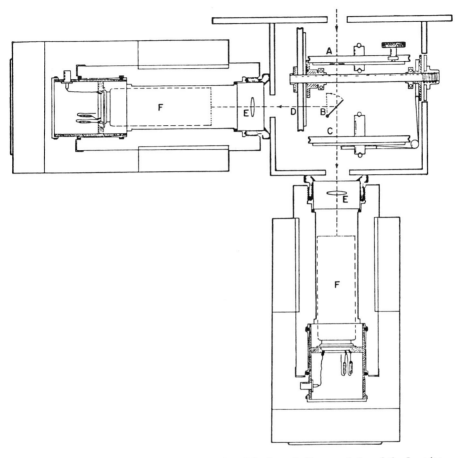

Figure 6.18 Single star photometer by Wood and Lockwood. (By permission of the Leander McCormick Observatory)

6.9d Multifilter Photometers

A multifilter photometer can also be classified as a coarse spectrometer. Light from a single star is broken into several beams that are measured individually. Dichroic beam splitters are not used because of their response curves; adequate separation of three or more colors is extremely difficult. Roberts[42] used aluminum beam splitters to obtain a three-channel photometer. Since aluminum is essentially neutral, three equal beams can be obtained with each passing through an appropriate

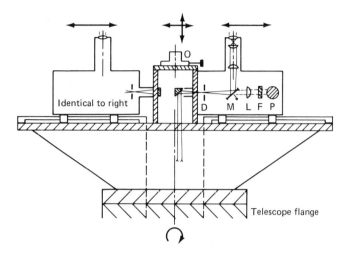

Figure 6.19 Schematic drawing of the Greyer photometer. *O* is the wide angle offset eyepiece, *D* the diaphragm wheel, *M* the periscope flip-flop mirror, *L* the Fabry lens, *F* the color filter wheel, and *P* the photomultiplier tube. (Courtesy of *Astronomy And Astrophysics*)

filter. However, each beam then contains only one-third of the light at any given wavelength so that the net result is the same as if you switched from one filter to another, in terms of throughput or net counts. However, the measurements are simultaneous.

A much more practical multifilter arrangement is the Walraven photometer.[43] Here a quartz prism is used to disperse the light, as in a spectrograph, and the resultant spectrum is sampled by five filter and detector combinations. It is impossible with such a combination to match the *UBV* system, as its wide-band response curves overlap each other. Medium- and narrow-band systems such as the Strömgren four-color and the Walraven five-color systems are ideally suited to multifilter photometers. The Mira group in California[44] have a 512-channel "photometer" covering the visible spectrum that they intend to use to acquire rapid spectrophotometry of the 125,000 stars of the Henry Draper Catalog visible in the Northern Hemisphere. As you can see, as the photometric instrumentation becomes more complicated, the dividing lines between types of instruments become very nebulous.

REFERENCES

1. Persha, G. 1980. *IAPPP Com.* **2**, 11.
2. De Lara, E., Chavarria K., Johnson, H. L., and Moreno, R., 1977. *Revista Mexicana de Astron. y Astrof.* **2**, 65.

3. Optec, Inc., 199 Smith, Lowell, MI 49331.
4. Burke, E. W. Jr., and Pippin, D. M. 1976. *Pub. A. S. P.* **88**, 561.
5. Edmund Scientific, 101 E. Gloucester Pike, Barrington, NJ 08007.
6. Jaegers, A., 691S Merrick Rd., Lynbrook, NY 11563.
7. EMI Gencom Inc., 80 Express St., Plainview, NY 11803.
8. ITT, Electro-Optical Products Division, 3700 E. Pontiac St., Fort Wayne, IN 46803.
9. RCA, Electro Optics and Devices, Lancaster PA 17604. RCA photomultipliers are available from electronics suppliers.
10. Fernie, J. D. 1974. *Pub. A. S. P.* **86**, 837.
11. Hamamatsu Corp., 420 South Ave., Middlesex, NJ 08846.
12 Perfection Mica Co., Magnetic Shield Division, 740 N. Thomas Drive, Bensenville, IL 60106.
13. Johnson, H. L. 1962. In *Astronomical Techniques,* Edited by W. Hiltner. Chicago: Univ. of Chicago Press, p. 157.
14. Pacific Precision Instruments, 1040 Shary Court, Concord, CA 94518.
15. EG & G Princeton Applied Research, P.O. Box 2565, Princeton, NJ 08540.
16. Products for Research, Inc., 88 Holten St., Danvers, MA 01923.
17. Corning Glass Works, Houghton Park, Corning, NY 14830. Corning filters may be ordered from Swift Glass Co., 104 Glass St., Elmira, NY 14902.
18. Schott Optical Glass Inc., 400 York Ave., Duryea, PA 18642.
19. EG & G Electro-Optics Division, 35 Congress St., Salem, MA 01970.
20. Dick, R., Fraser, A., Lossing, F., and Welch, D. 1978. *J. R. A. S. Canada* **72**, 40.
21. Photometry Committee. 1962. *Manual for Astronomical Photoelectric Photometry,* AAVSO, 187 Concord Ave., Cambridge, MA 02138.
22. Allen, W. H. 1980. *IAPPP Com.* **2**, 7.
23. Code, A. D. 1963. In *Photoelectric Astronomy for Amateurs.* Edited by F. B. Wood. New York: Macmillan.
24. Grauer, A. D., Pittman, C. E., and Russwurm, G. 1976. *Sky and Tel.* **52**, 86.
25. Stokes, A., 1980. In paper presented at IAPPP. Symposium, Dayton, OH.
26. *The Radio Amateur's Handbook.* Newington: The American Radio Relay League. Published yearly.
27. The Ted Dames Co., 308 Hickory St., Arlington, NJ 07032.
28. Kepco Inc., 131-38 Stanford Ave., Flushing, NY 11352.
29. Lambda Electronics Corp., 515 Broad Hollow Rd., Melville, NY 11747.
30. EG & G Ortec Inc., 100 Midland Rd., Oak Ridge, TN 37830.
31. Elkstrand, J. P. 1973. In *Linear Applications Handbook,* volume 1. National Semiconductor Corp. (A.N. 8).
32. Venus Scientific Inc., 399 Smith St., Farmingdale, NY 11735.
33. Taylor, D. J. 1980. *Pub. A. S. P.* **92**, 108.
34. Myrabø, H. K. 1978. *Observatory* **98**, 234.
35. Serkowski, K. 1970. *Pub. A. S. P.* **82**, 908.
36. Honeycutt, R. K. 1971. *Pub. A. S. P.* **83**, 502.
37. De Veny, J. B. 1967. *An Improved Technique for Photoelectric Measurement of Faint Stars.* Masters thesis, Indiana University.
38. Morbey, C. L., and Fletcher, J. M. 1974. *Pub. Dom. Ap. Obs.* **14**, 11.
39. Wood, H. J., and Lockwood, G. W. 1967. *Pub. Leander McCormick Obs.* **XV**, 25.

40. Geyer, H., and Hoffmann, M. 1975. *Ast. and Ap.* **38**, 359.
41. Reddish, V. C. 1966. *Sky and Tel.* **32**, 124.
42. Roberts, G. L. 1967. *Appl. Opt.* **6**, 907.
43. Walraven, T. and Walraven, J. H. 1960. *Bul. Ast. Inst. Neth.* **15**, 67.
44. Overbye, D. 1979. *Sky and Tel.* **57**, 223.
45. Nye, R. A. 1981. *Sky and Tel.* **62**, 496.

CHAPTER 7
PULSE-COUNTING ELECTRONICS

Pulse-counting systems are rapidly becoming comparable in expense to any other method of photoelectric photometry. A typical but very general layout of such a system is shown in Figure 7.1. The output from the photomultiplier is fed to the preamp, which amplifies the pulse, shapes it, and rejects noise pulses. This conditioned pulse is the input to the pulse counter, also known as a *frequency counter*. The pulse counter consists of three major parts: a counting circuit that counts every input pulse, a gate that allows pulses to reach the counter only for a specified time interval, and the timing circuit controlling gate.

The counts can be read directly from the counter or sent to a small computer through an interface. There the counts may be transformed to a crude magnitude scale. The data can be printed out on a teletype or displayed visually. It may be transferred to magnetic tape or disk to await further data reduction or for permanent storage.

Preamps and pulse counters are described in this chapter, along with some representative circuits. Some interfacing and testing procedures follow.

It should be emphasized that pulse counting cannot be performed with PIN diode photometers. Pulse counting requires hundreds of thousands of electrons for each incident photon. An incident photon produces only one electron-hole pair in a photodiode. Thus, with present photodiode technology, you must use DC methods.

7.1 PULSE AMPLIFIERS AND DISCRIMINATORS

The pulse amplifier increases the size and shapes the feeble pulse from the photomultiplier tube. The discriminator rejects pulses that are

168 ASTRONOMICAL PHOTOMETRY

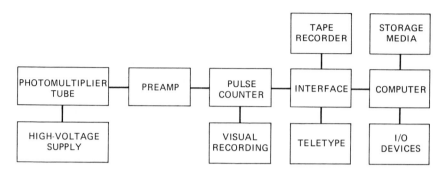

Figure 7.1. Block diagram of pulse counting system.

inherent to the photomultiplier tube itself and not from the source. The electronics that accomplish these two purposes is often in a single package, commonly called a *preamp*.

The amplification is necessary because each pulse contains on the order of a million (10^6) electrons, a current of only 10^{-12} A if averaged over 1 second. Most frequency counters require inputs of 100 mV (0.1 V) to count correctly. Therefore, using Ohm's law, we would have to use a series resistor of 10^{11} ohms to yield adequate counting voltage. This is a very difficult value to obtain.

If you look at the output of a typical photomultiplier tube at high time resolution, you would see something similar to Figure 7.2. Each pulse represents the output from a photon event, and the signal between pulses is the background or dark current of the tube. If you counted all of the pulses of a certain size occurring in a fixed time interval and

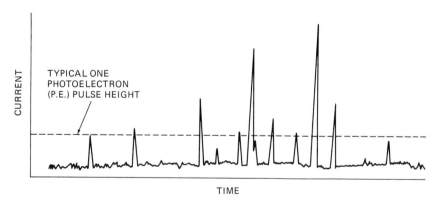

Figure 7.2. Typical photomultiplier output.

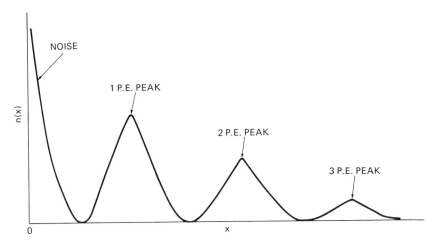

Figure 7.3. Pulse height distribution.

plotted your results, you would obtain a pulse height distribution. A theoretical pulse height distribution is shown in Figure 7.3. This shows that the number of background noise pulses decrease rapidly with the energy of the pulse. There are several peaks in the distribution, corresponding to the ejection of one or more electrons from the photocathode by the photon. Most events ejecting more than one electron are caused by cosmic rays and are few in number.

We do not want to amplify the noise pulses and count them along with photon events. Instead, we want to discriminate against them. This is usually achieved by setting a minimum threshold level below which no output pulse results. You can never eliminate all of the noise pulses because some arise on the photocathode itself and look like photon events, but by setting the threshold near the minimum between the noise and the one-photoelectron distribution, you will reject the maximum noise and accept the maximum signal. To be highly accurate, you would also reject the two and higher photoelectron pulses, because they are caused primarily by cosmic rays, and create a window discriminator. However, this trade-off is unnecessary because only a few pulses would be rejected with a large increase in circuit complexity.

A good pulse amplifier and discriminator should:

1. Have an output pulse no more than 50 nanoseconds wide, thereby providing a counting rate of about 20 MHz.

2. Have minimal temperature sensitivity.
3. Have stability and high noise immunity.
4. Be small, simple, and require only one operating voltage.
5. Be able to amplify a 0.5-mV pulse and provide a TTL-compatible output.

There are readily available commercial preamps. These include models from Princeton Applied Research,[1] Hamamatsu,[2] and Products for Research.[3] However, be prepared to spend several hundred dollars for one of these commercial devices. Amptek[4] has recently introduced hybrid charge-sensitive preamps that are the size of a dime. These could be mounted on the tube base and provide a very compact package. However, the current models are not sensitive enough for most astronomical applications. DuPuy[5] has recently published a circuit based on the MVL 100 single-chip amplifier. One limitation of this chip is that it may not have enough gain for some photomultiplier tubes.

7.2 A PRACTICAL PULSE AMPLIFIER AND DISCRIMINATOR

A preamp circuit that has been used by many observatories was described in 1973 by Taylor,[6] who recently revised the original circuit.[7] This enhanced preamp is presented in this section. The Taylor preamp was designed with simplicity and low cost in mind. The circuit is shown in Figure 7.4. It consists of nine 2N4124 transistors ($0.30 each) and a 1N3717 tunnel diode (about $10).

The first six transistors comprise the amplifier section and are connected as shunt-series feedback pairs, cascaded for an overall gain of about 1000. The tunnel diode monostable oscillator acts as the discriminator. When triggered, it generates a standard -0.5V pulse, which is buffered by an emitter follower, amplified and inverted, and fed to a second emitter follower to drive a 50-ohm cable to $+5$V. The shape of the output pulse is similar to the positive half of a sine wave with a base width of 20 nanoseconds. The discriminator level is adjusted by means of a current bias potentiometer. An IC regulator is included in the circuit to improve stability and eliminate the need for a separate zener regulator for the tunnel diode.

The circuit should be constructed on a single board. Point-to-point wiring is recommended, using a double-sided printed-circuit board as the chassis. Layout of the parts is not critical and no shielding between

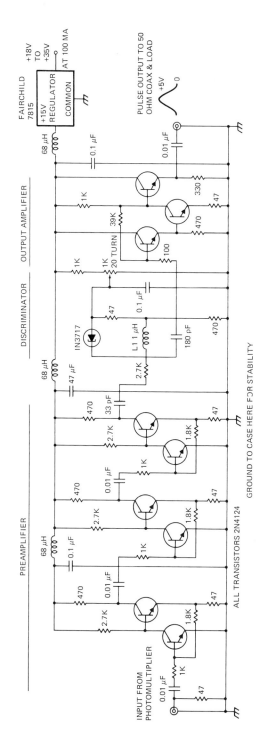

Figure 7.4. Schematic diagram of the improved pulse and amplifier circuit. All transistors are 2N4124's, L_1 is 1 μH (30 turns on a 1 Meg ½ watt resistor as a coil form). (Courtesy of the *Publications* of the Astronomical Society of the Pacific.)

stages is necessary, but a layout resembling the circuit diagram is suggested. After construction, the board should be mounted in a metal box for shielding. This can be a commercial box such as Pomona Electronics[8] model 3302 or constructed out of double-sided printed-circuit board material. The board should be grounded to the case in several places to prevent pulse doubling. BNC-type connectors should be used for the input and output.

The Taylor preamp is somewhat temperature-sensitive, losing sensitivity as the temperature decreases. The 1973 version had up to a 20 percent variation in the count rate with a 40 Celsius degree change. For this reason, obtaining accurate measurements requires thermostatting the circuit. Problems with temperature sensitivity can be minimized by using differential photometry and/or never observing near sunset when temperature variations are at their maximum.

With the pulse resolution of this preamp, dead-time corrections start to become important at around 100,000 counts per second. Use the methods discussed in Chapter 4 to correct for this error.

7.3 PULSE COUNTERS

A frequency or pulse counter for astronomical purposes has certain requirements:

1. Counting ability to 100 MHz or higher.
2. Selectable time base, with at least 1- and 10-second gating times for manual use, and 0.001- and 0.01-second gating for occultation observations.
3. Capability of external gate triggering and counter reset.
4. BCD or binary output for computer interfacing.

The last two items may not be necessary immediately when the counter purchase is contemplated, but should be considered for future applications. A rule of thumb is to be able to count 30 times faster than the most rapid anticipated nonuniform rate; 100 MHz gives a large margin of error in most cases. For astronomical purposes, a counting accuracy of 0.1 percent is entirely adequate. This is a condition met by all commercial frequency counters.

Examples of adequate commercial frequency counters are the Optoelectronics[9] 7010, Heathkit[10] SM-2420, and Hal-Tronix[11] HAL-600A;

all are 600-MHz counters using the newly released ICM 7216 frequency counter IC. Introduced by Intersil,[12] the 7216 has enormous potential for astronomical use because of its simplicity and low cost (under $25). The only additional major parts required for a 100-MHz counter with a selectable timebase are a 10-MHz crystal, a divide-by-ten prescaler (11C90 or 95H90), and a LED display of up to eight digits. Its only disadvantage is the very complicated interface for computer applications. Still, the ICM 7216 has allowed manufacturers to supply adequate frequency counters for manual use in the $100 to $200 price range.

If you want to build your own counter from scratch, consider the ICM 7216 and consult the data sheets supplied by Intersil. Other ICs are available to perform the major functions, such as a six-decade counter. The use of discrete ICs in building a counter makes for a cumbersome design, but has the advantage of easy computer interfacing as all signals are present continuously.

Normal quoted accuracies of the time bases for frequency counters are around 10 ppm/C°. This means that a change of 10 Celsius degrees causes a 0.01 percent change in the gating time, which is insignificant for overnight use. This error could become important if a less accurate time base were used. In addition, most commercial grade ICs quit working at 0°C (32°F), and the frequency response of both the pulse shaping input and the counter degrade as the temperature approaches zero. In other words, for best results the counter should be thermostatted to a constant temperature in summer and winter, just like the preamp. If you must operate without thermostatting, use the military (5400 series) instead of the commercial grade (7400 series) ICs.

7.4 A GENERAL-PURPOSE PULSE COUNTER

In this section, we describe a general-purpose frequency counter constructed from discrete integrated circuits. It contains 19 ICs and would cost around $100 to construct. Though more complicated than a counter using the ICM 7216 frequency counter IC, its main advantage is ease of computer control. All outputs are latched and can be brought out to a connector for computer input, and time base select and reset functions are simple to interface for computer control.

The counter is shown in Figures 7.5 through 7.7. The maximum count rate is 100 MHz, controlled by the 74S00 gate and the 74S196

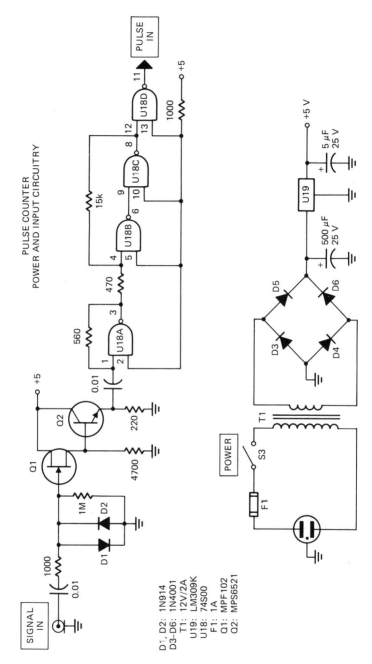

Figure 7.5. Power and input circuitry.

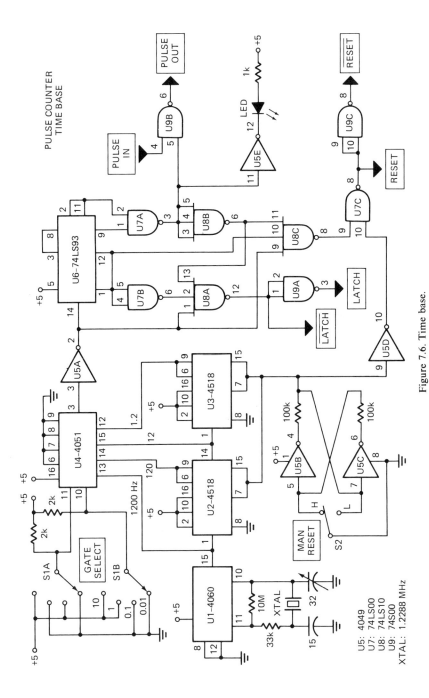

Figure 7.6. Time base.

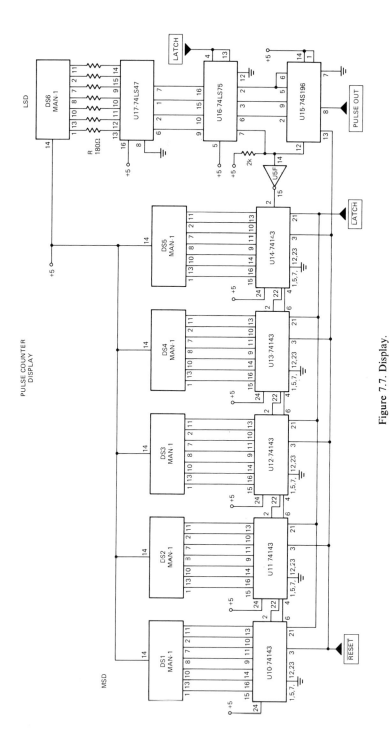

Figure 7.7. Display.

decade counter. If lower count rates are acceptable, the 74S196 can be replaced by a 74196 or another decade chip with some minor rewiring.

The input section amplifies the signal using a field-effect transistor (FET) and then conditions it into a square wave using device U18. This section can be eliminated if the counter's only use is for photometry, where there is a separate preamp in which the output pulse drives the gating circuitry directly. The pulse is routed through a timing gate and on to a series of decade counters. The 74143 is a combination counter, latch, and decoder-driver and would be used for all stages except that it has a low counting rate, 18 MHz. The 74S196 is negative-edge triggered, and its output must be inverted to drive the positive-edge triggered 74143. All LED displays are common-anode MAN-1 equivalents.

A CMOS 4060 IC is used as the oscillator. It divides the 1.2288-MHz crystal frequency by 2^{10} (1024) to provide a 1200-Hz square-wave output. This is further divided by the dual-decade 4518 counters, and four frequencies (1200, 120, 12, and 1.2 Hz) are fed to a 4051 demultiplexer. The desired frequency is selected by a DP4T rotary switch and routed to a divide-by-12 circuit. This opens the gate for 10 pulses, latches for one, and resets for one. Therefore, the final output gating times are 0.01, 0.1, 1, and 10 seconds, with 0.002, 0.02, 0.2, and 2 seconds, respectively, of dead-time between subsequent gatings. Note that the 4051 can accept up to 8 inputs, so that the second output stage of U3 (100 seconds) and the 2^7 output of the 4060 (1.2 milliseconds) could also be included in the gating selection with a larger switch.

The power supply is conventional with a full-wave bridge and an IC regulator. The bridge could be replaced by a single unit instead of four individual diodes. Be sure that the +5 V line on each board is bypassed by a 10-μF tantalum capacitor and that each counter IC is bypassed individually by 0.01-μF disc capacitors for noise immunity.

Our version of the counter was constructed on four boards: preamp, time base (and 74S196/74LS75), display, and power supply. Wire-wrap techniques were used and the final product was placed in a Radio Shack 270-270 cabinet.

To interface this counter to a computer, bring the 24 latched BCD lines to a back 25-pin connector along with a ground lead. The latch signal should also be available to the computer as you should not read the data while latching occurs. The computer should control the time base selection (add a DPDT toggle switch to change from manual to automatic time base select) and a reset signal to start counting (use one of the unused NAND gates in a similar manner to U7C).

7.5 A MICROPROCESSOR PULSE COUNTER

A high-speed pulse-counting board has been designed and built by Kephart.[13] This versatile board for the S-100 microcomputer bus satisfies the need for high time resolution (1 millisecond) for lunar occultation work and for moderate time resolution (0.1 second to several minutes) for multichannel applications.

Because of the complexity of the board and the fact that it is designed around a specific computer system (an 8080 with the S-100 bus), we do not give a complete schematic. Rather, Figure 7.8 shows a block diagram of the basic circuit in sufficient detail so that its logic can be implemented in other designs.

Two 21-bit counters are used as data counters. They each use one-half of a high-speed 74S112 J-K flip-flop for the least significant bit (LSB), with the pulse input routed to the J input and a control select signal to the K input. The remaining 20 bits are obtained from a 74197 and two 74393 binary counters. A third 16-bit counter using two 74393 ICs is used for interval timing.

For the millisecond time resolution application, the two data counters are used in a double-buffer mode to reduce the dead-time to a few nanoseconds of gate propagation time. One counter is disabled, read, and cleared while the other counter is acquiring data; at the end of the counting period the roles of the two counters are reversed. Dual-channel counting is performed by disabling, reading, and clearing both counters simultaneously, resulting in a dead-time of a few tens of microseconds per readout.

Communication with the microcomputer is through two parallel I/O ports (implemented on the pulse-counting board with Intel 8212 chips) and an interrupt instruction latch. One of the output ports from the microprocessor is used as a counting control latch. The other output port is used to set a comparison latch (with 7485 comparators) for the third onboard interval timer counter. An input port to the microcomputer is used to transfer any selected eight-bit byte from either 21-bit data counter to the CPU. The multiplexing is accomplished with 8T97 tri-state buffers. The CPU can be interrupted by the board, indicating to the CPU that service to the board is required.

The counting control latch byte from the CPU to the board is used to select functions of the pulse-counting board. From information written into the counting control latch, either interrupts or pulse counters

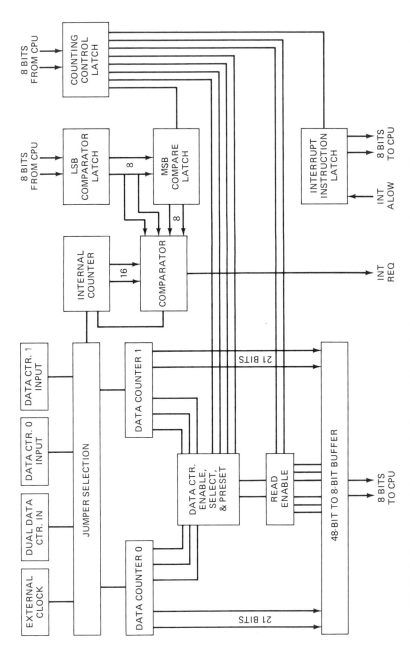

Figure 7.8. Block diagram of photon counting board. Published in the Proceedings of The Society of Photo-Optical Instrumentation Engineers, Volume 172, *Instrumentation in Astronomy III*, Bellingham, Washington.

or both can be disabled, counter selection (determining which counter is in the read mode) can be made, complete board reset or individual counter resets can be performed, byte selection between the MSB and LSB of the interval comparison latch, and byte selection for reading the 21-bit counters can be made.

The interval counter can be used to count either an external clock pulse or a signal from the S-100 bus. By setting the desired interval through software into the interval counter comparison latch, time resolution can be controlled by the user. The board creates an interrupt when the interval counter reaches the number held in the interval counter latch. This counter has 16-bit resolution allowing up to 65,535 clock pulses to be counted per interval. If a millisecond clock is used, then intervals from 0.001 to 65.535 seconds can be selected. This design allows the user to select the desired time resolution using interrupt control. The CPU can be used to display real-time data, reduce a previous observation, or perform any other desired task until an interrupt is issued from the pulse counting board, at which time the board is serviced and the CPU then returns to its previous task.

Figure 7.9. Microprocessor pulse counter.

Construction is straightforward, using wire-wrap techniques on a Vector prototype design board. As constructed, the board would cost about $125. Figure 7.9 shows the finished pulse counter. A second-generation counter would use the Intel 8255 triple-port I/0 chips to decrease chip density and power consumption. Such a microprocessor-controlled pulse-counting board demonstrates the versatility that can be achieved with a computer–pulse counting marriage.

7.6 PULSE GENERATORS

A useful piece of test equipment for the photometrist is the pulse generator. It produces pulses of known frequency, height, and duration that can be used to test frequency counters and preamps.

An example of a commercial pulse generator is the Continental Specialties Corporation[14] model 4001 (under $200). It has a frequency range of 0.5 Hz to 5 MHz, with pulses 100 mV to 10 V high and 100 nanoseconds to 1 second wide. Other generators are available from large manufacturers like Hewlett-Packard with tighter specifications and more ranges.

However, for testing photomultiplier preamps, pulses of −0.5 mV are desirable. A simple pulser satisfying this need is shown in Figure 7.10. Constructed at the Indiana University electronic shop, this pulser puts out a 0.5-mV, 500-nanosecond negative pulse to a 50-ohm load. This is a simple, inexpensive way to test a photomultiplier preamp. The

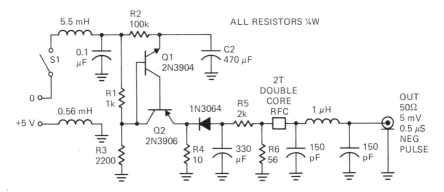

Figure 7.10. Simple pulse generator.

transistors Q1 and Q2 generate a ramp, with R2 and C2 controlling the ramp frequency. The R4/C2 pair provide the decay time of the ramp and D1 shapes the pulse. A voltage divider is formed by R5/R6, providing the 0.5-mV output pulse.

7.7 SETUP AND OPERATION

Pulse-counting systems are very sensitive to stray capacitances and noise. Stray impulses caused by heaters, motors, and relays turning on and off, along with other sources, are counted by the pulse counter and preamp just as if they came from the photomultiplier tube itself. To prevent this interference, connect the preamp solidly to the photomultiplier tube assembly. This keeps the interconnecting cable as short as possible and creates a common ground plane. Bypass all power leads with LC circuits to route all high-frequency interference to ground.

RG58 coxial cable should be connected between the tube and the preamp, and again between the preamp and the pulse counter. RG58 coax matches the input and output impedances of the preamp properly, and if connected to the 50-ohm input of a pulse counter will present proper termination to the preamp. If the coax is not terminated with 50 ohms, there will be a mismatch, and the fast pulses will be reflected back and forth giving rise to "ringing," where the pulse counter will count several pulses for each actual pulse from the preamp.

The discriminator level adjustment is one of trial and error. For an uncooled 1P21 tube, the final count rate for dark current should be about 200 counts per second. For a dry ice cooled tube, adjust the discriminator to allow about one or two counts per second. Taylor suggests that his preamp can be adjusted by attaching the preamp to a counter (but with no photomultiplier attached) and increasing the discriminator bias by the potentiometer until the discriminator oscillates. Then back off on the bias until first the oscillation and then the stray counts from amplifier-noise peaks cease. You still have to make final adjustments at the telescope to get optimum noise discrimination.

Setting the high voltage for an optimum signal-to-noise ratio is easier than in the case of DC amplifiers. First expose the photomultiplier tube to a constant light source (starlight for example) and then increase the voltage in steps of about 100 V. The observed count rate increases rapidly until a plateau is reached at at which the count rate from the source increases only slightly with increased voltage. Further increases in voltage serve no useful purpose; generally the dark current rises with no significant corresponding increases in signal count. For the 1P21 photometers in use at Indiana University, we have found that a high voltage around -900 to -950 V is optimum. To avoid dead-time effects during the voltage increase, pick a source that should eventually yield about 100,000 counts per second.

REFERENCES

1. EG & G Princeton Applied Research, P. O. Box 2565, Princeton, NJ 08540.
2. Hamamatsu Corp. 420 South Ave., Middlesex, NJ 08846.
3. Products for Research, Inc., 88 Holten St., Danvers, MA 01923.
4. Amptek, Inc., 6 De Angelo Dr., Bedford, MA 01730.
5. DuPuy, D. L. 1981. *Pub A.S.P.,* **93**, 144.
6. Taylor, D. J. 1972. *Pub. A. S. P.* **84**, 379.
7. Taylor, D. J. 1980. *Pub. A. S. P.* **92**, 108.
8. ITT Pomona Electronics, 1500 E. Ninth St., Pomona, CA 91766.
9. Optoelectronics Inc., 5821 N.E. 14th Ave. Ft. Lauderdale, FL 33334.
10. Heath Co., Benton Harbor, MI 49022.
11. Hal-Tronix, P. O. Box 1101, Southgate, MI 48195.
12. Intersil Inc., 10710 N. Tantau Ave., Cupertino, CA 95014.
13. Honeycutt, R. K., Kephart, J. E., and Henden, A. A., 1979. In *Instrumentation in Astronomy III*. Edited by D. L. Crawford. Society of Photo-Optical Instrumentation Engineers Proceedings, **172**, 408.
14. Continental Specialties Corporation, P. O. Box 1942, New Haven, CT 06509.

CHAPTER 8
DC ELECTRONICS

The photomultiplier tube is a high-gain current amplifier. For each electron released at the photocathode by a detected photon, about one million electrons are collected at the anode. A stream of incident photons generates a series of closely spaced bursts of current at the anode. In the pulse-counting mode, the goal is to count these bursts over a selected time interval. In the DC technique, the current is not resolved into bursts of current, but instead is averaged to give a continuous current. Despite the large amplification of the tube, the output current is extremely small and requires further amplification so that it can be measured easily. This current amplifier must be extremely linear because the photomultiplier's output current is directly proportional to the incident light flux. The 1P21 photomultiplier has a typical dark current of 10^{-9} A at room temperature. The amplifier should be capable of raising this to an easily measurable value, of about 1 mA (10^{-3} A). Thus, the amplifier should have a current gain of 10^6. This is easily achieved with some very simple electronics. On the other hand, the photodiode has an internal gain of unity and therefore requires an amplifier of much higher gain. This presents some special amplifier design problems, as discussed by Persha.[1] To date, very little has been published about amplifier designs for photodiodes used as astronomical detectors. For this reason, we restrict this chapter to an amplifier designed for use with a photomultiplier tube.

At this point, a very brief review of operational amplifiers (op amps) is needed. A more complete and lucid discussion of this topic can be found in the book by Melen and Garland.[2] The following discussion assumes a background in elementary electronics.

8.1 OPERATIONAL AMPLIFIERS

A few years ago, an *operational amplifier* was large, costly, fragile, and had a rather large power consumption. A modern op amp can be fabricated on a tiny silicon chip at a cost of a few dollars. Each chip may contain the equivalent of dozens of transistors, resistors, and capacitors. The details of the internal operation are not necessary for the present discussion. Figure 8.1 shows the symbol for an op amp. The op amp is a high-gain voltage amplifier. The "$-$" terminal is called the *inverting input* and the "$+$" terminal is called the *noninverting input*. An increasing voltage applied to the inverting input results in a decreasing voltage at the output (E_{out}). The same voltage applied to the noninverting input results in an increasing voltage at the output. If the same signal voltage was applied simultaneously to both inputs, the two amplified signals would be 180° out of phase and would cancel each other completely. The output is the amplified voltage difference between the two inputs. The output is unaffected by voltage changes that occur at both inputs; only the difference is amplified. It is sufficient for most of this discussion to consider the op amp to have "ideal" characteristics, namely infinite input impedance, infinite voltage gain, and zero output impedance. Figure 8.2 shows the op amp used as an inverting *voltage* amplifier. The *open-loop gain, A,* is defined as

$$A = \frac{E_{out}}{E_{in}}$$

where E_{in} and E_{out} are the input and output voltages, respectively. For an ideal op amp, A is infinite. For a practical op amp, A is 10^4 to 10^6. Real input impedances are typically 100 kΩ but the input impedance of op amps utilizing FETs can reach 10^{12} ohms or more. A typical output impedance is 50 ohms.

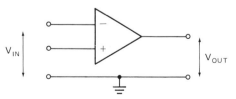

Figure 8.1. Op amp symbol.

186 ASTRONOMICAL PHOTOMETRY

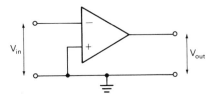

Figure 8.2. Voltage amplifier.

With the background developed above, it is now possible to discuss how the op amp is used as a *current* amplifier for DC photometry. Figure 8.3 shows a simplified current amplifier circuit. This particular type of circuit is *not* recommended. It is illustrated here as an example of a circuit design to be avoided. This type of circuit has been used in astronomical photometry in the past without a general appreciation of its inherent inaccuracy. An example of this type of circuit can be found in Wood.[3] We do not discuss the operation of this circuit except to point out the source of the problem. An input current from the photomultiplier tube, I_i, flows through R_L to ground instead of entering the higher-impedance amplifier. This elevates the potential at point B to $I_i R_L$. This means that the potential difference between the anode of the photomultiplier and ground has been changed slightly. The amount of change depends on I_i, which in turn depends on the brightness of the star. Young[4] has shown that this can result in a nonlinear tube response of a few tenths of a percent. This is a small error, but it need not be tolerated because there is a very simple solution. Circuits that use an anode load resistor should be avoided. The anode should always see ground potential directly.

Figure 8.4 illustrates the necessary circuit. In this type of circuit point a, the input seen by the anode, is always very nearly at ground potential. To see this, suppose a small positive external voltage, E_t, is applied at

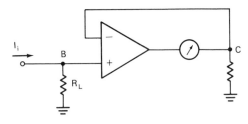

Figure 8.3. Current amplifier (not recommended).

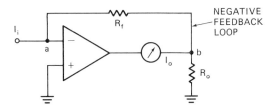

Figure 8.4. Current amplifier utilizing virtual ground.

the inverting input. This results in a negative output voltage, E_o. The output is connected, via the feedback loop, to the inverting input. The total input voltage at point a, E_a, is then

$$E_a = E_i + E_o. \tag{8.1}$$

The output voltage, E_o, is related to the input, E_a, and the voltage gain, A, by

$$E_o = -AE_a.$$

The minus sign results from the use of the inverting input. Combining the above two equations yields

$$E_i = E_a(1 + A).$$

By Equation 8.1,

$$E_o = E_a - E_i$$
$$= E_a - E_a(1 + A).$$

Thus the input and output voltage are related by

$$\frac{E_o}{E_i} = \frac{E_a - E_a(1 + A)}{E_a(1 + A)} = \frac{-A}{1 + A}.$$

As long as A is a large number, then

$$E_o \simeq -E_i. \tag{8.2}$$

Combining Equations 8.1 and 8.2, we obtain

$$E_a \simeq 0.$$

Point *a* is said to be at *virtual ground* because it is essentially at ground potential, but current does not flow to ground at this point. The anode of the photomultiplier always sees ground potential when connected to this circuit and the tube linearity is not affected.

Now consider specifically how this circuit is used as a current amplifier. An input current from the photomultiplier flows through resistors R_f and R_o to ground. Only a negligible amount of current enters the amplifier because of its very high input impedance. Because R_o is always made very small compared to R_f, point *a* would seem to be at a potential of $I_i R_f$ with respect to ground. This cannot happen, by our discussion above. By Equation 8.2, the voltage at point *b* must be $-I_i R_f$. This potential results in current flowing between the amplifier output, resistor R_o, and ground. This output current, I_o, must be related to the input current by

$$I_i R_f = -I_o R_o$$

or

$$I_o = -I_i \frac{R_f}{R_o}. \tag{8.3}$$

There is a linear amplification value that depends on the ratio of these two resistors and not on the characteristics of the op amp itself. While this is strictly true only for an ideal amplifier, it does show that a circuit that is very insensitive to changes within the op amp can be built. It would be very difficult to find an op amp with amplification stable enough for photometry, especially with the large temperature changes found in an observatory, if the feedback loop were not utilized. Even so, it is advisable to purchase a high-quality op amp to insure stability.

8.2 AN OP-AMP DC AMPLIFIER

Figure 8.5 shows a practical DC amplifier circuit. The op amp used is a Sylvania ECG 940 which has an FET input with an impedance of

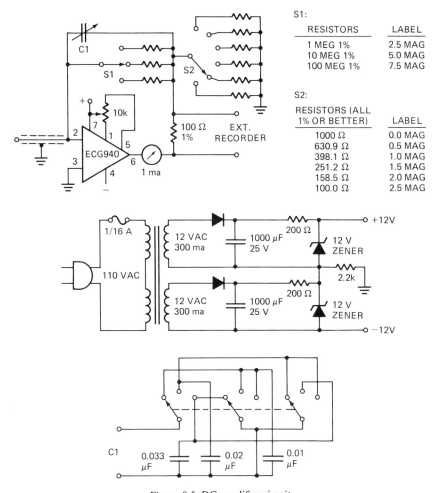

Figure 8.5. DC amplifier circuit.

about 10^{12} ohms and an open-loop gain of 10^6. This amplifier has a specified operating range from 0° to 70°C (32° to 158°F), which means that the amplifier must be kept warm in the winter. This is not a big disadvantage because it is always a good idea to operate the electronics at a constant temperature to avoid drift. Other op amps with a wider temperature range, such as the Analog Devices AD523K can be used. This latter device has a range from −55° to +125°C, but costs twice as much.

The resistors R_f and R_o have been replaced by switches that allow various combinations of resistors, and hence, various combinations of current gain to be selected. To make the gain large, the resistors on switch S1 (R_f) must be made large and those on switch S2 (R_o) rather small. The 1-megohm resistor of switch S1 is used for the brightest stars and the and the 100-megohm resistor is used for the faintest stars. As discussed in Section 6.2, the 1P21 shows fatigue effects if the tube current exceeds 10^{-6} A. Because our amplifier does not give a direct readout of the tube current, it would be advantageous to have some safety mechanism to let us know when this level is reached. The simplest approach is to design the amplifier so that a tube current of 10^{-6} A yields a full-scale deflection on the amplifier meter when the lowest gain setting is used. Because a full-scale reading on the meter is 10^{-3} A, the lowest current gain should be 10^3. This sets the value of the largest R_o resistor at 1000 ohms by Equation 8.3, since the lowest R_f is 10^6. The remaining resistors in switch S2 decrease in 0.5 magnitude steps; that is each is smaller than its predecessor by a factor of 0.6310. The resistors in switch S1 change by a factor of 10, yielding 2.5 magnitude steps. With the highest current gain ($R_f = 10^8$, $R_o = 10^2$), an input current of 10^{-9} A, which is the dark current of the 1P21 at room temperature, gives a full-scale deflection.

If you intend to use this amplifier with a cooled photomultiplier tube, one or two more resistors should be added to switch S1 with values of 1000 and 10,000 megohms, respectively. These resistors are required to take advantage of the reduction (by a factor of 100) in dark current, which allows much fainter stars, and hence, much lower currents to be measured. Unlike the idealized case, an actual op amp draws a small amount of input current during operation. This is referred to as the *input bias current*. This current itself can be a noise source just like the dark current from the photomultiplier. The op amp used in this circuit has an input bias current of 10^{-10} A, which is 10 times less than the dark current from an uncooled 1P21 and is therefore negligible. If you use a cooled tube, however, the bias current will become the dominant noise source when measuring faint stars. If you plan to use a cooled tube, the op amp in Figure 8.5 should be replaced by one with an input bias current of 10^{-12} A or less. Such op amps are available but are more expensive.

The feedback resistors have large values to achieve high amplifier gain and to minimize noise. Thermal (or Johnson) noise in these resistors varies with the square root of the resistance. Large values of R_f make this noise small compared to the current to be measured. The feedback resistors, R_f, should be accurate to 1 percent or better. Victoreen Instrument Company[5] can supply high-megohm resistors in glass encapsulation for the required accuracy. The R_o resistors have much lower values of resistance, which makes them inherently more stable, and they need not be glass-encapsulated. However, they should be accurate to 1 percent or better, because precision resistors are more stable than ordinary carbon resistors. It is impossible to find 1 percent resistors that equal the listed values exactly, so the values should be matched within 10 percent. A perfect match is not necessary because the amplifier is calibrated after construction.

The 10 kΩ potentiometer in Figure 8.5 is used to balance the circuit so that a zero input current produces a zero output current. An ordinary one-turn pot can be used but a five-turn pot makes circuit balancing much easier, especially at the high-gain settings. The purpose of selector switch C1 is to add a time constant that helps to smooth variations resulting from atmospheric scintillation and tube noise. A disadvantage of this circuit is that the time constant varies with the S1 switch position. The time constant, which is the product of R_f and C1, is negligible at the lowest gain setting and becomes significant only on the highest setting. Table 8.1 lists the time constant for each combination of R_f and C1. The capacitor values in this circuit can be changed to obtain other time constants. The variation in time constant with R_f is not a serious problem because larger time constants are preferred when higher gain settings are in use. It would be more convenient to be able to use the same time constant for all gain settings. The circuit designed by Oliver[6] avoids this problem by using a second op amp, of unity gain, connected to the output of the first op amp. This second amplifier drives the meter and any external recorder. It also has its own adjustable RC time constant in its feedback loop, which is independent of R_f in the first amplifier.

Selector switch C1 is wired so that each capacitor is shorted when

TABLE 8.1. Amplifier Time Constants

R_f \ C1	0.01 μF	0.02 μF	0.03 μF
1 Meg	0.01 sec	0.02 sec	0.03 sec
10 Meg	0.1	0.2	0.3
100 Meg	1.0	2.0	3.0

Figure 8.6. The DC amplifier, front view (top), and rear view (bottom).

not in use. This prevents some unwanted voltage spike from appearing at the amplifier input when a new capacitor is switched into the circuit. The 100-ohm resistor in the output circuit produces a voltage drop for an external chart recorder. An output of 1 mA produces a 100-mV drop, which produces a full-scale reading on a 100-mV chart recorder. The value of this resistor may be changed for recorders with different full-scale sensitivity.

The power supply circuit is taken from Stokes.[7] He found this simple zener-regulated design adequate for an earlier DC amplifier design. Both the amplifier and power supply can be built into one small chassis. Figure 8.6 shows a photograph of the completed unit, which is small enough to be mounted directly on the telescope if so desired.

There are two important comments to be made about the operation of this amplifier. First, FET devices, such as the op amp in this circuit, are damaged easily by an electrostatic charge at the input. As a precaution, always turn the amplifier on before connecting the signal cable. This prevents damage from any charge accumulated by the cable. Also, be careful when handling the op amp in a dry-air environment. Always "discharge" yourself by touching a ground, such as a water pipe, before handling the op amp. The second comment concerns zero-point drift. During the first 20 minutes of operation, there is substantial drift on the high-gain settings. It is therefore a good idea to turn on the electronics at least one-half hour before observing.

The circuit presented here is inexpensive and very simple to build. A more advanced circuit has been designed by Oliver.[6] As already mentioned, this circuit handles the time constant problem nicely. It has some other valuable features such as an internal constant current source for calibration and sky background cancellation. Although these extra features increase the cost, this circuit should be seriously considered by the advanced observer.

8.3 CHART RECORDERS AND METERS

Once your amplifier, high-voltage supply, and photometer head are built, you are ready to begin making some measurements. The question then arises as to the method of recording the data. The simplest technique is to read the amplifier meter and record the measurement with pencil and paper. To achieve good photometric accuracy, you must be able to read the meter to an accuracy of at least 1 percent. This is obviously not possible with the tiny edge meter seen in Figure 8.6. This meter was intended to be used only to monitor the functioning of the amplifier. If you plan to "meter read," you must invest in a larger meter of good quality. The minimum size should be 7.5 centimeters (3 inches) or preferably larger. The meter should have a quoted accuracy of 1 percent or better at full scale. Meters with a mirrored scale are preferred because they minimize parallax. The meter can be mounted in the amplifier chassis or in its own separate chassis connected to the amplifier with a cable. The later arrangement is convenient if you plan to mount the amplifier on the telescope. If you plan to take measurements with a meter, invest in a good one.

The disadvantages to meter reading are obvious. Atmospheric scintillation causes the needle position to jitter, making estimates difficult. The problem is complicated by observer fatigue, especially after 3 a.m.! Above all, there is no permanent record of the actual meter output. If the meter reading is written incorrectly, the observation is lost forever. Any DC observer who can afford it quickly invests in a *strip-chart recorder*. A strip-chart recorder consists of a device that drives a long roll of chart paper under a pen whose transverse displacement across the width of the paper is proportional to the input voltage. This device reduces the amount of effort required at the telescope. The observer needs only to write comments on the paper occasionally such as the amplifier gain or the time. Because the chart paper advances at a constant rate, the time of any observation can be found by interpolation. The chart record can be studied by an alert observer the following day. A permanent record exists that can be checked if an observation fails to reduce properly. Figure 9.9 illustrates the appearance of a chart recording for a series of observations.

There are a number of characteristics you should look for in a chart recorder to be used for photometry. There are a number of small and inexpensive models on the market. Unfortunately, these units must use narrow paper. This limits the accuracy to which the pen tracing can be read. It is best to buy a recorder that uses chart paper at least 25 centimeters (10 inches) wide. The recorder should be a DC voltmeter type with a range from zero to a few hundred millivolts. The latter value is not critical because the resistor in the output circuit of Figure 8.5 can be changed. A 100-mV input is used commonly by chart recorder manufacturers. The accuracy of the recorder at full scale should be at least one percent. Recorders with 0.5 percent accuracy are readily available. Finally, the chart speed must be considered. Most recorders have an adjustable speed. Experience has shown that a speed of 1 to 5 centimeters per minute is a good choice for most kinds of photometry. Be sure that the recorder you consider has a speed in this range.

There are many companies that manufacture chart recorders to meet the above criteria. Examples are the Markson Science Incorporated[8] model 5740, the Cole-Parmer Instrument[9] models C-8386-32 and C-8373-00, the Hewlett-Packard[10] model 7131A, and the Heath[11] model

IR-18M. This is just a very short list and the inclusion or omission of a company's name does not constitute an endorsement or criticism of their products. This list is merely a starting place for the would-be chart recorder owner. Unfortunately, all of these recorders, with the exception of the Heath IR-18M, cost over $700. This price is certainly beyond a limited budget. Consequently, most amateur astronomers have turned to the less expensive $230 Heath recorder. This recorder is not so strong mechanically as the other recorders but it does meet all of the selection criteria listed above. Experience has shown that this recorder, as most others, does not function well in the cold, winterlike environment of an observatory. A small heated enclosure solves the problem nicely. If you plan to do a lot of DC photometry, a strip-chart recorder is a very worthwhile investment.

8.4 VOLTAGE-TO-FREQUENCY CONVERTERS

One of the advantages of pulse counting over DC is the digital output, which frees the observer from making amplifier gain adjustments and calibrations. There is another approach to DC photometry that has these same attributes. The traditional DC amplifier is replaced by a voltage-to-frequency converter (VFC) circuit. The basic idea is to convert the current output of the photomultiplier to voltage which can serve as the input of a voltage-controlled oscillator. This oscillator has a frequency that varies linearly with the input voltage. The output of this oscillator is fed to a frequency counter in exactly the same way the output of a pulse amplifier is when pulse counting (see Sections 7.3 and 7.4).

The VFC design of Dunham and Elliot[12] is shown in Figure 8.7. The current from the photomultiplier tube is converted to a voltage by the first op amp. If the signal is weak, a second op amp is switched into the circuit for additional amplification. This voltage is applied to the input of a voltage-controlled oscillator. This single-chip device produces an output frequency of 10^5 Hz per volt at the input. The frequency counter is not shown in Figure 8.7. Before constructing this circuit, you should consult Dunham and Elliot for valuable commentary. The purpose for showing this circuit is to emphasize its simplicity. It is also possible to use a voltage-controlled oscillator and frequency counter with a conventional DC amplifier. This is certainly better than meter reading and may be less expensive than a chart recorder.

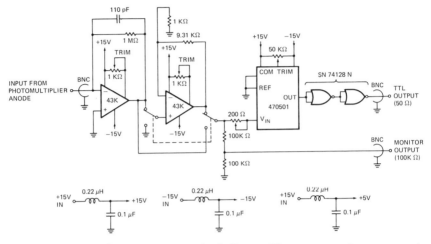

Figure 8.7. Voltage to frequency converter circuit diagram. The current to voltage converter is at left followed by the optional gain of 10.3 amplifier. The 470501 is the voltage controlled oscillator and the NOR gates on the right are line drivers. The filters at the bottom are lowpass power supply filters. Courtesy of the *Publications* of the Astronomical Society of the Pacific.

The VFC approach has some definite advantages over the conventional DC amplifier. There are fewer gain switches to adjust, which is especially valuable when light levels change very rapidly as occurs during occultation photometry. This also does away with the need for frequent amplifier gain calibration. Unlike pulse counting, there are no dead-time corrections to be made, and the digital output does away with the tedium of reading chart recorder tracings. However, unless you have some recording device such as a minicomputer with a disk drive and/or a printer, there is no permanent record of an observation. One approach that has been used by McGraw et al.[13] is to record the output frequency on magnetic tape as an audio signal. Then, as with a chart recorder, observations can be reviewed the following day. Finally, it must be kept in mind that just because you get a digital output from a VFC, this does not mean that you are photon counting. This is still DC photometry and the conclusion of Section K.5b still applies; the signal-to-noise ratio of DC photometry is inferior to pulse counting.

8.5 CONSTANT CURRENT SOURCES

The most common method of calibration of a DC amplifier requires a constant input current. In principle, the photomultiplier tube could be

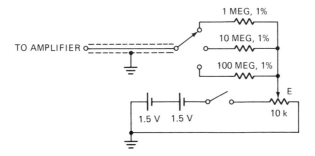

Figure 8.8. Constant current source.

used by exposing it to a constant light source, but in practice such a source is difficult to find. For instance, the brightness of an ordinary light bulb is very sensitive to changes in line voltage. Furthermore, an uncooled tube introduces noise that limits the accuracy of the calibration. A much better approach is to build a constant current source. Figure 8.8 shows such a circuit, which is extremely simple to build. The rotary switch and the potentiometer are used to adjust the current. The resistors used on this switch are the same value as those used on switch S1 of Figure 8.5. The rotary switch steps the current by factors of 10 just as switch S1 steps the amplifier gain by factors of 10. The 10 kΩ potentiometer is for fine adjustment of the current. Because the amplifier input is a virtual ground, the calibration current is given by E/R, where E is the voltage at the potentiometer wiper, 0 to 3 V, and R is the value of the rotary switch resistor. This circuit draws very little current, so two ordinary 1.5-V batteries comprise a sufficient power supply. We now describe how this circuit is used to calibrate the amplifier.

8.6 CALIBRATION AND OPERATION

The current gain of a DC amplifier is established by the resistors associated with its gain switches. The accuracy of your photometry depends, in part, on an accurate knowledge of the gain differences between switch positions. The actual sizes of these gain steps must be measured for two other reasons. First, it is very difficult to find resistors that exactly match the required values for 0.5 magnitude steps. Second, resistors tend to change value with time. This is especially true for the feedback resistors with values exceeding 10^8 ohms. This means that the calibration of the amplifier must be checked regularly. The frequency of these calibrations depends upon the quality of the resistors and their storage environment. Initially, you should plan to do a calibration every few months. You may find this is too frequent if the calibration appears to change little. On the other hand, you may find this is not frequent enough if significant calibration changes are seen. If calibration drifts seem to be associated with just one switch position, you may wish to replace that resistor with a more stable one.

There are usually two approaches to the amplifier calibration. The first measures the resistances of the amplifier resistors with a laboratory Wheatstone bridge. It is best to make these measurements with the resistors actually in place in the circuit. The process of cutting the resistor leads and soldering them to the switch may alter their values slightly. As a rule, most commonly found Wheatstone bridges cannot measure the large megohm resistors found in the feedback loop. Unfortunately, it is these large resistors that tend to be the most unstable. The second approach uses a constant current source. This has the advantage that you actually measure the amplifier gain directly and both gain switches can be calibrated by this process. We now describe the calibration using a constant current source in detail.

The actual calibration procedure is quite simple. Turn the amplifier on at least 30 minutes early to minimize drift while measurements are being taken. The fine gain steps are calibrated first. Place the coarse gain switch to its lowest position (2.5) and connect the constant current source to the amplifier input. Turn the fine gain switch to its lowest position (0.0) and adjust the rotary switch of the current source to its highest position, that is, the 1-megohm position. Turn the current source on and adjust its potentiometer to obtain an amplifier meter deflection of about one-half of full scale. Record the reading, turn the current source off, and record the zero-point level. Turn the current source on and repeat the process. Once a half dozen measurement pairs have been taken, increase the fine gain setting by one step (0.5 magnitude) and repeat the entire process. The ratio of the net deflections at these two switch positions yields the actual magnitude difference.

With the fine gain set at 0.5 magnitude, adjust the current source to reduce the meter reading to about half scale and take another series of readings at 0.5 and 1.0 magnitude. The entire process is repeated until measurements have been made at every fine gain switch position. In Table 8.2, we have listed a set of calibration readings. To save space, only one measurement per switch position is shown.

The next step is to subtract the zero-point readings to obtain the net

TABLE 8.2. Data for Fine Calibration

Gain Position	Deflection	Zero Point	Net	Magnitude Difference
0.0	48.6	7.1	41.5	
0.5	74.8	6.3	68.5	0.544
0.5	57.3	6.3	51.2	
1.0	86.9	5.3	81.6	0.506
1.0	53.8	5.2	48.6	
1.5	82.4	4.3	78.1	0.515
1.5	58.9	4.3	54.6	
2.0	90.4	3.8	86.6	0.501
2.0	58.1	3.8	54.3	
2.5	90.5	3.8	86.7	0.508

deflection for each measurement. Finally, the magnitude difference between each switch position is calculated by

$$\Delta m = 2.5 \log (d_H/d_L),$$

where d_H and d_L are the net deflection at the high and low gains, respectively. These values are listed in the last column in Table 8.2. You have about a half dozen such values for each switch position pair. If you determine an average and compute the standard deviation of the mean, you will have the best estimate of the gain difference and its error. You should strive to obtain a standard deviation of less than 0.005 magnitude. With the amplifier in Figure 8.5 and the current source of Figure 8.6, a standard deviation of 0.002 magnitude was easily obtained in laboratory tests.

The next step is to use these magnitude differences to construct a gain table. This table is used during data reduction. It allows the researcher to find the gain difference between any two switch positions at a glance. Table 8.3 shows a gain table constructed from the magnitude differences listed in Table 8.2. The horizontal and vertical axes are the gain positions. For example, to find the gain difference between the 1.5 and the 0.5 positions we simply look to where the "1.5 row" intersects the "0.5 column" and read 1.021 magnitudes. The entries in this table were determined by simply summing the magnitude differences of Table 8.2 between each combination of switch positions.

TABLE 8.3. Gain Table for Fine Adjustment Switch

	0.0	0.5	1.0	1.5	2.0
2.5	2.574	2.030	1.524	1.009	0.508
2.0	2.066	1.522	1.016	0.501	
1.5	1.565	1.021	0.515		
1.0	1.050	0.506			
0.5	0.544				

Once the fine gain switch positions have been calibrated, the coarse positions can be calibrated with respect to them. The rotary switch of the constant current source is placed in the next position (10-megohm resistor) with both the coarse and fine gain set to 2.5. Again the series of measurements are made. The fine gain is then reduced to 0.0 and the coarse gain increased to 5.0. Another set of measurements is then made. In Table 8.4, we list a sample measurement. If the amplifier had a perfect set of resistors, these two sets of measurements would be identical because the total gain of the two switches is the same (5.0). The rotary switch of the current source is moved to the last position (100-megohm resistor) and the procedure is repeated. The first set of measurements is taken with the coarse gain at 5.0 and the fine gain at 2.5. For the second set, the coarse and fine gains are set to 7.5 and 0.0, respectively.

Once the net deflections have been calculated, the first step is to correct for the fact that the fine gain difference is not exactly 2.5 magnitudes. According to Table 8.3, the gain difference between the 0.0 and 2.5 position is actually 2.574. Because this gain is larger than it should be, the deflections taken when the fine gain was set to 2.5 need to be corrected downward. If D_H is the net deflection, then the corrected deflection, D_H^*, is given by

$$D_H^* = D_H \, 10^{-0.4(\Delta f - 2.5)},$$

where Δf is the actual magnitude difference of the fine gain control (2.574). The results appear in column 6 of Table 8.4. Finally, the true coarse gain differences, Δm, can be calculated by

$$\Delta m = 2.500 - 2.5 \log (D_H^*/D_L)$$

TABLE 8.4. Data for Coarse Gain Calibration

Gain Coarse	Fine	Deflection	Zero Point	Net	D_H^*	Δm
2.5	2.5	72.2	14.8	57.4	53.6	
5.0	0.0	75.4	21.2	54.2		2.512
5.0	2.5	83.4	15.2	68.2	63.7	
7.5	0.0	83.2	21.4	61.8		2.467

where D_L is the net deflection obtained when the fine gain is set to 0.0. The coarse gain differences appear in the last column of Table 8.4.

The operation of the amplifier is straightforward, requiring only a little care and common sense. This is a sensitive device that should be used only to measure the output of the photomultiplier tube or the constant current source. As mentioned earlier, the amplifier should be turned on before connecting the signal cable as a precaution against damaging the FET input of the op amp. Finally, when measuring an unknown star for the first time, begin at the lowest gain setting. Gradually increase the gain until the desired deflection is reached. This procedure avoids possible damage to your meter or chart recorder if a bright star is measured with a gain setting that is too high.

REFERENCES

1. Persha, G., 1980. *IAPPP Com.* **2**, 11.
2. Melen, R. and Garland, H. 1971. *Understanding IC Operational Amplifiers.* Indianapolis: Howard W. Sams and Co.
3. Wood, F. B. 1963. *Photoelectric Astronomy for Amateurs.* New York: Macmillan, p. 70.
4. Young, A. T. 1974. In *Methods of Experimental Physics: Astrophysics.* vol. **12A**. Edited by N. Carleton. New York: Academic Press, p. 52.
5. The Victoreen Instrument Company, 10101 Woodland Ave., Cleveland, OH 44104.
6. Oliver, J. P., 1975. *Pub. A. S. P.* **87**, 217.
7. Stokes, A. J., 1972. *J. AAVSO* **1**, 60.
8. Markson Science Inc., 565 Oak St., Box 767, Del Mar, CA 92014.
9. Cole-Parmer Instrument Co., 7425 N. Oak Park Ave., Chicago IL 60648.
10. Hewlett-Packard Co., 5201 Tollview Dr., Rolling Meadows, IL 60008.
11. Heath Co., Benton Harbor, MI 49022.
12. Dunham, E., and Elliot, J. L. 1978. *Pub. A. S. P.* **90**, 119.
13. McGraw, J. T., Wells, D. C., and Wiant, J. R. 1973. *Rev. Sci. Inst.* **44**, 748.

CHAPTER 9
PRACTICAL OBSERVING TECHNIQUES

Chapters 1 through 5 present the foundation for understanding photometry and starlight in general, along with the rudiments of data reduction. Chapters 6 through 8 show how to construct or buy the necessary equipment, set it up, and perform the necessary calibrations. We are now ready to discuss using your photometer. This chapter explains in more detail how to perform photometric measurements, from selecting comparison stars and making a finding chart, through the actual acquisition of data. It ends with some comments about sources of error external to your equipment with which you must contend.

No book can replace actual experience with the equipment at hand. We can give you some practical advice and try to guide you past some of the pitfalls that we found, but you must learn much of photometry by trial and error. One suggestion we would like to make is to pick one variable that is bright, short period, and very well observed as your first trial. In this manner, you can be sure that your data compares favorably with previous results.

9.1 FINDING CHARTS

Sirius, Polaris, and other bright stars are easy to find in the sky. Fainter stars become increasingly difficult to find, not only because they are harder to see but also because there are more of them. With care, stars fainter than those visible by eye through a telescope can be measured by photoelectric photometry. Fainter stars, being more difficult to locate, require the use of a good finding chart.

The usual method of identifying program stars is through the preparation of a finding chart: a sketch or photograph of the region of the sky containing the object. You can prepare a finding chart from various atlases, from your own photographs of the area, or by obtaining previously prepared charts from published sources. Each of these methods is described below.

9.1a Available Positional Atlases

Positional atlases are drawn from catalogs of star positions. In many cases, stars are omitted for lack of data or were positioned incorrectly. Still, they can contain more information about the stars than a photograph. Several atlases include stars brighter than eighth magnitude. Most of these atlases have been reviewed by Larson[1] and are readily available. For objects brighter than ninth or tenth magnitude, three atlases are commonly available. They are:

1. *Bonner Durchmusterung (BD) and Córdoba Durchmusterung (CD) Atlases.*[2,3] The BD was produced by Argelander and Schönfeld in the period 1859–1886, covering the northern sky, and the CD was published between 1892 and 1932, covering the southern sky. Together, they contain approximately 580,000 stars to a limiting visual magnitude of 10 and have been the mainstay for almost a century. These catalogs are available at existing libraries and observatories. New copies are available in magnetic tape form only. Epoch 1855 coordinates are used and must be precessed.
2. *The Smithsonian Astrophysical Observatory (SAO) Atlas.*[4] These charts contain approximately the same stars as the BD and CD atlases, but the charts are smaller and stars are plotted closer together on a smaller scale. Most variables brighter than ninth magnitude are marked. All stars are identified in the accompanying catalog, available from the U. S. Government Printing Office. Transparent overlays allow the location of stars with arc minute accuracy. The coordinates are for epoch 1950.
3. *Atlas Borealis, Eclipticalis, and Australis.*[5] These charts by Becvar cover the sky to approximately tenth magnitude and identify variables by their variable star designations. One of the nice features of these charts is the color coding of spectral type. Epoch 1950 coordinates are used with transparent overlays. This atlas set is widely used by amateur astronomers and is relatively inexpensive.

9.1b Available Photographic Atlases

For stars fainter than ninth magnitude, photographic atlases must be used because of the large number of stars involved. These atlases consist of either photo-offset charts from original plates or actual photographic prints. Ingrao and Kasperian[6] review early photographic atlases. The major photographic atlases are listed below.

1. *Photographic Star Atlas (Falkau Atlas).*[7] This atlas used plates that were blue-sensitive and covers the entire sky in two volumes. The limiting magnitude is 13, the scale is 1 millimeter = 4 arc minutes, and each chart is about 10° on a side.
2. *Atlas Stellarum 1950.0.*[8] This atlas covers the entire sky in three volumes using blue-sensitive plates. The limiting magnitude is 14.5, the scale is 1 millimeter = 2 arc minutes, and a complete set of extremely useful transparent overlay grids is included. This atlas costs $225 in 1980.
3. *True Visual Magnitude Photographic Star Atlas.*[9] This atlas is very similar to Atlas Stellarum in that it covers the entire sky in three volumes with the same scale. The limiting magnitude is 13.5, and a green-sensitive emulsion has been used. For finding charts, green sensitivity is a great advantage as it closely matches the response of the eye.
4. *Lick Observatory Sky Atlas (North)*[10] *and Canterbury Sky Atlas (South).*[11] Rather than using the photo-offset methods of the previously listed photographic atlases, these two atlases are actual prints of blue plates. The scale is 1 millimeter = 3.88 arc minutes, the limiting magnitude is 15, and each print covers about 18° on a side. No overlays exist and copies are no longer available except at existing libraries and observatories.
5. *National Geographic–Palomar Observatory Sky Survey (POSS).*[12] This survey with the Palomar 48-inch Schmidt is the Rolls Royce of the astronomical atlases. Both red and blue plates were used, with the atlas consisting of positive prints with a scale of 1 millimeter = 1.1 arc minutes and a limiting magnitude of 20 (red) or 21 (blue). Each print is 6.6° on a side and a sequence of overlays exists, though less useful than most as no fiducial marks are found on the prints. The POSS is complete to −24° declina-

tion, with a red-plate extension to $-45°$. The plates were taken in the early 1950s, and Palomar intends to redo the survey starting sometime in 1985. A complete set of prints costs several thousand dollars.

6. *The European Southern Observatory (ESO)/Science Research Council (SRC) Atlas of the Southern Sky.*[13] In a similar manner to that of the POSS, the southern sky is presently being photographed by the two large Schmidt telescopes in the Southern Hemisphere.[14] The ESO 1 meter at La Silla is taking red plates and the SRC 1.2 meter at Siding Spring is taking the blue survey plates, both with a scale of 1 millimeter = 1.1 arc minutes and a limiting magnitude of 22. The atlas covers the sky from $-90°$ to $-17°$ declination with plate centers at $5°$ spacing. This atlas is being released in limited quantities (150 copies) only on 36-centimeter (14-inch) Aerographic Duplicating Film.

9.1c Preparation of Finding Charts

The goal of a finding chart is to allow easy identification of the program object at the telescope. Generally, two charts are prepared. A small-scale chart matching the field of view of the main telescope, typically 15 arc minutes square, should be prepared carefully, including stars two to three magnitudes fainter than the variable. Mark the program object, any nearby comparison stars, and an area with no stars to be used for sky background measurements. Many observers prefer these charts to match exactly the view of the telescope, that is, reversed and/or inverted. Cardinal directions should be indicated as well as the chart scale, perhaps by an angular measurement grid. A large field chart roughly matching the finder can provide pointing information for the main telescope, and at the same time identify photoelectric comparison stars and readily identifiable patterns to help locate the field.

The best charts are Polaroid copies of photographic atlases, negative copies of atlas prints, which can then be enlarged, or prints made from your original negatives of the sky. Direct tracings of atlases or xeroxes may be acceptable provided that the limiting magnitude near the program object, comparison stars, and sky measurement position is sufficiently faint.

9.1d Published Finding Charts

In most cases, earlier observers have published finding charts for your object of interest. These charts may lie in obscure journals or suffer from poor quality. However, it is highly recommended to search for published finding charts of variables fainter than ninth magnitude before preparing your own.

The major source for finding charts is the General Catalog of Variable Stars (GCVS).[15] Its extensive reference list contains chart references for the vast majority of identified variable stars. However, many of these finding charts are published in Russian journals, which are identified only in the Cyrillic alphabet rather than the English transliterated names under which they are cataloged in most major libraries.

There are several collections of finding charts available if the GCVS itself or its referenced charts are not available. These are listed below.

1. *AAVSO Variable Star Atlas.*[16] There are 178 charts in this atlas, measuring 11 × 14 inches with a scale of 15 millimeters per degree. These charts contain all of the American Association of Variable Star Observers' program stars and all variables that reach 10.5 visual magnitudes or brighter at maximum. The charts are very similar to the SAO Atlas.
2. *The Sonneberg charts.*[17] Several thousand variables were discovered at Sonneberg during the early part of the twentieth century. The charts contained in reference 17 cover 3600 of the Sonneberg variables.
3. *The Odessa charts.*[18] This reference contains light curves and finding charts for 266 stars.
4. *Atlas Stellarum Variabilium.*[19] These charts cover all variables brighter than tenth magnitude at minimum that were known by 1930. The entire sky was covered, with each chart measuring 20° on a side and plotting stars to a limiting magnitude of 14.
5. *Charts for Southern Variables.*[20] The 400 charts in this collection were sponsored by the International Astronomical Union and cover all long-period variables brighter than thirteenth magnitude at maximum and south of $-30°$ declination. Later volumes in the series have photoelectric sequences for nearby stars.
6. *Atlas of Finding Charts of Variable Stars.*[21] This work is hard to find in Western libraries.

Goddard Space Flight Center (GSFC) in Greenbelt, Maryland is the distributor for the magnetic tape version of the GCVS. The tape also includes a cross-reference list for the Sonneberg variables, which is extremely helpful. In addition, GSFC can generate POSS Atlas overlays to identify faint variable stars. Write to Code 680 of GSFC for more details.[22]

9.2 COMPARISON STARS

Now that you have located your program star, you need to decide whether or not to use a comparison star in your measurements. These are stars that are near to your variable and are observed immediately before and/or after your program star. Usually, measurements are reported as differential observations, star A minus star B. This difference in the magnitudes is less likely to vary, as conditions that affect one star usually affect the other. Use of comparison stars has several advantages as noted below.

1. Slow atmospheric variations are eliminated, as the variation affects both stars equally.
2. First-order extinction corrections are nearly equal for both stars and becomes unimportant for differential measurements. Extinction measures are not necessary except to place the comparison star on the standard system.
3. Zero-point differences between you and other observers are removed if common comparison stars are used.
4. If a comparison star of similar color is used, errors in the transformation equations will have less effect.
5. If a comparison star of a similar magnitude is used, differential dead-time corrections become unimportant in pulse counting, and gain correction errors are removed if the same gain setting is used for both stars in DC photometry.
6. The comparison star can serve a double purpose as an extinction star if several measurements over wide air mass differences are made. In addition, if advantage 4 holds, second-order extinction is automatically taken into account.
7. For high accuracy and long-term reproducibility, errors in differential measurements with respect to a comparison star can approach the equipment short-term accuracy, less than $0^m.01$,

instead of the night-to-night transformation variation of $0^m.02$ or more. This is very important for small amplitude variables.

You can observe variable stars without using comparison stars. Many extensive surveys of variables have been carried out in the past without these aids, especially in the southwestern United States where photometric skies are common. However, for accurate low-amplitude photometry they are almost a necessity.

9.2a Selection of Comparison Stars

Make every effort to use the same comparison star as previous observers of the variable star. This should be done to minimize systematic differences between data sets of various observers. However, if this cannot be done (because no one has observed this star before or previous observers did not specify their comparison star), you will need to select your own.

Comparison stars should meet five criteria: (1) less than 1° from the program star, (2) of similar color, (3) equal in magnitude, (4) nonvarying, and (5) not red in color. Red stars are almost always variable, and are quite likely flare stars. Rules 2 and 3 are not rigid, as few stars will be near the variable and be the same brightness and color. But make every effort to enforce rules 1 and 4! Pick a brighter comparison star rather than one which is fainter than the variable, as better statistics can be obtained in a shorter time.

Selection of similarly colored stars can be difficult. For stars brighter than tenth magnitude, the Becvar atlases[5] indicate spectral class by the color of the dot representing the star. A more exact method is to use the Henry Draper (HD) Catalog,[23] which lists accurate spectral types for over 200,000 stars. This catalog is available at most colleges. A microfiche version is available from the University of North Carolina at Greensboro,[24] and a magnetic tape version from GSFC. Modern spectral classification yielding more accurate spectral types for the HD stars is being carried out by the University of Michigan[25] and by the Mira group in California.[26]

Nearby comparison stars of similar brightness can be found by examining the prepared finding charts for the variables. If you have a choice, pick a star listed in one of the major catalogs: BD, CD, HD, or SAO. This allows easy publication of your results without the necessity of including a finding chart for your comparisons.

If all else fails, a few minutes of observation in the field surrounding your variable will usually find a nearby comparison star candidate of the same magnitude. These observations can be performed quickly, as you are looking for a star with similar deflections or count rate to those of your variable and can easily eliminate stars that differ widely.

9.2b Use of Comparison Stars

Two comparison stars for each variable are usually chosen. One is considered to be the actual comparison star. The other is called a *check star*, and is used to test the stability of the comparison star. Differential measurements between the comparison and check stars should remain constant to within the nightly errors, usually $0^m.02$ or less with good skies. If variations larger than this consistently occur, either the comparison or the check star is variable, a fairly frequent occurrence! Which one is the culprit can be decided by comparing the nightly standardization of the comparisons or by deciding which star gives a light curve for the variable that has the smallest scatter. The check star is observed once or twice a night; the remaining measurements use only the comparison star.

The recommended observing sequence for each variable star observation is:

1. Sky
2. Comparison
3. Variable
4. Comparison
5. Variable
6. Comparison
7. Sky
8. Check (optional)

Each step in turn is comprised by the U, B, V, red leak, and perhaps dark-current measurements as discussed in Section 9.3. If the variable is faint, so that steps 3 and 5 would take several minutes each to reach 1 percent accuracy, it is much better to perform several sequences with less accuracy and average them than to risk sky transparency changes during the variable and comparison star observations.

If the comparison star is to be placed on the standard *UBV* system, it should be done on several occasions: at the beginning of the program, in the middle, and at the end. This can be done through differential measurements of nearby standard stars, or by using the transformation equations and nightly determination of the extinction.

9.3 INDIVIDUAL MEASUREMENTS OF A SINGLE STAR

An individual star observation really consists of four separate determinations in UBV photometry: U, B, V, and red leak. If you are doing only BV photometry, the U and red-leak measurements are removed. Red leak is unimportant in blue stars because little of their energy falls in the red. Sometimes observers also measure the dark current. This is usually negligible for bright stars or a dry-ice photomultiplier tube, and is subtracted automatically during sky background subtraction.

For pulse counting, the usual measurement sequence is V, B, U, red leak, and dark current. This is because V is the most commonly listed magnitude and gives a check for proper star selection and equipment operation. If you are using a filter wheel, you will end on the dark filter position, allowing movement to the next star, and then will be ready to perform the V observation with the next rotation of the wheel. The time of the observation is the average of the starting and stopping times of the sequence.

For DC photometry, the usual sequence is V, B, U, red leak, U, B, and V. This sequence allows checking for instrumental drift effects during the sequence.

Achieving one percent accuracy in a star measurement always requires a signal-to-noise ratio (S/N) of 100. How this is determined is different for pulse counting and DC photometry.

9.3a Pulse-Counting Measurements

Remember from Chapter 3 that 10,000 total source counts are necessary from a star to approach one percent or $0^m.01$ accuracy. Figure 9.1 shows the typical count rate for telescopes of 20-, 40-, and 75-centimeter diameters. For example, an $11^m.7$ star produces roughly the count rates shown in Table 9.1. For faint stars, measuring the U magnitude is always a problem. For example, the count rate for cepheids in U is ten times less than in V. You must decide in these cases if having the $(U - B)$ color index is essential to your program and make allowances for the increased time if it is.

PRACTICAL OBSERVING TECHNIQUES 211

TABLE 9.1. Pulse-Counting Rates

Telescope Size (centimeters)	Rate (counts/second)	Integration Time to Achieve 1% Accuracy (seconds)
20	100	100
40	380	26
75	1400	7

The times quoted in Table 9.1 are only for measuring the star in a single color. However, the sky must also be measured accurately to subtract its contribution from the star measurement. The ideal ratio of time spent measuring the sky background to time spent on star and sky is

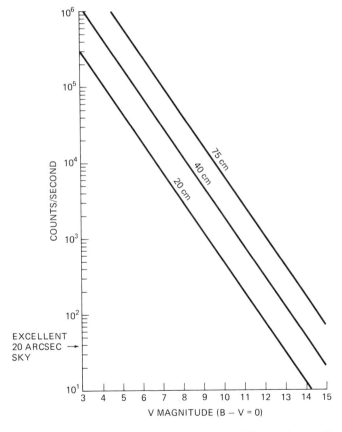

Figure 9.1. Count rate versus apparent magnitude for three different telescope diameters.

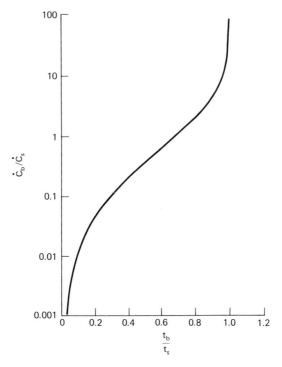

Figure 9.2. The optimum fraction of observing time to be spent on sky background.

found by taking the derivative of Equation K.41 with respect to the source time. The result is

$$\frac{t_b}{t_s} = \sqrt{\frac{\dot{C}_b/\dot{C}_s}{\dot{C}_b/\dot{C}_s + 1}} \qquad (9.1)$$

where the subscript s refers to source measurements and the subscript b refers to background measurements, $\dot{C}$ is the count rate, and t is the time per measurement. This ratio is plotted in Figure 9.2. From a known count rate ratio the corresponding time ratio can be read directly from this figure.

Example: $11^m.7$ star produces roughly 100 counts per second in the 20-centimeter telescope. Assume the sky background gives roughly 50 counts per second in the chosen diaphragm. How long do we have to observe the star and the background?

$$\frac{t_b}{t_s} = \sqrt{\frac{50/100}{(50/100) + 1}} = 0.577$$

Or, $\dot{C}_b/\dot{C}_s = 50/100 = 0.5$, and reading from Figure 9.2, the corresponding time ratio of 0.56. Because we need to observe the star for 100 seconds for 1 percent accuracy, we need to observe sky alone for at least 56 to 58 seconds.

9.3b DC Photometry

When pulse counting, there is a fairly simple rule to follow to achieve an accuracy of 1 percent. This requires an S/N of 100, which means that a total of 10,000 counts must be accumulated. If a star produces only 1000 counts in one second, you simply observe it for at least 10 seconds. The guidelines are not so simple in DC photometry. If you were to watch the output of a DC amplifier on a chart recorder as starlight strikes the detector, you would see the pen rise and then jitter about some mean level.

Figure 9.3 shows the chart recorder tracing for the first observation of UZ Leonis in Table 4.1. The mean level of the pen represents the signal from the star and the jitter is the noise. It is noise that prevents us from determining the stellar signal with perfect accuracy. Unlike

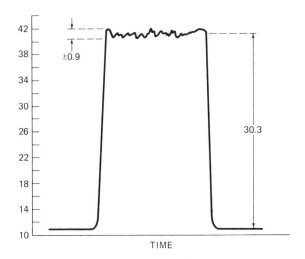

Figure 9.3. Chart recorder tracing.

pulse counting, the S/N does not improve when we observe the star longer, the chart tracing just gets longer and continues to look much the same. The same is true if we watch the amplifier meter. Increasing the observing time improves the photometric accuracy to some extent because the longer chart tracing makes it easier to estimate the mean level using a straightedge. However, once the tracing is several centimeters long, continuing the deflection brings diminishing returns (unlike pulse counting). The problem is that in DC photometry the integration time is set by the RC time constant at the amplifier input. The obvious way to improve the S/N is to increase this time constant. Indeed, the tracing does become smoother when the time constant is increased. However, there is a trade-off when using a capacitor to smooth the signal. When the detector is exposed to light, the current entering the amplifier must charge the capacitor. The rate at which the capacitor charges depends on the RC time constant. As the capacitor charges, the pen makes an exponential rise to its final value. For the pen to reach 99 percent of its final value requires a period of 4.6 time constants. If a large time constant is used, a significant amount of observing time is spent waiting for the pen to reach its final level. For this reason, DC amplifiers seldom use time constants that exceed a few seconds. The DC amplifier in Chapter 8 has an adjustable time constant. For bright stars that have a large S/N, a small time constant is used to save observing time, while for fainter stars a longer time constant is used to improve the S/N.

For many stars, a time constant of less than a second is not enough to achieve a S/N of 100. Figure 9.3 shows a case with a 0.5 second time constant. We can estimate the S/N by the amount of jitter about the mean. The mean signal level is 30.3 units on the chart paper. The noise causes variations of 0.9 unit to either side of the mean. If the sky background is large, it would be necessary to subtract this from the star to obtain the net signal. In this particular example, the sky background is low so that the S/N is very nearly $30.3/0.9 = 34$. By Equation K.25, we see that the S/N increases with the square root of the total integration time, t. Therefore, we would need to increase the amplifier time constant by a factor of nine to achieve an S/N of 100. A time constant of this length is not available for this amplifier. Then the procedure is to take several deflections (each many time constants in duration) and form an average to make a single observation. A rough estimate of the necessary number, n, of such observations is given by

$$n = \left(\frac{100}{\text{S/N}}\right)^2 \tag{9.2}$$

The time between each of these deflections can be spent recentering the star in the diaphragm (if necessary) or making a deflection in another filter. For the chart tracing in Figure 9.3, the S/N implies that a single measurement would have an error of 0.03 magnitude. The formula above implies that nine observations should be averaged for a 0.01 magnitude error. In fact, the actual standard deviation from the mean of nine observations that night was 0.012 magnitude.

This example points out a disadvantage of DC compared to pulse counting. The nine observations required would take several minutes of observing time. With a pulse-counting system, if we obtained a S/N of 34 in 0.5 second (the time constant used that night), we would need only to integrate nine times longer (because S/N $\propto t^{1/2}$) or 4.5 seconds for a S/N of 100. For many observing projects, this difference in observing time is unimportant. But there are rapid variable stars and short-period binaries that have measurable changes of brightness in just a few minutes. We must observe the light curve on more nights to obtain the same quality of data as that obtained with a pulse-counting system. Alternatively, you can retain the time resolution by not averaging as many deflections but you obtain a noisier light curve.

Note that this disadvantage of DC photometry disappears if the chart recorder is replaced by a voltage-to-frequency converter and a counter. A very small amplifier time constant can then be used to integrate on the star until the desired S/N is reached. However, the S/N analysis has a further complication. Unlike pulse counting, the number that appears on your counter is not equal to the number of detected photons. Instead, it corresponds to some level of current flowing in the feedback loop of the amplifier. This in turn depends on both the brightness of the star and the amplifier gain. In this case, an empirical method is the simplest way to determine S/N. Take a series of short test integrations and calculate the standard deviation from the mean. If $\bar{c}$ is the mean counts and s.d. is the standard deviation, then

$$\text{S/N} = \frac{\bar{c}}{\text{s.d.}} \tag{9.3}$$

To obtain an S/N of 100, the required integration time, T, is

$$T = \left(\frac{100}{\text{S/N}}\right)^2 t. \tag{9.4}$$

where t is the total time of all the test integrations.

The amount of observing time spent on sky measurements can be estimated by Equation 9.1 just as it is for pulse counting. The count rates in that equation are simply replaced by net pen deflections. In the example above, the sky background was $\frac{1}{25}$ of the stellar signal. Equation 9.1 then tells us that 20 percent of our observing time should be spent on sky background.

There are also some differences in the observing procedure between pulse counting and DC photometry. These are discussed in Section 9.7.

In Sections 9.3a and 9.3b, much emphasis has been placed on the S/N as an indicator of the quality of an observation. However, a high S/N is a necessary but not a sufficient condition for an accurate observation. There are many other factors that can come into play. For instance, electronic drift or the slow passage of cirrus clouds are not obvious in the noise level in a single measurement. However, they become apparent when measurements of the same object, such as the comparison star, fail to repeat. Discrepancies that exceed the noise levels are indicators of a problem. Even if a single measurement has a very high S/N, never assume that it will be reproducible; always take at least two. There is no substitute for an alert, experienced observer who can tell when "things are not quite right."

9.3c Differential Photometry

Differential photometry is the simplest and potentially the most accurate of photometric techniques. The basic idea is to compare the brightness of the variable star to that of a nearby and constant comparison star. However, simple as it sounds, certain observing procedures must be followed strictly if differential photometry is to be done properly.

The golden rule of differential photometry is: *interpolate, never extrapolate*. To illustrate the meaning of this rule, consider what happens as we observe our two stars during the night. Suppose that early in the evening the variable and comparison star are near the eastern horizon. We measure the comparison star and then, for some reason, delay measuring the variable for 20 minutes. During those 20 minutes, the stars have risen higher and the extinction is considerably less than it was when the comparison star was measured. The result is that the variable looks too bright with respect to the comparison star measurement. This is an example of extrapolation; we took a comparison star measurement and assumed it was valid 20 minutes later. Obviously, a better procedure is to measure the comparison, variable, and then the comparison star again. We can then interpolate to estimate the apparent brightness of the comparison star at the time of the variable star measurement. Our golden rule can be restated: *always sandwich the variable star measurements between comparison star measurements.*

If the variable star is faint or varies slowly in brightness, the observing sequence in Section 9.2b is recommended. However, if you are observing a star that varies rapidly, such as an eclipsing binary with a half-day orbital period, a slightly different observing pattern is preferred. If we let C and V represent an observation through each filter of the comparison and variable star respectively, then the observing sequence might look like the following.

$$CVV \ldots VVCVV \ldots VVCVV \ldots VVCVV \ldots$$

The brackets mark a data group that we refer to as a *block*. Each block begins and ends with a comparison star measurement. The number of variable star measurements in a block depends on three factors. First, the required number of measurements needed so that when combined, a single observation with a S/N of at least 100 is produced. (This was discussed in Sections 9.3a and 9.3b.) Second is the speed with which the variable changes. Obviously, if the star only varies by 0.1^m during the entire night, you need not look at it as often as one that changes by the same amount in 30 minutes. In the latter case, it would be desirable to obtain several observations per block (i.e., spend a higher percentage of the observing time on the variable). The third factor is zenith distance. When the air mass is large, variations in extinction can have a large impact. Therefore, the comparison star must be observed more frequently. There is no simple rule on how long to make a block, but of course, there is no substitute for experience. However, it is certainly advisable to observe the comparison star as frequently as possible. Experience with short-period eclipsing binaries observed through the somewhat variable skies of the midwestern United States suggests that the comparison star should be observed at intervals of 20 minutes or less. If the observing sequence must be interrupted at any point, it is important to end with a comparison star measurement. If observing resumes later, you should begin with a comparison star measurement.

The block structure outlined above does not indicate sky background measurements. The reason is that the amount of time spent measuring the sky depends on the relative brightness of the star and the sky background. The method of Section 9.3a can be used to estimate the percentage of observing time spent monitoring the sky background. If, for instance, it turns out that 25 percent of your time should be spent on the sky, then every fourth measurement in the block should be of the sky. If you suspect the sky background is changing rapidly (for example, if the moon is rising), then you should measure the sky more frequently.

Note that our block structure does not contain separate measurements of the dark current. Some authors recommend measuring the dark current frequently. However, our experience has been that with well-designed amplifiers and fairly stable photomultiplier tube temperatures, the dark current is very constant. Every time the sky background is observed, we actually measure sky plus dark current (plus any zero-point shift if a DC system is used). When this is subtracted from the stellar measurement, the dark current (plus any zero-point shift) is subtracted automatically. There is no need to measure and subtract the dark current specifically from all the measurements. Therefore, the dark current need be measured only occasionally as a check on the stability of the photometer.

9.3d Faint Sources

Photometry of faint sources can be a time-consuming and exasperating project that should only be undertaken by the experienced observer. By faint sources we mean objects that are comparable to the sky background in brightness, or objects near the visual limit of the telescope. There are several points to consider when observing faint sources.

First, pulse counting with a cooled photomultiplier tube is the most practical method of observing faint sources. DC methods yield chart recorder deflections that are not significantly greater than the random fluctuations in the sky background. A smoother trace can be obtained by increasing the time constant, thereby integrating over a longer period. However, the time required to reach a constant level is also increased. Eye measurement of a star plus sky trace that is only a few percent greater than the sky trace alone is very difficult. Long integrations with pulse counting are very easy, requiring only the selection of a longer gating time on the pulse counter.

Second, time is limited by the accuracy of your telescope drive and by sky conditions. Generally, never integrate on one star for more than 5 minutes, including time to measure *all* colors. If you have insufficient counts with the 5-minute limitation, move to the comparison star or background and then return to the program object for another 5-minute observation.

Third, never observe except under optimum conditions. This includes using only moonless nights, observing near the zenith, and with the best possible seeing conditions. Under these conditions, you can use the smallest diaphragm to reduce the sky background and increase the contrast between the star and the sky.

For stars that approach the sky background in intensity, the time spent on observing the star and observing the sky should be about equal (see Figure 9.2). This means that you should alternate 5-minute integrations between star and sky. Always cycle through all filters on one object before moving the telescope to look at sky or a comparison source.

If you are observing one source for a significant amount of time, say 30 minutes or more, plot the sky values versus time. You may find a significant trend because of a brighter sky near the horizon or slowly varying sky brightness that allows you to interpolate between adjacent sky readings to give a better sky value at the time of observation of your program star.

Stars near the visual limit or fainter can be measured photoelectrically, but are very difficult to place in the diaphragm. The usual procedure is to have the guiding or finding eyepiece on a stage with X and Y movements and offset to a brighter star in the same field. This requires the ability to measure the amount of offset in both axes and the knowledge of the plate scale of the telescope. For simple systems, the first requirement can be met by counting screw turns between two stars in the field with known positions. To perform offset photometry, first position the cross hairs on the center of the diaphragm, and then move the eyepiece in X and Y the distance between the object to be measured and the nearby bright star. The positioning of the bright star on the cross hairs places the source to be measured in the diaphragm.

You must find a region near the star where no stars within five magnitudes of the program star's brightness exist. For example, a ninth magnitude variable must have a sky reading with no stars brighter than fourteenth magnitude in the diaphragm. Otherwise, the sky background reading gives a sky value significantly higher than the sky reading at the star itself, and the measurement for the star becomes fainter than it actually is when you subtract the incorrect brighter sky value from it. This becomes troublesome particularly around tenth magnitude for the program object, as there are many fifteenth magnitude stars within an average 30 arc second area. It is difficult to find a clear region for the sky reading. Also, there is a large chance of including a faint companion near the star itself. A secondary problem is finding charts that go faint enough; even using Atlas Stellarum with its fourteenth magnitude limit, you cannot observe any stars fainter than ninth magnitude without the chance of significant sky background subtraction errors.

9.4 DIAPHRAGM SELECTION

Chapter 6 presented the practical details of constructing the photometer head and the diaphragms. However, there are some basic considerations to be made when using these diaphragms in your observing program.

The most obvious reason for having more than one diaphragm is to prevent unwanted stars from contributing to the light entering the diaphragm. In addition, a smaller diaphragm allows less sky background radiation to pass through, while permitting most of the program star's light to pass through unhindered.

It is tempting when using a photometer with several diaphragms to use a small one on one star, then to use a larger one on another star, because it is brighter or has no companions. There are several reasons for *not* doing this, and these are presented in this section.

*9.4a The Optical System

The image of a point source created by a telescope is not a point source but rather a very complicated distribution. Because the telescope is not infinite in size, only a portion of the total wave front of light from the star is intercepted by the telescopic mirror or lens. The mirror is then called the *entrance aperture* of the optical system.

As the light is focused by this aperture, the path length of light rays from a given wave front is different for different parts of the image. Some of these rays add together in phase, resulting in *constructive interference;* others add together 180° out of phase, resulting in *destructive interference.* The resultant image of a point source by a circular aperture looks like a bright central disk, the *Airy disk,* surrounded by fainter concentric rings of light. This image can be seen with good optics on a good night and high magnification, and actually can be used to collimate the optical system. The important point to remember is that the image is *not* concentrated at a point even if the source appears pointlike. The size of the central disk and the brightness of the secondary rings are inversely proportional to the diameter of the telescope.

An additional problem arises when a secondary mirror is used, because part of the circular aperture is obscured and an annular entrance aperture results. A theoretical description of the resultant image of a point source is discussed by Young.[27] The major change from a strictly uniform, circular aperture is an increase in the amount of light in the secondary rings.

If a diaphragm is placed in the focal plane of the telescope, a certain amount of the energy from the star will be removed, the amount depending on how many of the secondary rings are larger than the diaphragm. A good approximation to the excluded energy is

$$X(d) = \left[\frac{82,500\lambda}{d\,D(1-t)} \right] \quad (9.5)$$

where $X(d)$ is the fractional excluded energy as a function of diaphragm diameter, λ is the wavelength of light, D is the diameter of the lens or mirror, d is the diameter of the diaphragm in arc seconds, and t is the fractional obscuration of the primary mirror by the secondary mirror. The variables D and λ must be in the same units. The total energy included by a diaphragm is then given by

$$I(d) = 1 - X(d) \quad (9.6)$$

For a 40-centimeter (16-inch) telescope with a 40 percent central obscuration and a wavelength of 5000 Å (5×10^{-5} centimeters),

$$I(d) = 1 - \frac{82{,}500(5 \times 10^{-5})}{40d(1 - 0.4)}$$

$$I(d) = 1 - \frac{0.1718}{d} \tag{9.7}$$

Figure 9.4 shows the result of Equation 9.7. The approximation breaks down for small diaphragms, which is why the included energy from the figure does not approach zero for small diaphragms. The point to note, however, is that for a 40-centimeter telescope, changing from a 10 arc second diaphragm (2 percent excluded energy) on one star to a 20 arc second diaphragm (1 percent excluded) on another star causes at least a 1 percent error. However, the difference between a 20 arc second and a 30 arc second diaphragm is small enough to neglect unless

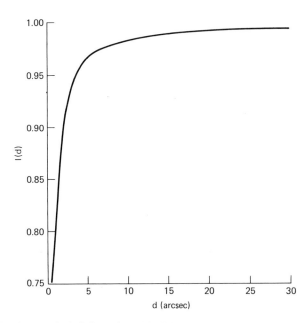

Figure 9.4. Total energy included as a function of diaphragm size, for a 40-centimeter aperture telescope.

precision greater than ± 0.01 magnitude is desired. But reducing the size of the telescope to 20 centimeters would then make the 20 to 30 arc second change as large as the 10 to 20 arc second change with the larger telescope. Therefore, use as large a telescope as possible to eliminate this error; many observers even do not use diaphragms smaller than 20 arc seconds with telescopes smaller than about 20 centimeters in diameter.

We have not included other effects that result from diffraction spikes from the secondary supports or the optical aberrations caused by the mirror itself, all of which increase the amount of energy outside of the central disk. In other words, consider the above estimates to be lower limits on the errors involved from the optical system itself.

9.4b Stellar Profiles

Just as the light from the sun scatters, making the sky blue, the light from any star scatters over the entire sky. The profile of a stellar image on the sky is therefore not strictly pointlike, but rather spread out by refraction, diffraction, and scattering in the atmosphere and diffraction and scattering within the telescope. The profile concentrates heavily towards the center, producing a "seeing disk" typically 2 arc seconds across and then decreases rapidly outside that diameter. The seeing disk or stellar profile is *not* constant for a given instrument. A hazy night can broaden the image greatly.

Figure 9.5 shows a typical stellar profile for a $m_v = 0$ star based on the results by King[28] and Picarillo.[29] Note that, although the intensity falls off rapidly to a 10 arc second radius, the decreasing intensity soon approaches an inverse square law drop. You might think because at a 10 arc second radius, equivalent to a 20 arc second diaphragm, the starlight is ten magnitudes fainter than near the profile center that the remaining radiation could be neglected. However, the total light entering the diaphragm is the product of the intensity per unit area and the total area of the diaphragm, which increases as the square of the radius. The resultant total light pattern is shown in Figure 9.6. Here you can see that a 30 arc second radius circle, 60 arc second diaphragm, includes most of the light of a star, but that using a 20 arc second diaphragm on one star and a 10 arc second diaphragm on another can cause significant error.

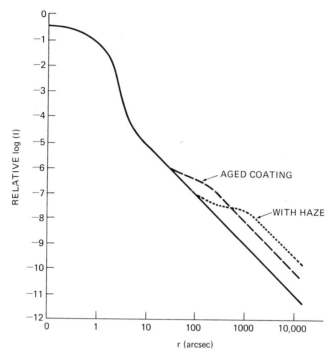

Figure 9.5. Typical stellar profile.

9.4c Practical Considerations

The fact that a stellar image is not concentrated at one point has several direct consequences to photometry:

1. Changing diaphragms between a comparison and the variable star causes different fractions of each star's total light to reach the photometer.
2. On hazy nights, even observations with the same diaphragm size can be in error as the haze scatters a changing percentage of the starlight out of the diaphragm, depending on observation time or altitude.
3. A miscentered image increases the chance for observational error greatly, because a significant fraction of one star's light may be misplaced out of the diaphragm.
4. A misfocused image enlarges the seeing disk and allows less light to reach the photomultiplier tube and increases the chance of mis-

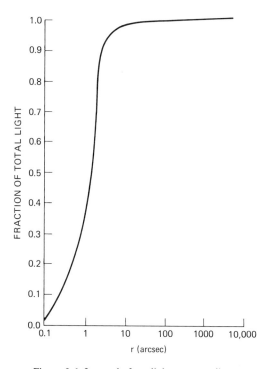

Figure 9.6. Integral of starlight versus radius.

centering errors. This error is common when comparing early and late evening observations if the telescope has not been refocused for temperature changes.

5. Never change diaphragms between star and sky measurements. Whereas the stellar contributions through a 20 or a 60 arc second aperture is the same to within 1 percent, the 60 arc second aperture allows nine times more skylight to pass through. At times it is difficult to find a clear patch of sky for the background measurement. It is then the observer's discretion whether to make the sky measurement further from the program star, or to use a smaller diaphragm for all measurements.

6. There is a limit on the smallness of the diaphragm because of the stellar profile size. If apertures much less than 10 arc seconds are used, a significant fraction of the star's light is rejected. One reason the 2.1-meter Space Telescope (to be launched from the space shuttle by NASA in 1986) is expected to outperform the largest earth-based telescopes is that with no atmospheric seeing, a dif-

fraction-limited star profile is obtained, where 70 percent of the light is concentrated within a 0.1 arc second circle. Then diaphragms of 0.4 and 1 arc second are easily used, removing much more of the sky background from the measurement than is possible from the ground.
7. The diameter of the seeing disk varies with altitude above the horizon because of the differing air mass. Near the zenith the disk may be 1 or 2 arc seconds in diameter, but near the horizon it has expanded to perhaps 10 arc seconds. Therefore, even using the same diaphragm on stars at differing altitude, differing amounts of the total starlight are admitted. Young[27] gives a good review of this and other seeing and scintillation effects.

At all costs, try to use one diaphragm size for all observations that are to be compared on a given night. You do not have to measure 100 percent of the light from a star, or even 95 percent, to get accurate results. What you must try to do is measure the same *fraction* of light from every star you want to compare.

9.4d Background Removal

Because the sky background acts like an extended source, a larger diaphragm will admit more background radiation since a larger area of the sky is seen by the photometer. However, few more star photons are acquired as they come from a nearly pointlike image amply covered by the diaphragm. The light from the sky passing through the diaphragm cannot be distinguished from the light of some faint star as the photometer knows only that a certain *number* of photons have reached it, not the *origin* of those photons in the area covered by the diaphragm. Therefore, we can compute the magnitude of an equivalent star that produces the same number of photons as produced by the sky. This is shown in Figure 9.7 for a typical site where the sky brightness is $22^m.6$ per square arc second at the zenith. You can see that if a 30 arc second diaphragm is used, as much light reaches the photometer from the sky as if a $15^m.5$ star were in the aperture. As Equation 9.1 indicates, the larger the sky-to-star ratio becomes, the longer is the time required to complete an observation. Therefore, you should use the smallest diaphragm feasible when measuring faint sources. A size of 15 to 20 arc seconds is usually considered the minimum size for amateur telescopes,

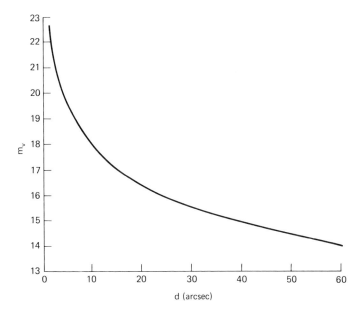

Figure 9.7. Sky brightness as seen through different diaphragms assuming a surface brightness of $22^m.6$ per square arcsec.

though diaphragms of 4 to 7 arc seconds have been used on the Hale 5-meter (200-inch) telescope with good seeing and the advantage of its superior drive.

The full moon can easily increase the brightness indicated in Figure 9.7 by three or more magnitudes. Also, because of the increased scattering near the horizon, the sky is two to three times brighter there than near the zenith. Faint star observations are therefore usually only performed under moonless conditions near the zenith and with good seeing.

9.4e Aperture Calibration

There are two main reasons why an accurate calibration of the aperture size might be necessary. When photometry of extended objects such as galaxies or comets is intended, the measurements must be reduced to magnitudes per square arc second. When you want to compare measurements taken with different apertures, both measurements must be reduced to a common aperture size.

Because of the small size involved, not many methods are available to measure the diameter or area. A direct measurement of a 0.01 inch hole to 0.1 percent accuracy requires a precision of ± 10 microns, not available from your typical ruler! Usually this is performed with a large instrument called a *measuring engine,* and must be done at an astronomy department possessing such an instrument. An indirect measurement of the area can be obtained (and the diameter from $A = \pi D^2/4$) by measuring a uniform object like a white card on a dome through each diaphragm and noting the intensity ratios. After the relative aperture sizes have been obtained, the size of one aperture can be obtained by measuring an object with known surface brightness, such as a constant light source, or by direct measurement of the largest diaphragm.

9.5 EXTINCTION NOTES

Chapter 4 gave the types of stars to be used for extinction measurements. Listed below are some of the methods for obtaining these observations during a program of all-sky photometry (i.e., not differential photometry).

1. Use more than one star to determine extinction. This reduces errors due to the non-uniform sky.
2. Use stars near the celestial equator so that they move through air mass values (X) quickly. Also, measure them several times while they are near the horizon, as X varies quickly there and you want to fill in the plot with more than one point at large X. Do not get too close to the horizon. Refraction and particulate problems are much more severe at low altitudes. A practical limit is $X < 5$.
3. If you are in a dry climate or are under an extended high-pressure system, you can average values from several consecutive nights to obtain mean extinction coefficients.
4. Observe in the following sequence: extinction stars, program stars, extinction stars, etc., spending about 80 percent of time on your program stars but obtaining extinction measurements throughout the night.
5. To check on the temporal, that is, variation with time, consistency of the sky, look at some of the North Polar Sequence stars from Appendix E throughout the night, as the zenith distance of these stars (and therefore X) changes very little. Temporal variations occur most frequently near sunrise, sunset, and when there is an approaching atmospheric frontal zone.

6. Always interpolate, never extrapolate. In other words, make sure that you have extinction measurements at air masses larger and smaller than those for your program stars.

9.6 LIGHT OF THE NIGHT SKY

Even when the sun disappears beneath the horizon, our world is filled with light. It may be fainter than daylight or present in the infrared and therefore undetectable by the eye, but is a major contributor to the errors of photometric observations. Roach and Gordon[30] give a good review of the light of the night sky, which should be studied if more details are needed. In this section we will consider only the natural sources of night sky light, neglecting man-made light pollution.

There are six general contributors to the night sky brightness: (1) integrated light from distant galaxies; (2) integrated starlight from within our galaxy; (3) zodiacal light; (4) night airglow; (5) aurora; and (6) twilight emission lines. Night airglow, aurora, and twilight emission lines are results of a planet with an atmosphere and magnetic field. Zodiacal light is a result of being within a solar system. The remaining two contributors would be present anywhere within our galaxy. We discuss only the spectral region covered by the *UBV* system with some digression into the near-infrared.

Background light from faint stars and galaxies is probably the limiting factor in photometry of faint sources. Miller[31] notes that the sky background can vary over distances of a few arc minutes. This means that if you observe a star and then offset to another location to measure only the sky, the two sky values may not agree. However, in most cases the light from these background stars and galaxies will not be important unless the program object is very faint and difficult to observe even if you are using a very large telescope. In addition, these contributors are static; if you always offset to the same location to measure sky, the contribution from these faint sources will always be the same.

Zodiacal light is caused by sunlight reflecting off dust in the plane of the solar system. It increases in brightness as the observer looks closer to the sun, and is always confined to the ecliptic plane. Zodiacal light may or may not be important as a background source in your observations depending on the location of your source with respect to the sun and the ecliptic. For instance, within 50° of the sun the zodiacal light in the ecliptic is brighter than the brightest part of the Milky Way, and it is brighter than integrated starlight over most of the sky. The zodiacal light is relatively uniform in the range of arc minutes, follows the solar spectrum, and is highly polarized like the blue sky.

Twilight emission lines are only important for a very short time after sunset and seldom interfere with astronomical observations. A layer in the upper atmosphere containing sodium atoms is illuminated by the sun after sunset as seen from the earth's surface. This illumination excites the atoms, causing them to emit the sodium D lines (5892 Å). However, only at a *solar depression angle* (the distance at which sun is found below the horizon) of 7° to 10° is this emission observable. If the sun is closer to the horizon, scattering overpowers the emission; below 10°, the layer is no longer illuminated. A similar case is noted for the red lines of oxygen atoms (6300, 6364 Å). Both of these effects contribute to errors in the V magnitude, but are important for less than an hour. Note, though, that observing in twilight carries its disadvantages.

All of the effects discussed in detail so far are minor contributors that cause errors only for short intervals or when very faint stars, those whose brightness is comparable to the sky brightness, are to be observed.

The final two terrestrial contributors can cause larger errors, both spatial and temporal. *Night airglow* is the fluorescence of the atoms and molecules in the air from photochemical excitation. It occurs primarily in a layer about 100 kilometers above the earth and is variable, depending on sky conditions, local time, latitude, season, and solar activity. There is a component that is present at most wavelengths, called the continuum, primarily caused by nitrous oxide and other molecules, but the major component is caused by distinct emission lines. Both components are always present, tend to increase in brightness near the horizon, and are not strongly affected by geomagnetic activity. The primary lines in the airglow are atomic oxygen (5577 Å), sodium (5892 Å), molecular oxygen (7619, 8645 Å), and hydroxyl, OH^- (mostly in the near-infrared). All of these emissions can be fairly strong, with some observers seeing the 5577 Å structure with the unaided eye at dark locations. Peterson and Kieffaber[32] present photographs of the hydroxyl near-infrared emission showing the mottled structure of the emission. As observed with infrared sensors, the night airglow can look like bands of cirrus clouds and move across the sky. Therefore, at least in the V filter and certainly in any filter redward of this, the airglow is a variable that always reduces the consistency of the measurements during any given night.

The *aurora* again occurs with the same mechanisms and altitudes as airglow, but varies with the solar cycle. The primary excitation mechanism is incoming charged particles from the sun. These particles become trapped in the geomagnetic field and spiral towards the poles, where they excite the atoms and molecules in the air. These polar auroras differ from airglow primarily in their strength, being up to several hundred times brighter, and their higher degree of excitation. The main auroral lines are atomic oxygen (5577, 6300, 6364 Å), hydrogen (6563 Å), and the red molecular nitrogen bands. These lines vary according to three factors: (1) solar activity, during which aurora occur more often near sunspot maxima when flares are more common; (2) latitude, because the particles concentrate near the poles, most of the radiation occurs there; and (3) time of year, peaking in March and October. Our suggestion is not to observe if an aurora is visible at your site. Though photometry can be accomplished, auroras vary rapidly in intensity and direction. Special techniques are necessary to achieve accurate results. This is one of the reasons few observatories are located above the Arctic Circle or in auroral zones. Myrabø[33] gives an example of the possible results when a chopping technique is used to remove the effects of a bright aurora that can vary five magnitudes within 1 minute.

The conclusions of this section are threefold. First, the light of the night sky is not constant in time or space, and therefore will always limit the accuracy of your measurements. Do not expect $\pm\ 0^m.001$ results! Second, do not observe during an aurora, near the horizon, or at twilight. And last, measurements redward of the *V* filter are strongly affected by the varying night sky and should be avoided until experience is gained with the *UBV* system. Except for auroras or while observing very faint stars, the night sky variations will probably never be noticeable in your photometry, but knowledge of the possibility of these errors should be filed in your mind for later reference.

9.7 YOUR FIRST NIGHT AT THE TELESCOPE

For the newcomer to photometry, the previous sections may seem helpful but somehow cloud the answer to the question "What do I do at the telescope?" In this section, we attempt to pull together the concepts of earlier chapters to outline some procedures to follow at the telescope.

It is important to be prepared before going to the telescope. Unless you plan to observe some very bright objects, you should prepare a set of finding charts in order to identify your "targets" for the night. It is important to mark the charts to indicate the orientation and the size of the field of view of your guide telescope or wide-field eyepiece of the photometer. This can save hours of frustration when you try to identify the star fields at the telescope. If you are doing differential photometry, be sure the comparison and check stars are marked on the chart in addition to the variable star. If you are using a pulse-counting system, you should determine the dead-time coefficient as outlined in Section 4.2. For a DC photometer, you should calibrate the gain settings as described in Section 8.6.

It is a good idea to arrive at the telescope early to uncap the tube and let the optics adjust to the outdoor temperature. All the electronics should be turned on at least 60 minutes prior to observing. This is necessary to allow the electronics time to stabilize. DC amplifiers, for example, tend to drift rapidly for the first few minutes after they are turned on. The high voltage should also be applied to the photomultiplier tube (with the dark slide closed) because the tube tends to be noisier than normal during the first few minutes of operation. If your photometer uses a cooled detector, this is the time to turn on the cooling system or to add dry ice to the cold box.

If you are using a photomultiplier tube, the next step is very important. In order for the photometer's optical elements to be in the proper location within the telescope's light cone, it is important that the diaphragm be positioned in the focal plane of the telescope. To do this, first aim the telescope at some bright background (such as the observatory wall, or the twilight sky) so that the outline of the diaphragm can be seen clearly in the diaphragm eyepiece. This can also be accomplished with a diaphragm light. This eyepiece is then focused to make the diaphragm as clear as possible. *Do not adjust this eyepiece focus again the rest of the night.* Next, a bright star is centered in the diaphragm and focused sharply using the *telescope focus*. Now both the star and the diaphragm appear sharp when viewed in the diaphragm eyepiece. The diaphragm is now in the focal plane of the telescope.

This procedure is unnecessary for a photodiode photometer. In this case, the photometer head does not have a diaphragm. The head is constructed so that when the stellar image appears focused in the eyepiece, it is also focused on the photodiode when the flip mirror is removed from the light path. The set-up procedure is concluded by choosing the diaphragm to be used (see Section 9.4), setting the telescope coordinates, and setting the observatory clock.

There are some important comments to be made about the art of data taking at the telescope. Making a stellar measurement with a pulse-counting system is the simplest. The star is centered in the diaphragm, the dark slide is opened, and the counts are recorded. For this system, use the advice of Section 9.3a to estimate the total observing time through each filter. Measurements through the red-leak filter should have the same duration as the U filter. If you are using a DC system, some additional advice is in order. The DC amplifier zero point may drift slightly during the night. This drift may be either positive or negative. Most chart recorders and meters do not follow a negative drift off the bottom of the scale. For this reason, it is a good idea to put the zero point at about 10 percent of full scale. To do this, simply close the dark slide (with the high voltage on) and adjust the "zero adjustment" knob on the DC amplifier. Chart recorders and meters are most accurate at their maximum (full-scale) reading. Therefore, whenever possible, adjust the amplifier gain so that a stellar measurement gives a nearly full-scale reading.

There are three additional rules to follow when doing DC photometry. When you measure a star for the first time, be sure the amplifier gain is at its lowest setting. You then increase the gain to achieve the desired meter deflection. This procedure is designed to protect the meter and/or chart recorder from damage if too high a gain setting is used. If this happens, the pen or meter needle will slam beyond full scale. With a little practice, you can estimate a safe gain setting simply by observing the apparent brightness of the star in the eyepiece. The second rule is to use the same gain settings for the sky measurements as was used for the stellar measurements. A much higher gain may be used for the U measurements than is used for B or V. The U sky background measurement also must be made at this higher gain setting. This is obviously necessary if we are to subtract the sky measurement from the stellar measurement directly. However, this is also necessary because it is possible for the amplifier zero point to change slightly with different gain settings. The third rule is to record the coarse and fine gain settings separately and not their total. The reason is that each gain position requires a calibration correction as discussed in Section 8.6. More than one combination of switch positions can yield the same total. If your records only contain the total gain, you will not be able to determine the proper gain correction without ambiguity.

The observing pattern used at the telescope depends strongly on the

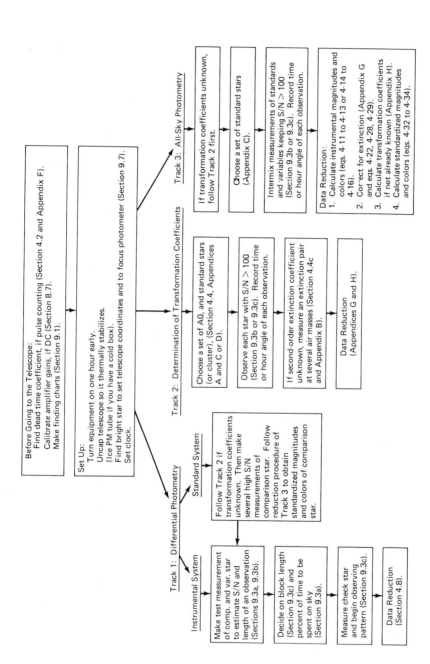

Figure 9.8. Observing flowchart.

research program and the habits of the observer. For example, one might begin the night with simple differential photometry of a rapid variable but after an hour or two start a program of observing cepheid variables on the *UBV* system scattered all over the sky. Clearly, this calls for a combination of observing techniques. However, on any given night, the observing pattern can be broken into three categories. The first category is differential photometry either on the standard or the instrumental photometric system. The second is a sequence of observations designed to yield the transformation coefficients to the standard system. The third category is what we call "all-sky" photometry in which many objects are observed throughout the night, located at various positions in the sky. The cepheid observations mentioned above fall into this category. We must stress that there are several other forms of photometry, such as occultation or galaxy surface photometry, that do not fit these categories directly. These categories are merely the most common types and they are defined to help the novice to see how things are done. Figure 9.8 illustrates possible observing patterns with the three categories labeled as tracks 1, 2, and 3. This figure is a flowchart that can be followed from observing preparations to data reductions. For your reference, we have labeled the appropriate sections where a discussion or a worked example can be found.

A final comment about data recording is in order for those using DC photometers. If you are using the amplifier meter to make your measurements, it is a good idea to follow a simple pattern. As you watch the needle, it appears to jitter above and below some mean value. Rather than watching the meter for a long stretch of time, it is better to watch the meter for, say, 10 seconds and then estimate and record the mean value. This procedure should be repeated several times and averaged to make a single measurement. When using a chart recorder, the jitter of the needle is transformed to the jitter of a pen recording on strip-chart paper. The main advantage of the chart recorder is that you have a permanent record of the actual observations and you can transform the pen tracings to numbers the next day. Figure 9.9 shows a sample chart tracing to illustrate the technique of extracting the data. Each measurement is estimated by drawing a straight line through the middle of the jitter. The two sets of sky measurements illustrate that our golden rule of differential photometry really applies to all photometry. By drawing a line connecting the sky measurements, we can interpolate the sky background to the time of each stellar measurement. In this exam-

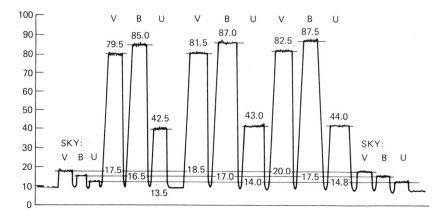

Figure 9.9. Typical chart tracing.

ple, the sky background was increasing because of a rising moon. Note that if we had made only the first sky measurement, it would have appeared that the star was brightening steadily. With the two sets of sky measurements and the graphical interpolation, you can see that the star was essentially constant once the sky background is subtracted.

We wish you good luck on your first night of photometry.

REFERENCES

1. Larson, W. J. 1978. *Sky and Tel.* **56**, 507.
2. Argelander, F. W. A. and Schönfeld, E. 1859–1886. *Astronomiche Beobachten Sternwarte Konigl.* **3, 4, 5, 8**.
3. Thome, J. M. and Perrine, C. D., 1892–1932. *Resultados Observatorios Nacional Argentino* **16, 17, 18, 21**.
4. Staff, 1969. *The Smithsonian Astrophysical Observatory Star Atlas.* Cambridge: MIT Press.
5. Becvar, A., 1964. *Atlas Borealis, Eclipticalis, and Australis.* Cambridge: Sky Publishing Co.
6. Ingrao, H. C. and Kasperian, E. 1967. *Sky and Tel.* **34**, 284.
7. Vehrenberg, H. 1963. *Photographic Star Atlas.* Düsseldorf: Treugesell-Verlag KG.
8. Vehrenberg, H., 1970. *Atlas Stellarum 1950.0.* Düsseldorf: Treugesell-Verlag KG.
9. Papadopoulos, C. 1979. *True Visual Magnitude Photographic Star Atlas.* Elmsford, NY: Pergamon.
10. Lick Observatory 1965. *Lick Observatory Sky Atlas.* Mt. Hamilton, California.
11. Doughty, N. A., Shane, C. O., and Wood, F. B. 1972. *Canterbury Sky Atlas (Australis).* New Zealand: Mount John University Observatory.

12. Palomar Observatory, 1954. *National Geographic–Palomar Observatory Sky Survey.* Pasadena: California Institute of Technology.
13. ESO/SRC. *Atlas of the Southern Sky.* In preparation.
14. Overbye, D. 1979. *Sky and Tel.* **58**, 30.
15. Kukarkin, B. V., Kholopov, P. N., Efremov, Yu. N., Kukarakina, N. P., Kurochkin, N. E., Medvedera, G. I., Peruva, N. B., Fedorovich, V. P., and Frolov, M. S. 1969–1974. *General Catalog of Variable Stars.* Moscow: Academy of Sciences of the U.S.S.R.
16. Scovil, C. E. 1980. *AAVSO Variable Star Atlas.* Cambridge: Sky Publishing Co.
17. Hoffmeister, C. *Mitteilungen der Sternwarte zu Sonneberg,* No. 12–22 (1928–1933), No. 245–330 (1957).
18. Anon., 1954. *Communications of the Observatory of the University at Odessa* **4**, 1–3.
19. Hagen, J. G. 1934. *Atlas Stellarum Variabilium.* Vatican.
20. Bateson, F. M., Jones, A. F., et al. 1958–1977. *Charts for Southern Variables.* Wellington.
21. *Odessa Universitet Observatoriia Izvestiia,* 1953, 1954, 1955, vol. **IV**, parts I, II, III.
22. Code 680, Goddard Space Flight Center, Greenbelt, MD 20771
23. Cannon, A. J., and Pickering, E. C. 1918–1924. *Henry Draper Catalog and Extensions,* Harvard Annals 91–100.
24. Danford, S., and Muir, R., 1978. *Bul. Amer. Astr. Soc.* **10**, 461.
25. Houk, N. and Cowley, A. P. 1975. *University of Michigan Catalogue of Two Dimensional Spectral Types for the HD Stars.* Ann Arbor: Univ. of Michigan Press, vol. 1.
26. Overbye, D. 1979. *Sky and Tel.* **57**, 223.
27. Young, A. T. 1974. In *Methods of Experimental Physics: Astrophysics.* vol. **12A**. Edited by N. Carleton. New York: Academic Press.
28. King, I. R. 1971. *Pub. A. S. P.* **83**, 199.
29. Picarillo, J. 1973. *Pub. A. S. P.* **85**, 278.
30. Roach, F. E., and Gordon, J. L. 1973. *The Light of the Night Sky.* Boston: D. Reidel.
31. Miller, R. H. 1963. *Ap. J.* **137**, 1049.
32. Peterson, A. W. and Kieffaber, L., 1973. *Sky and Tel.* **46**, 338.
33. Myrabø, H. K. 1978. *Observatory* **98**, 234.

CHAPTER 10
APPLICATIONS OF PHOTOELECTRIC PHOTOMETRY

By now, you probably have built or purchased a photoelectric photometer and have measured the magnitudes of a few bright stars. You may rightfully wonder what to do next! Happily, this is the least of your worries, as there are thousands of interesting stars and projects within the range of most amateur and small college telescopes.

In this chapter, we present a few of these projects. By no means is this an exhaustive compendium and we invite you to branch out and pursue other topics. However, the ideas mentioned here should give you a feeling of the breadth and diversity of the field of photometry. Not only can it be as fun and interesting as photography or visual observing, but you can help in the advancement of scientific knowledge.

10.1 PHOTOMETRIC SEQUENCES

Are you unselfish and like to share your work? Then determining photometric sequences should be right up your alley! The object is to determine the magnitudes of several standard stars in a field surrounding or near some interesting source. These sequences have to be determined carefully, as many observers following in your footsteps will be using your data as the foundation for their own research. For the purposes of this section, we identify three types of comparison star sequences: for visual observers, for photoelectric observers, and for calibrating photographic plates.

Visual variable star observers have two primary advantages over photoelectric observers: they are less influenced by light cloud cover because the comparison is made almost instantaneously, and they can work faster because the object's light does not have to be centered in a diaphragm. For these reasons, among others, visual photometry still flourishes in the ranks of amateur astronomers. However, the comparison star sequences around visual variables are generally not very accurate. The stars should cluster uniformly in space and brightness around the variable, appear similar in color, and have their magnitudes measured in a filter bandpass approximating the response of the unaided eye. This latter requirement has been difficult to achieve. Landis[1] mentions that the V filter does not match the eye's response, as blue stars look fainter and red stars brighter in V than with the eye. Stanton[2] experimented with filter responses and found that a Schott GG4 filter along with an EMI 9789B (S-11 cathode) matched the eye's response best. Each visual observer's eye has a slightly different response because of inherent and dark adaptation variations. We suggest that you cooperate with the American Association of Variable Star Observers (AAVSO)[3] and help them in determining photoelectric sequences in poorly defined fields and in finding good filter-photometer combinations for visual magnitude sequences.

Photoelectric observers need fewer comparison stars per variable than do visual observers, but the selection of suitable comparisons can be very tedious. The color (especially) and magnitude should match as closely as possible, and the star should be within 1° of the variable. Unlike visual observers, where 0.1 magnitude accuracy is acceptable, photoelectric observers need to determine their variables to 0.01 magnitude or better. This means the comparison stars must be at least this stable. Two stars should be located and measured on several occasions to ascertain whether they are truly constant in magnitude. It is difficult to obtain comparison stars for red variables because a large fraction of cool stars are variable, often with long periods, so that they may be constant during a few night's observations. Try to determine comparison stars for a group of variables, such as bright flare stars, cepheids with 5-day periods, etc. Another project is to reobserve comparison stars used in the literature to see if they remain constant or are long-period variables. Check with a local professional astronomer or the AAVSO for additional or more specific ideas.

Photographic photometry is still valuable today because of its great multiplexing advantages. By calibrating a cluster plate accurately, for example, you can measure the brightness of many hundreds of stars in a few hours. In crowded fields, photographic photometry may be the only practical method of determining magnitudes. It is also much faster in measuring faint stars than photoelectric methods. Calibrating photographic plates is impossible unless a good photoelectric sequence appears in the plate field. This sequence should cover as wide a magnitude range as possible, and be obtained in the filter bandpass of the plate, which is usually blue. The plate calibration of the large surveys such as the National Geographic–Palomar Sky Survey is particularly important, as the Space Telescope will require accurate magnitudes over the entire sky, obtainable only from such surveys. Argue et al.[4] have gathered the known photoelectric sequences, which are usually around galactic and globular clusters. Complete coverage of the sky is lacking, though, and amateur observations are particularly needed here to take some of the burden from the professional observatories. There are over 1000 fields in the National Geographic–Palomar Sky Survey alone, each requiring 10 to 20 sequence stars to fifteenth visual magnitude for the fine guidance system of the Space Telescope. You can see that there is plenty of opportunity here for your photometer to feel needed!

10.2 MONITORING FLARE STARS

In 1924, the astronomer E. Hertzsprung[5] was photographing an area of sky in the constellation of Carina. He discovered that on one photograph a star was two magnitudes brighter than it had been on previous photographs. The rate at which the star brightened was much too fast to be either a nova or any kind of intrinsic variable then known. Hertzsprung tried to explain the event as the result of an asteroid falling into a star. This is the first known observation of what we now call a *flare star*. Later, this same star was seen to flare many more times and was given the variable star designation DH Car. In the years that followed, other M-type dwarfs were seen to flare but they did not receive recognition as a new class of variable stars until 20 years after Hertzsprung's initial observation.

In 1947 the American astronomer Carpenter (Luyten[6]) was photographing a red dwarf. This star was known to have a large proper motion, which implied that it was nearby. Carpenter was attempting to measure its parallax. On a single plate, he took a series of five 4-minute exposures, moving the telescope slightly after each one. When the plate was developed, he expected to see five equally bright images. But the second image was much brighter than the first. The star faded in the later images, returning to its normal brightness. The star had become 12 times brighter in about 3 minutes. This rate of increase is even faster than a supernova, though the total luminosity is much less. This star is now called UV Ceti and flare stars are now commonly referred to as *UV Ceti variables.*

The first photoelectric recording of a flare was obtained by Gordon and Kron[7] in 1949. They were observing a late-type eclipsing binary with the 0.9-meter (36-inch) refractor at Lick Observatory. They were measuring the comparison star when the needle on the DC amplifier went off the top of the scale. After several frantic minutes of experimentation, they realized their equipment was not malfunctioning and the comparison star had flared. They were able to observe the star slowly return to its normal brightness. This star is known today as AD Leo. This story simultaneously points out the sudden nature of flare star eruptions and one of the dangers of using a red comparison star!

Flare stars are usually M-type dwarfs that have emission lines in their spectra. K-type stars occasionally have been seen to flare but they do so much less frequently. Flare stars are typically one-twentieth to one-half the mass of the sun and only 10^{-5} times as luminous. They are so intrinsically faint that they must be nearby to be seen at all. Their sudden increases in luminosity have been likened to solar flares, only thousands of times stronger. These are very remarkable events for such small stars. Flare amplitudes can range from a few tenths of a magnitude or less to several magnitudes. There are two general types of flares. A spike flare increases suddenly with a rise time of only a few seconds and then fades in a few minutes. A slow flare brightens gradually over an interval of minutes to tens of minutes and then fades at about the same rate.

However, these classifications should not be taken too literally. Many flares are a combination of both types and there is great variability from flare to flare even in the same star. A more complete description of these interesting objects is beyond the scope of this section. The interested reader is referred to the work of Moffett,[8] Lovell,[9] Gershberg,[10] and Gurzadyan.[11]

The monitoring of flare stars can only be recommended to the beginner with some reluctance. Catching a flare photoelectrically can be every bit as exciting today as it was for Gordon and Kron in 1949. However, the observer must be prepared for possibly spending many tens of hours at the telescope and recording nothing but a constant star. Flares cannot be predicted, and catching one is a matter of looking in the right place at the right time. For those who cannot resist the chance of actually seeing a flare, we supply a list of flare stars, brighter than twelfth visual magnitude, in Table 10.1. Finder charts for some of these stars are available from the AAVSO at a modest cost.

Besides the requirement of patience, there are certain requirements of your equipment. Unlike normal photometry, in flare star monitoring you measure the same star for hours at a time, interrupted occasionally by a measurement of the sky background. It is therefore important that your telescope can track well and keep the star centered in the diaphragm as long as possible. Because the flare may occur very rapidly and at some unknown time, the observations are usually made in a single filter. Flares are brightest in the ultraviolet, so an ultraviolet or blue filter is preferred. Pulse-counting photometers cannot be used unless a large digital memory that can store hundreds or thousands of measurements is available. Such a system has been described by Warner and Nather[12] and Nather.[13] This instrumentation is fairly elaborate and a much simpler system for the beginner is a DC amplifier and a strip-chart recorder. If the telescope tracks well, the equipment can be left unattended for small intervals of time and the chart recorder will preserve a record of any flare activity. It is important to use the shortest possible amplifier time constant. This is necessary so the amplifier can follow any rapid flickering during the flare event. Unlike the normal procedure, it is a good idea to adjust the amplifier gain so that the star does *not* read nearly full scale. After all, if the star flares it would be helpful if the amplifier and chart recorder do not go off the top of the scale.

As mentioned above, catching a flare involves a lot of luck. One of the authors (RHK) recalls accumulating over 23 hours of monitoring of a flare star only to collect what seemed like miles of chart paper with a constant ink line. In utter frustration, the telescope was moved to the flare star EV Lac. Within the first hour of monitoring the spike flare shown in Figure 10.1 was recorded. Note that this flare lasted only 82 seconds. It would have been very unfortunate if during those 82 seconds the sky background was measured instead!

TABLE 10.1. Flare Stars

Star	RA (1950.0)	Dec. (1950.0)	RA (1985.0)	Dec. (1985.0)	Vis. Mag.	Comments
BD +43° 44	00^{h}15.5^m	43° 44.4	00^{h}17.3^m	43° 56.0	8.1	Visual double
CQ And	00 15.5	43 44.4	00 17.3	43 56.0	11.0	
BD +66° 34	00 29.3	66 57.8	00 31.3	67 09.3	10.5	
Butler's Star	00 58.1	−73 13.4	00 59.2	−73 02.0	10.6	
LPM 63	01 09.9	−17 16.0	01 11.6	−17 04.9	11.6	
CC Eri	02 32.5	−44 00.6	02 33.8	−43 51.4	8.7	
40 Eri C	03 13.1	−07 44.1	03 14.8	−07 36.3	11.2	
Ross 42	05 29.5	09 47.3	05 31.4	09 48.8	11.5	
V371 Ori	05 31.2	01 54.8	05 33.0	01 56.2	11.7	
BD −21° 1377	06 08.5	−21 50.6	06 10.0	−21 51.0	8.1	
Ross 614	06 26.8	−02 46.2	06 28.5	−02 47.6	11.1	
PZ Mon	06 45.8	01 16.6	06 47.6	01 14.2	10.8	
AC +38° 23616	07 06.7	38 37.5	07 09.1	38 34.1	11.5	
YY Gem	07 31.6	31 58.8	07 33.6	31 54.2	9.1	
YZ CMi	07 42.1	03 40.8	07 43.9	03 35.7	11.2	
BD +33° 1646B	08 05.7	32 56.0	08 07.9	32 49.8	11.0	
AD Leo	10 16.9	20 07.3	10 18.8	19 56.7	9.4	
SZ UMa	11 17.5	66 07.0	11 19.6	65 55.5	9.3	
Ross 128	11 45.2	01 01.0	11 47.0	00 49.3	11.1	
DT Vir	12 58.3	12 38.7	13 00.0	12 27.4	9.8	
EQ Her	13 32.1	−08 05.1	13 33.9	−08 15.8	9.3	
V645 Cen	14 26.3	−62 28.1	14 29.0	−62 37.4	11.1	
DM +16° 2708	14 52.1	16 18.3	14 53.7	16 09.8	10.2	
DM +55° 1823	16 16.0	55 23.8	16 16.8	55 18.7	10.0	
V1054 Oph	16 52.8	−08 14.7	16 54.7	−08 18.0	9.8	Visual Companion, Ross 867, also flare star (12.9 mag.)
Ross 868	17 17.9	26 32.8	17 19.3	26 30.7	11.4	
BY Dra	18 32.7	51 41.0	18 33.5	51 42.7	8.6	
V1216 Sgr	18 46.8	−23 53.5	18 48.9	−23 51.1	10.6	

TABLE 10.1.. Flare Stars (continued)

Star	(1950.0) RA	Dec.	(1985.0) RA	Dec.	Vis. Mag.	Comments
V1285 Aql	18 53.0	08 20.3	18 54.7	08 26.0	10.1	
WOLF 1130	19 20.1	54 18.2	19 20.9	54 22.2	11.9	
AT Mic	20 38.7	−32 36.6	20 40.9	−32 29.1	10.8	
AU Mic	20 42.1	−31 31.1	20 44.2	−31 23.5	8.6	
AC +39°57322	20 58.1	39 52.7	20 59.4	40 00.9	10.3	Visual Companion, DO Cep, also flare star (13.3 mag.)
BD +56°2783	22 26.2	57 26.8	22 27.5	57 37.5	9.9	
L717-22	22 36.0	−20 52.8	22 37.9	−20 41.9	11.5	
EV Lac	22 44.7	44 04.6	22 46.2	44 15.5	10.2	
DM +19°5116	23 29.5	19 39.7	23 31.2	19 51.3	10.4	
BD +1°4774	23 46.6	02 08.2	23 48.4	02 19.9	9.0	

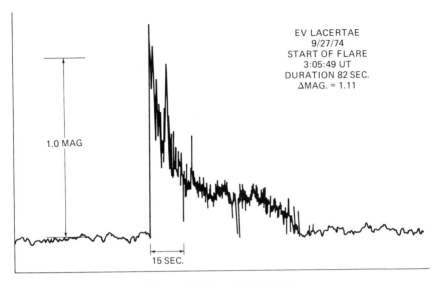

Figure 10.1. Flare of EV Lacertae.

10.3 OCCULTATION PHOTOMETRY

Occultations are among the oldest astronomical observations. With the advent of modern photoelectric photometers, chart recorders, and accurate timekeeping via WWV, amateur astronomers now have the means to make high-quality observations of occultation events. As more and more observers couple microcomputers to their equipment, the limit to complex occultation projects will depend only on the ingenuity of the astronomer. Space does not permit a thorough coverage of the methods of observing occultations. Instead, the reader is urged strongly to study the appropriate references in this section for a more in-depth treatment.

This section describes several types of occultation observations. Occultation observers routinely seek information on the figure of the lunar limb, the separation of otherwise unresolvable binary stars, the diameter of an asteroid, or the angular size of a star. No matter what sort of occultation project is undertaken, one theme underlies each of these observations: occultations yield very high angular resolution. Observers can resolve angles as small as a few arc milliseconds. Measuring, say, 5 arc milliseconds is like determining the angular size of an orange at a distance of 3200 kilometers (1920 miles). The reason for this greatly enhanced resolution over the Dawes limit is that you no longer use the telescope's optics to do the resolving. Instead, you take advantage of the geometry of the occultation by using the separation between your earth-bound location and the occulting body as a sort of "interplanetary optical bench" in probing the object of interest.

Requirements for observing occultations vary with the complexity of the project. For many observations, only a strip-chart recorder and an accurate timepiece are needed. In order to obtain the high resolution afforded by occultations, observations must be made at high time resolution. Instrumentation such as that described by McGraw et al.[14] provides a great deal of versatility and high time resolution. Integrations must be less than 1 second and should be somewhat less for some of the projects described. This may be accomplished by running the chart recorder at high speed or for those with microcomputers, reading photometer counts into a "circulating" memory at a rate controlled by the computer. By use of microcomputers, a time resolution of 0.001 second is not difficult to obtain. This high time resolution is necessary for angular diameter studies of stars.

Grazing lunar occultations have long been of interest to amateur astronomers. Amateurs have banded themselves into groups that pack up their portable equipment, travel to "graze lines," and through cooperative efforts obtain lunar limb profiles. Many graze observers obtain very good results with little more than a stopwatch and telescope. Use of photometry for the quick disappearances and reappearances characteristic of grazes yields an unbiased, more accurate recording of grazes. Harold Povenmire[15] has devoted considerable effort to the observation of grazing occultations.

From time to time, the planets occult stars. During such events, observers are afforded the chance to obtain accurate diameters of these objects or information on their upper atmospheres. Rather startling discoveries have recently come from occultations of stars by planets. The discovery of the Uranian rings in 1977 by Elliot et al.[16] was made by occultation. Almost all information on those rings continues to be from occultation observations. In April 1980, Pluto nearly occulted a star. Such an observation would be extremely valuable as a measure of Pluto's diameter; no such occultation has yet been reported. During that 1980 appulse, Pluto's moon Chiron did occult the star as seen by one observer, A. R. Walker[17] in South Africa. An upper limit of 1200 kilometers for the diameter was deduced from the observation. Occultations of stars by planets is still fertile ground for new discoveries.

In the past few years, astrometric techniques have been pushed toward the goal of providing better predictions of occultations by asteroids. Accurate predictions are important because the path of observability is equal to the diameter of the occulting body, and asteroids are generally less than a few hundred kilometers across. Results from these measurements are impressive. Often, diameter determinations are accurate to within 1 percent.

The observations just described do not require microcomputer control of the occultation event. However, as more observers couple computers to their telescopes, it is probably only a matter of time before an intrepid amateur attempts to determine a stellar angular diameter. During a lunar occultation, the light from a star is diffracted by the lunar limb. What the photometer "sees" during the occultation is the passage of the fresnel diffraction fringes: an oscillation of the stellar flux just prior to occultation. By analysis of the diffraction fringe spacing and height, the angular size of the occulted star may be determined. References concerning aspects of lunar occultations may be found in a two-article sequence by Evans.[18,19] A more rigorous development of the topic may be found in Nather and Evans,[20] Nather,[21] Evans,[22] and Nather and McCants.[23] The fresnel diffraction pattern is only seen at high time resolution, that is, integrations of around 1 millisecond.

Related to occultations are mutual phenomena of planetary satellites, that is the mutual occultations, eclipses, and transits of satellites during the time the nodes of their orbital planes align with the earth. Accurate timing of these events can give accurate shapes and sizes of these satellites.

Predictions of the described occultations may be found from several sources such as *Sky and Telescope,* the *Astronomical Almanac,*[24] or the *Observer's Handbook,*[25] to name a few. Ambitious astronomers may wish to try their hand at predicting occultations for their particular location. These readers are referred to Smart's *Spherical Astronomy,*[26] where an excellent treatment may be found. Amateurs interested in any part of occultation observing are urged to join the International Occultation Timing Association.[27] IOTA provides excellent predictions, information, and hints on observing all types of occultations. Furthermore, this organization serves as a clearing house for observations of almost all types of occultations.

[Note: Section 10.3 was contributed by T. L. Mullikin, Space Operations and Satellite Systems Development, Rockwell International.]

10.4 INTRINSIC VARIABLES

Intrinsic variables are those stars that vary in brightness because of internal changes. Normally, this is evidenced by pulsational behavior: the star periodically shrinks and expands. Sometimes the light variations are highly regular, as in the case of cepheids and RR Lyrae variables; sometimes it is quite erratic, as for Mira and RV Tauri variables.

A thorough description of the characteristics and pulsational mechanisms for these stars would occupy more space than this text. Therefore, we refer the interested reader to Glasby,[28] Strohmeier,[29] and Kukarkin[30] for more detail. The major point of interest in this section is the fact that there are thousands of intrinsic variables, each one unique though they are placed in general categories. Astronomers have neither the manpower nor the telescope time to investigate even a majority of these stars with the thoroughness they deserve. The energies of the amateur variable star observer should be channeled in the following directions:

1. Determining standard comparison stars and preparing finding charts for those variables where no charts exist.
2. Obtaining light curves for stars with missing or incomplete curves.
3. Observing a variable star at an epoch several years removed from previous measurements, in order to detect temporal variations.
4. Investigating stars mentioned in the *Catalog of Suspected Variables*[31] or other publications to determine their nature and period.

Two rules should be followed for accurate data: use comparison stars for all measures, and obtain as many observations as possible, covering the entire light curve.

In this section, several tables of stars are presented. In each case, the 1950 coordinates are as accurate as could be found. Those variables for which coordinates are less accurate can be identified by their 1950 declination, where the arc seconds are multiples of six. The magnitudes are a general range, with the particular color indicated with a letter: *P* for photographic, *B* for the *B* filter of the *UBV* system, and *V* for the visual. Each list contains approximately 30 bright variables in each period range, as culled from the magnetic tape version of the *General Catalog of Variable Stars*. For those of you with access to large computer facilities, we highly recommend acquiring a copy of this magnetic tape through Goddard Space Flight Center in Greenbelt, Maryland.[32] Another good source of bright variables is the *Atlas of the Heavens Catalog*,[33] which has 13 pages of variables with 1950 coordinates.

Note that as the periods of the stars mentioned in this section increase, the accuracy with which they are determined decreases. This is because photoelectric data are missing on most of the long-period variables. They have not been observed for nearly as many periods as those variables with short cycles. The use of the terms *epoch* and *period* is explained in Section 10.5.

10.4a Short-Period Variables

As with the other classifications, as soon as researchers find more than one star with similar characteristics, they invent a new group of variables. The short-period intrinsic variables have therefore been divided into four major groups:

1. δ *Scuti stars*. These have low amplitude, sinusoidal light curves. Periods are less than 0.3 day.
2. *Dwarf cepheids*. These have larger amplitudes (though less than one magnitude), varied light curve shapes (either sinusoidal or asymmetric with a faster rise to maximum brightness than the ensuing decline), and periods also less than 0.3 day.
3. *RR Lyrae variables*. Named after their prototype, these are similar to dwarf cepheids but with periods ranging between 0.3 and 1.0 day. They are commonly called *cluster variables* because of their affinity to globular clusters.
4. *Cepheids*. Again similar to dwarf cepheids, these variables have periods ranging between 1 and approximately 50 days.

Eggen[34] and Tsesevich[35] give more information on these stars. For the amateur, δ Scuti stars are very difficult to observe as the amplitude is seldom greater than 0.10 magnitude in the visual. Therefore, we concentrate on the latter three groups.

Dwarf cepheids brighter than twelfth visual magnitude and with a ΔV of 0.5 or less have been listed by Percy et al.[11] The coordinates that they give are for 1900.0 and will have to be precessed to be usable. Fourteen stars are included on their list. A complete list (but without coordinates) is given by Eggen,[34] including all δ Scuti and dwarf cepheids known to date that are brighter than visual magnitude 11.5 at maximum. Coordinates can be found in Eggen's references or by using the BD or HD catalogs.

No exclusive catalog exists for RR Lyrae variables. Tsesevich[35] presents older visual observations for several hundred RR Lyrae stars and this work should be reviewed by the interested reader. A complete catalog of the several thousand Magellanic Cloud RR Lyrae variables has been published by the Gaposchkins,[37,38] but these stars are too faint and in overcrowded fields for photoelectric work by all but the large professional telescopes. Sturch[39] lists photoelectric observations on 100 field RR Lyrae stars.

Table 10.2 lists the magnitudes and coordinates for selected very short-period (less than 1 day) variables. Table 10.2 is by no means complete, but can be used as a stepping stone to more difficult objects.

Because of their relative rarity and larger luminosity, cepheid variables have been cataloged more carefully than the other variables. Schaltenbrand and Tammann[40] give a complete list of those cepheids with photoelectric data. Henden[41,42] gives an extensive list of short-period (less than 5 days) cepheids that are visible from the Northern Hemisphere. Table 10.3 lists a representative sample of cepheid variables with periods less than 10 days.

10.4b Medium-period Variables

These stars are classified primarily as *RV Tauri stars* or *long-period cepheids*. Eggen[43] gives a review of these stars whose periods range from 30 to 100 days or so. Most of the long-period cepheids are found in the Small Magellanic or Large Magellanic Cloud. Very little data exist on these variables because of their rarity and long periods. These and the longer period Miras are prime candidates for dedicated amateurs as their periods are too long for good coverage from national observatories. RV Tauri light curves are unstable and show irregularities in both shape and period. Like RR Lyrae stars, they belong to the old galactic halo star population type. The long-period cepheids have not been studied extensively, and in some cases, such as RU Cam, they have quit pulsating for extended periods of time. Representative members of both of these classes with periods between 10 to 100 days are shown in Table 10.4.

TABLE 10.2. Intrinsic Variables, $0.1 < P < 1.0$ Days

NAME	R.A. (1950.0)	DEC.	R.A. (1985.0)	DEC.	TYPE	MAG.	EPOCH	PERIOD
BS AQR	23 46 11.6	- 8 25 24.5	23 47 59	- 8 13 44	RRS	9.4-10.0 B	28095.330	0.19782278
DN AQR	23 16 28.5	-24 33 28.7	23 18 19	-24 21 59	RRAB	10.0-10.5 P	28425.284	0.63464
X ARI	3 5 48.0	10 15 23.8	3 7 41	10 23 25	RRAB	9.2-10.5 B	37583.568	0.651139
VZ CNC	8 38 9.5	10 10.4	8 40 3	9 52 41	RRS	7.5- 8.2 P	33631.8461	0.17836376
AD CMI	7 50 11.7	1 43 39.1	7 52 0	1 38 12	RRS	9.1- 9.4 B	36601.8228	0.122974
V743 CEN	13 25 17.7	-51 43 58.5	13 27 26	-51 12 50	RRS	8.3- 8.5 P	0.104	
RZ CEP	22 37 27.9	64 35 42.6	22 38 40	64 46 38	RR	9.5-10.3 B	38207.938	0.308645
RR CET	1 29 34.0	1 5 12.0	1 31 21	1 15 59	RRAB	9.3-10.3 P	17501.4421	0.5530253
XZ CET	1 57 52.6	-16 35 15.2	1 59 33	-16 25 5	RR	8.5- 9.2 P	—	0.451
XZ CYG	19 31 27.4	56 16 47.2	19 32 10	56 21 20	RRAB	9.1-10.5 B	36933.981	0.466579
DX DEL	20 45 5.0	12 16 42.0	20 46 44	12 24 26	RRAB	9.5-10.3 V	30950.506	0.47261673
SU DRA	11 35 6.8	67 36 27.0	11 37 6	67 24 49	RRAB	9.2-10.2 V	20605.7569	0.66041926
XZ DRA	19 9 24.4	64 46 33.1	19 9 37	64 50 2	RRAB	9.6-10.6 V	27985.648	0.4764944
RX ERI	4 47 28.5	-15 44 35.3	4 49 3	-15 45 59	RRAB	9.2-10.1 V	21692.479	0.58724622
SV ERI	3 10.9	-11 32 36.0	3 11 7	-11 22 24	RRAB	9.6-10.2 V	28398.200	0.7137590
CS ERI	2 35 10.9	-43 10 47.3	2 36 30	-43 1 41	RRC	8.7- 9.2 V	—	0.311331
SS FOR	2 5 36.1	-27 6 5.9	2 7 11	-26 56 8	RRAB	9.5-10.6 V	38668.951	0.495432
RS GRU	21 39 48.2	-48 25 6.5	21 42 5	-48 15 30	RRS	7.9- 8.5 V	34325.2931	0.14701147
V IND	21 8 11.0	-45 16 42.0	21 10 30	-45 5 6	RRAB	9.1-10.5 V	27993.567	0.479591
TT LYN	8 59 49.0	44 47 6.0	9 2 9	44 38 47	RRAB	9.5-10.2 V	36651.358	0.5974379
RR LYR	19 23 52.1	42 41 11.9	19 24 59	42 45 24	RRAB	7.2- 8.6 B	38241.460	0.5668054
TY MEN	5 40 38.1	-81 30 19.1	5 37 37	-81 37 14	RRS	7.7- 7.9 P	38314.493	0.18747
V429 ORI	4 53 43.0	- 3 36 30.0	4 55 27	- 3 33 12	RRS	9.0-10.0 V	28876.413	0.5017
DH PEG	22 12 55.0	6 34 12.0	22 14 40	6 44 39	RRC	9.3- 9.8 V	38251.872	0.255510
RU PSC	1 11 42.0	24 9 6.0	1 13 36	24 20 12	RRC	10.0-10.4 V	24057.945	0.3903174
V440 SGR	19 29 20.0	-23 57 36.0	19 31 26	-23 53 6	RRAB	9.8-11.3 B	37526.324	0.477474
V703 SCO	17 39 0.6	-32 30 0.0	17 41 17	-32 31 0	RRS	7.8- 8.5 B	37186.365	0.11521789
MT TEL	18 58 30.8	-46 43 27.3	19 1 6	-46 40 26	RRC	8.7- 9.3 V	38479.332	0.316899
TU UMA	11 27 9.7	30 20 37.7	11 29 1	30 9 2	RRAB	9.3-10.2 V	38510.756	0.557659
AI VEL	8 12 26.2	-44 25 21.8	8 13 35	-44 31 46	RRS	6.4- 7.1 V	—	0.11157396

TABLE 10.3. Intrinsic Variables, $1.0 < P < 10.0$ Days

NAME	R.A. (1950.0)	DEC.	R.A. (1985.0)	DEC.	TYPE	MAG.	EPOCH	PERIOD
FF AQL	18 56 1.2	17 17 32.4	18 57 34	17 20 24	CDEL	5.8- 6.4B	41576.428	4.470916
V1162 AQL	19 49 35.2	-11 29 46.2	19 51 31	-11 24 20	CEP	8.6- 9.3P	25803.400	5.3761
ETA AQL	19 49 55.5	0 52 33.2	19 51 42	0 57 59	CDEL	4.1- 5.4B	32926.749	7.176641
V CAR	8 27 42.5	-59 57 17.9	8 28 25	-60 4 20	CDEL	7.8- 8.8B	35612.16	6.69638
GI CAR	11 11 48.0	-57 38 18.1	11 13 20	-57 49 44	CEP	8.8- 9.2B	34521.40	4.43061
TU CAS	0 23 36.7	51 0 13.5	0 25 30	51 11 51	CW	7.4- 8.9B	36792.94	2.139292
V381 CEN	13 47 22.5	-57 19 58.4	13 49 43	-57 30 22	CW	7.9- 9.0B	34932.29	5.07878
V419 CEN	11 28 34.2	-56 37 22.1	11 30 12	-56 48 57	CEP	8.5- 9.2B	34906.43	5.50746
V553 CEN	14 43 32.2	-31 57 42.0	14 45 38	-32 6 30	CEP	8.7- 9.3P	34235.65	2.06119
V659 CEN	13 28 12.8	-61 19 29.7	13 30 32	-61 30 18	CEP	7.1- 7.4P	30049.637	5.621605
DELTA CEP	22 27 18.5	58 9 31.8	22 28 36	58 20 17	CDEL	3.9- 5.2B	42756.458	5.366270
AX CIR	14 48 29.9	-63 36 17.8	14 51 21	-63 44 55	CEP	5.6- 6.1V	38199.325	5.2734
V1334 CYG	21 17 22.4	38 1 31.8	21 18 46	38 10 25	CEP	5.8- 6.0V	41760.900	3.333020
TX DEL	20 47 42.0	3 27 54.8	20 49 27	3 35 45	CW	9.8- 9.5V	42947.033	6.165907
BETA DOR	5 33 11.3	-62 31 20.2	5 33 29	-62 29 58	CDEL	4.0- 5.1B	40905.26	9.84200
W GEM	6 32 5.6	-15 22 16.4	6 34 31	-15 20 35	CDEL	7.3- 8.5B	42755.172	7.913779
GH LUP	15 20 56.5	-52 44 38.9	15 23 31	-52 48 5	CEP	7.8- 8.2P	38202.145	9.285
V526 MON	6 59 21.1	-1 3 29.9	7 1 7	-1 6 32	CEP	9.0- 9.4P	40286.290	2.674985
AU PEG	21 21 40.4	-18 3 49.4	21 23 18	-18 12 51	CW	9.0- 9.4V	41739.439	2.40525
16 PSA	23 0 44.1	-35 12 12.7	23 2 39	-34 49 53	CEP	5.2- 5.5P	38260.250	7.975
AP PUP	7 56 1.0	-39 59 14.8	7 57 14	-40 4 56	CDEL	7.1- 7.5V	40689.21	5.084310P2
S SGE	19 53 44.9	16 30 4.0	19 55 20	16 35 40	CDEL	5.9- 7.0B	42678.783	8.382086
V636 SCO	17 19 5.4	-45 34 1.1	17 21 39	-45 36 1	CEP	7.2- 8.0B	34906.47	6.79663
ST TAU	5 42 13.3	13 33 23.8	5 44 12	13 34 15	CW	8.5- 9.6B	41761.963	4.034229
SZ TAU	4 34 20.2	18 26 35.2	4 36 22	18 30 48	CDEL	7.1- 7.7B	41659.194	3.148380
U TRA	16 2 51.2	-62 46 36.6	16 5 58	-62 52 15	CEP	7.7- 9.1B	19722.284	2.568438
ALFA UMI	1 48 48.8	89 1 43.7	2 16 16	89 11 48	CW	1.9- 2.1V.	39253.23	3.96978
AH VEL	8 10 25.6	-46 29 36.8	8 11 31	-46 35 56	CEP	5.5- 5.9V	40742.14	4.22713
BG VEL	9 6 39.1	-51 14 0.0	9 7 46	-51 22 31	CDEL	7.4- 7.9V	41053.30	6.92357
T VUL	20 49 20.8	28 3 43.7	20 50 49	28 11 37	CDEL	5.4- 6.1V	41705.121	4.435462
U VUL	19 34 26.5	20 13 12.3	19 35 58	20 17 55	CDEL	6.8- 7.5V	42526.328	7.990629

TABLE 10.4. Intrinsic Variables with $10 < P < 100$ Days

NAME	R.A. (1950.0)	DEC.	R.A. (1985.0)	DEC.	TYPE	MAG.	EPOCH	PERIOD
DS AQR	22 50 36.0	-18 51 30.0	22 52 28	-18 40 19	RV	10.3-11.6P	30972.6	78.30
V341 ARA	16 53 2.0	-63 7 54.0	16 56 18	-63 11 11	CEP	10.9-11.3V	34237.6	11.95
RU CAM	7 16 20.4	69 45 53.4	7 20 7	69 41 58	CW	10.3-10.4B	37356.9	22.055
TW CAM	4 16 39.0	57 19 18.0	4 19 32	57 24 19	RVA	10.4-11.5P	28647.	85.6
TW CAP	20 11 40.0	-13 59 30.0	20 13 37	-13 53 6	CW	10.3-12.0B	35664.8	28.5578
U CAR	10 55 45.6	-59 27 50.8	10 57 17	-59 39 5	CDEL	6.7- 8.5B	41118.2	38.7681
IW CAR	9 25 42.9	-63 24 16.0	9 26 32	-63 33 52	RVB	7.9- 9.6P	29401.	67.5
L CAR	9 43 52.4	-62 16 36.4	9 44 50	-62 26 18	CDEL	4.3- 5.5B	40736.44	35.5330
V420 CEN	11 37 24.0	-47 41 6.0	11 39 6	-47 52 44	CW	9.9-11.6B	25350.67	24.7678
RS COL	5 13 33.0	-28 48 24.0	5 14 55	-28 46 4	CEP	9.4P	27809.	14.66
X CYG	20 41 26.6	35 24 23.9	20 42 48	35 31 59	CDEL	6.6- 8.4B	25739.90	16.3866
SS GEM	6 5 33.5	22 37 32.8	6 7 48	22 37 12	RV	9.3-10.7P	34365.	89.31
ZETA GEM	7 1 8.6	20 38 43.4	7 3 13	20 35 35	CDEL	3.7- 4.2V	34416.78	10.15082
AC HER	18 28 9.0	21 49 53.1	18 29 37	21 51 21	RVA	7.4- 9.7B	35052.	75.4619
AP HER	18 48 13.0	15 52 42.0	18 49 47	15 55 18	CW	10.4-11.2V		10.408
FW LUP	15 19 7.2	-40 44 53.3	15 21 25	-40 52 23	CEP	9.2- 9.6P	38197.3	16.73
T MON	6 22 33.2	-7 6 51.8	6 24 26	-7 5 39	CDEL	5.6- 6.0V	40838.59	27.0205
U MON	7 28 24.3	-9 40 15.4	7 30 4	-9 44 41	RVB	6.1- 8.1P	30347.	92.26
Y OPH	17 49 57.8	-6 7 59.0	17 51 50	-6 8 26	CEP	7.1- 7.9B	34921.49	17.12326
EI PEG	23 19 15.0	12 19 18.0	23 21 21	12 30 48	SRA	10.0-11.5P	30961.0	61.15
RS PUP	8 11 9.0	-34 25 35.6	8 12 29	-34 31 57	CDEL	6.5- 7.6V	35734.426	41.3876
ST PUP	6 47 12.0	-37 13 6.0	6 48 24	-37 15 31	CW	9.7-11.5B	35617.45	18.8864
AR PUP	8 1 10.0	-36 27 18.0	8 2 27	-36 33 13	RVB	8.7-10.9P		75.
LS PUP	7 56 58.0	-29 16 18.0	7 58 22	-29 16 24	CEP	9.8-10.8P	38376.500	14.1464
LX PUP	8 6 6.0	-16 34 48.0	8 7 41	-16 40 56	CEP	9.5-10.0P	37314.25	13.88
R SGE	20 11 46.7	16 26.2	20 13 22	16 40 49	RVB	9.5-11.5B	23627.0	70.594
AL VIR	14 8 26.8	-13 4 32.9	14 10 20	-13 14 25	CW	9.1-10.7V	37462.5	10.3040
W VIR	13 23 26.9	-3 7 8.8	13 25 15	-3 18 3	CW	9.5-10.7V	32697.783	17.2736
V VUL	20 34 24.0	26 25 48.0	20 35 53	26 33 7	RVA	8.1- 9.3V	14871.1	75.72
SV VUL	19 49 27.8	27 19 52.8	19 50 53	27 25 17	CDEL	6.7- 7.8V	38268.9	45.035

10.4c Long-period Variables

The *Mira* type of variable has regular periods of 150 to 450 days. They differ from the *semiregular* (SR) variables primarily in their amplitude of pulsation. Miras vary by more than 2.5 magnitudes in the visual, whereas SR variables are defined to have smaller amplitudes than this. Miras are red giants, and are named after their prototype Mira Ceti, the first intrinsic variable star ever discovered. Most long-period variables show bright hydrogen emission lines and titanium oxide bands in their spectra. Table 10.5 lists the red long-period variables whose periods range from 100 to 1000 days.

Landolt has one of the longest series of *UBV* observations on long-period variables. The papers in his series are listed below.

I: 1966, *Pub. A.S.P.* **78**, 531.
II: 1967, *Pub. A.S.P.* **79**, 336.
III: 1968, *Pub. A.S.P.* **80**, 228.
IV: 1968, *Pub. A.S.P.* **80**, 450.
V: 1968, *Pub. A.S.P.* **80**, 680.
VI: 1969, *Pub. A.S.P.* **81**, 134.
VII: 1969, *Pub. A.S.P.* **81**, 381.
VIII: 1973, *Pub. A.S.P.* **85**, 625.

From these papers, it is immediately obvious that one of the problems of observing this class of variables is: they are very red, with typical $(B - V)$ indices of 1.5 to 4.0 magnitudes. Therefore, when a long period variable is near minimum at V equals 10, it may be fourteenth magnitude at B!

10.4d The Eggen Paper Series

Eggen was one of the pioneers in observing variable stars photoelectrically, and has been classifying variables systematically for many years. His series of papers should be consulted for information concerning any variable class, and contain many observations that are useful in setting up your own program. The papers are somewhat technical and may be difficult for the amateur to read, but are listed below.

I: "The Red Variables of Type N," 1972. *Ap. J.* **174**, 45.
II: "The Red Variables of S and Related Types," 1972. *Ap. J.* **177**, 489.

TABLE 10.5. Intrinsic Variables, 100 < P < 1000 Days

NAME	R.A. (1950.0)	DEC.	R.A. (1985.0)	DEC.	TYPE	MAG.	EPOCH	PERIOD
VX AND	0 17 15.0	44 26 0.0	0 19 6	44 37 39	SRA	7.8- 9.3V	25558.	369.
THET APS	14 0 23.3	-76 33 24.7	14 3 50	-76 43 29	SRB	6.4- 8.6P	28625.	119.
V AQR	20 44 17.7	2 15 11.9	20 46 3	2 22 54	SRB	7.6- 9.4V	34275.	244.
RV BOO	14 37 9.3	32 45 15.2	14 38 37	32 36 13	SRB	7.9- 9.6P	—	137.
RW BOO	14 39 6.2	31 41 6.6	14 40 35	31 38 8	SRB	8.0- 9.5P	—	209.
RV CAM	4 26 31.9	57 18 12.7	4 29 26	57 22 46	SRB	8.2- 9.0P	28861.	101.
V CVN	13 17 17.1	45 47 22.4	13 18 48	45 36 21	SRA	8.2- 8.8V	34930.	191.88
RU CEP	0 14 23.7	84 52 10.1	1 19 5	85 8 12	SRD	8.2- 9.4V	—	109.
T CET	0 19 14.5	-20 20 6.2	0 21 0	-20 8 27	SRB	6.6- 7.7P	36460.	159.
OMI CET	2 16 49.0	-3 12 13.4	2 18 35	-2 2 34	M	2.0-10.1V	38457.	331.65
W CYG	21 34 8.2	45 9 0.4	21 35 27	45 18 25	SRB	6.8- 8.9P	30684.	130.85
RS CYG	20 11 34.5	38 34 36.4	20 12 50	38 40 59	SRA	6.5- 9.3V	37930.	418.0
R DOR	4 36 10.4	-62 10 31.8	4 36 35	-62 6 21	SRB	5.9- 6.9V	—	338.
UX DRA	19 23 22.4	76 27 41.7	19 22 0	76 31 49	SRA	5.9- 6.5V	—	168.
AH DRA	16 47 24.0	57 53 59.2	16 48 57	57 50 21	SRB	8.5- 9.3P	30520.	158.
TV GEM	6 8 50.9	21 52 51.6	6 10 57	21 52 52	SRC	8.7- 9.5P	—	182.
UW HER	17 12 39.0	36 25 20.7	17 13 52	36 23 4	SRB	8.6- 9.5P	—	100.
AK HYA	8 37 35.7	-17 23 4	8 39 12	-17 14 50	SRB	7.8- 8.2P	—	112.
T IND	21 16 52.5	-45 14 3.6	21 19 10	-45 5 10	SRB	7.7- 9.4P	—	320.
T MIC	20 24 52.8	-28 25 39.2	20 27 0	-28 18 42	SRB	7.7- 9.6P	30100.	347.
RY MON	7 31 31.0	-1 28 48.0	7 6 12	-1 32 5	SRA	7.7- 9.2V	23743.	466.
X OPH	18 35 57.6	8 47 19.7	18 37 37	8 49 11	M	5.9- 9.2V	38475.	334.22
TW PEG	22 1 43.2	28 6 20.4	22 3 21	28 16 31	SR	7.0- 9.2V	38370.	956.4
L2 PUP	7 12 0.7	-44 33 26.4	7 13 4	-44 37 4	SRB	2.6- 6.0V	36946.	140.83
U PYX	8 27 49.1	-30 9 25.9	8 29 14	-30 16 29	SR	8.6- 9.4P	—	345.
BM SCO	17 37 42.8	-32 11 21.1	17 39 59	-32 12 25	SRD	6.8- 8.7P	—	850.
CE TAU	5 29 16.8	18 33 32.0	5 31 19	18 35 2	SRC	6.1- 6.5P	—	165.
RY UMA	12 18 4.0	61 35 13.3	12 19 44	61 23 34	SRA	8.4- 9.3P	34670.	311.2
VW UMA	10 55 38.0	70 15 25.5	10 58 70	70 4 10	SR	8.4- 9.1P	—	125.
SS VIR	12 22 40.0	-1 23 48.0	12 24 27	-2 51 10	M	6.0- 9.6V	38890.	354.66
SW VIR	13 11 29.7	-2 32 32.6	13 13 17	-2 43 39	SRB	8.2- 9.4P	—	150.

255

III: "Calibration of the Luminosities of Small-Amplitude Red Variables of the Old Disk Population," 1973. *Ap. J.* **180**, 857.
IV: "Very-Small-Amplitude, Very Short-Period Red Variables," 1973. *Ap. J.* **184**, 793.
V: "The Large-Amplitude Red Variables," 1975. *Ap. J.* **195**, 661.
VI: "The Long-Period Cepheids," 1977. *Ap. J. Supl. Ser.* **34**, 1.
VII: "The Medium-Amplitude Red Variables," 1977. *Ap. J.* **213**, 767.
VIII: "Ultrashort-Period Cepheids," 1979. *Ap. J. Supl. Ser.* **41**, 413.
IX: "The Very Short-Period Cepheids," *Ap. J.* In press.

10.5 ECLIPSING BINARIES

At least 50 percent of all stars are members of systems in which two or more stars orbit around their common center of mass. Many of the stars that appear single are in fact unresolved binaries. Some of these systems have orbital planes that lie nearly along our line of sight. As a result, the two stars take turns blocking each other from our sight during each orbital period. The apparent single point of light seen on earth fades and recovers as a star goes through eclipse. There are two eclipses each orbital period. The amount of light lost depends on the temperature of the two stars. The greatest light loss, called *primary eclipse,* occurs when the hotter star is blocked from view. The shallower, *secondary eclipse* occurs when the cooler star is blocked from view. These stellar systems are referred to as eclipsing binary systems. Their light curves contain valuable information about the stellar sizes, shapes, limb darkening, mass exchange, and surface spots, to name but a few items. As a result, the study of these systems forms an important part of stellar research.

It has been traditional to classify an eclipsing binary system as one of three types based on the shape of its light curve. When binaries are classified in this way, many systems of diverse physical structure and evolutionary state are grouped together. Furthermore, there are systems for which light curves defy classification in this scheme. It has become clear in recent years that this classification system is now obsolete. However, there is no new classification scheme that is universally accepted. For this and historical reasons, we describe the three traditional classification types.

Algol or β Persei is a naked-eye star that was discovered to be a variable star by ancient Arab astronomers. This was the first known variable star. It is now the prototype of the *Algol class* of eclipsing binaries, which now numbers several hundred. Typically, these systems contain an early-type (B to A) star that is brighter and more massive than its late-type (G to K) companion. The primary eclipse is deep because of the loss of light from the hot early-type star. On the other hand, the secondary eclipse, because of the light loss of the cooler star, is very shallow and difficult to detect. The orbital periods of Algol systems range from about 1 day to more than a month. The light curve in Figure 10.2a is from a typical Algol system. Table 10.6 lists a few of the brighter systems.

The light curve of Figure 10.2b is that of a typical β Lyrae type eclipsing binary. In these systems, the apparent magnitude changes con-

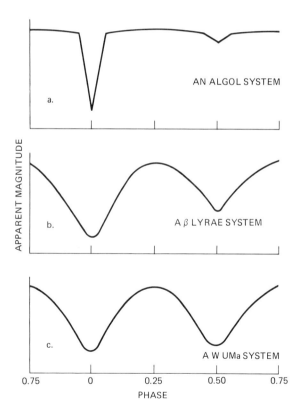

Figure 10.2. Light curves of eclipsing binaries.

TABLE 10.6. Algol-Type Binary Systems

Name	(1950.0) RA	(1950.0) Dec.	(1985.0) RA	(1985.0) Dec.	Mag.	Epoch	Period (Days)
XZ And	01^{h}53^{m}49^s	41°51′26″	01^{h}55^{m}57^s	42°01′41″	10.02–12.99 p	2441954.340	1.357293
V889 Aql	19 16 34	16 09 30	19 18 09	16 13 22	8.7–9.3 p	27210.596	11.12071
RZ Cas	02 44 23	69 25 36	02 47 33	69 34 21	6.38–7.89 p	39025.3025	1.1952499
TV Cas	00 16 36	58 51 40	00 18 30	59 03 19	7.3–8.39 p	36483.8091	1.8126130
V636 Cen	14 13 40	–49 42 48	14 15 58	–49 52 32	8.7–9.2 v	34540.340	4.28398
U Cep	00 57 46	81 36 25	01 00 56	81 47 43	6.80–9.10 V	42327.7697	2.493083
U CrB	15 16 09	31 49 42	15 17 35	31 42 04	7.04–8.35 p	16747.964	3.45220416
SW Cyg	20 05 24	46 09 21	20 06 30	46 15 27	9.24–11.83 V	38602.6009	4.573116
AI Dra	16 55 09	52 46 30	16 55 57	52 43 15	7.05–8.09 V	37544.5095	1.19881520
TW Dra	15 33 07	64 04 23	15 33 38	63 57 24	8.2–10.5 p	38539.4457	2.8068352
S Equ	20 54 44	04 53 10	20 56 29	05 01 16	8.0–10.08 V	37968.345	3.436072
CD Eri	03 45 21	–08 46 04	03 47 03	–08 39 37	9.5–10.2 p	29910.567	2.876766
TT Hya	11 10 46	–26 11 36	11 12 29	–26 23 02	7.5–9.5 p	24615.388	6.9534124
AU Mon	06 52 22	–01 18 42	06 54 09	–01 21 24	8.2–9.5 v	32888.554	11.11306
β Per (Algol)	03 04 55	40 45 50	03 07 12	40 53 53	2.12–3.40 V	40953.4657	2.8673075
IZ Per	01 28 56	53 45 42	01 31 08	53 56 30	7.8–9.0 p	25571.360	3.687661
RW Per	04 16 48	42 11 41	04 19 14	42 16 43	9.7–11.45 V	39063.684	13.19891
Y Psc	23 31 53	07 38 52	23 33 40	07 50 28	9.0–12.0 v	41225.473	3.765859
BH Pup	08 06 34	–41 52 55	08 07 46	–41 59 05	8.4–9.1 p	26100.891	1.915854
U Sge	19 16 37	19 31 06	19 18 09	19 34 58	6.58–9.18 V	40774.4638	3.3806260
λ Tau	03 57 54	12 21 04	03 59 50	12 26 58	3.3–3.80 p	35089.204	3.952955
RW Tau	04 00 49	27 59 24	04 02 58	28 05 10	8.02–11.59 V	40160.3771	2.7688425
X Tri	01 57 43	27 38 48	01 59 43	27 48 57	8.9–11.5 p	40299.296	0.9715330
AC UMa	08 51 36	65 09 45	08 54 37	65 01 44	9.2–10.2 p	42521.58	6.85493
TX UMa	10 42 24	45 49 47	10 44 27	45 38 44	7.06–8.76 V	39193.310	3.063243

tinuously even outside of eclipse because of the distorted shapes of the stars. The stars in these systems are nearly touching, and their mutual gravitational attraction has distorted them into elongated, egg-like shapes whose long dimensions always point toward each other. As the stars move in their orbits, their projected surface area as seen by the observer varies. The light curve then peaks when both stars are seen "broadside" and then fade as they turn. These systems usually contain early-type stars and often show complications in their spectra indicating gas streams and shells. Their orbital periods usually exceed 1 day. Table 10.7 is a list of some bright β Lyrae-type systems.

The curve in Figure 10.2c belongs to a typical member of the *W Ursae Majoris (W UMa) class.* These systems too are highly distorted by gravitational fields. The two component stars have nearly identical surface temperatures, which results in near equality in the depths of the two eclipses. The similarity in temperature is believed to be the result of a common atmosphere that surrounds both stars. These stars really are touching! Unlike the β Lyrae systems, these stars belong to spectral classes F, G, and K. The orbital periods are less than 1 day. Table 10.8 is a list of some brighter W UMa-type systems.

There is a newly recognized type of eclipsing binary, *RS Canun Venaticorum (RS CVn) systems*, which shows one of the shortcomings of the old classification system. In the past, members of this group had been classified with the Algols, yet they clearly contain stars that are very different from those found in Algol systems. The hotter star is type F or G and the cooler is late G to early K. They have orbital periods of a few days. The curious feature of their light curve is a broad depression of about $0^{m}.1$ which slowly migrates in phase. Several of these systems are now known to flare in the radio and x-ray regions of the spectrum. The depression in the light curve is believed to be the result of large starspots that migrate slowly in longitude on the surface of the cooler star. These spots are like sunspots, only they cover a much higher percentage (about 10 percent) of the stellar surface. The x-ray and radio emissions are believed to be related to strong coronal activity around the spotted star. A more complete review can be found in an article by Zeilik et al.[44] Table 10.9 contains a partial list of some RS CVn-type systems. Anyone seriously considering observing these systems would do well to contact Douglas Hall,[45] who is coordinating the photometric work of amateur and professional observers. When monitoring these systems, it is important to follow the light curve distortions and to cor-

TABLE 10.7. β Lyrae-Type Binary Systems

Name	RA (1950.0)	Dec. (1950.0)	RA (1985.0)	Dec. (1985.0)	Mag.	Epoch	Period (Days)
AN And	23ʰ16ᵐ01ˢ	41°29′59″	23ʰ17ᵐ41ˢ	41°41′28″	6.0–6.16 p	2421060.326	3.219565
σ Aql	19 36 43	05 16 58	19 38 27	05 21 48	5.18–5.36 B	22486.797	1.95026
LR Ara	16 49 07	−61 30 08	16 52 17	−61 33 37	10.0–10.6 p	28004.430	1.519304
TT Aur	05 06 15	39 31 24	05 08 40	39 34 03	8.3–9.2 p	21242.2564	1.33273365
SZ Cam	04 03 24	62 11 59	04 06 29	62 17 37	7.0–7.29 B	27533.5191	2.6984378
CX CMa	07 19 57	−25 46 44	07 21 23	−25 50 46	9.9–10.6 p	28095.601	0.954608
X Car	08 30 11	−59 03 27	08 30 57	−59 10 35	8.0–8.7 p	15021.114	1.0826310
AO Cas	00 15 04	51 09 22	00 16 56	51 21 02	5.96–6.11 B	24002.579	3.523487
LW Cen	11 35 12	−63 04 18	11 36 50	−63 15 56	9.3–9.6 p	24824.462	1.0025674
LZ Cen	11 48 05	−60 30 59	11 49 49	−60 42 40	8.0–8.6 p	26096.384	2.757717
AH Cep	22 46 04	64 47 49	22 47 20	64 58 55	6.9–7.12 p	34989.3702	1.7747274
V366 Cyg	20 43 06	53 55 10	20 44 05	54 02 49	10.0–10.46 p	34489.593	1.0960183
V548 Cyg	19 55 47	54 39 51	19 56 37	54 45 32	8.9–9.72 p	38972.1706	1.805244
V836 Cyg	21 19 21	35 31 25	21 20 47	35 40 22	8.59–9.30 B	26547.5224	0.65341090
u Her	17 15 29	33 09 13	17 16 47	33 06 59	4.6–5.28 p	27640.654	2.0510272
β Lyr	18 48 14	33 18 13	18 49 32	33 20 41	3.34–4.34 V	36379.532	12.93016
η Ori A	05 21 58	−02 26 27	05 23 44	−02 24 34	3.14–3.35 B	33420.215	7.98926
VV Ori	05 30 59	−01 11 24	05 32 46	−01 09 58	5.14–5.51 p	40545.899	1.48537769
AU Pup	08 15 56	−41 33 05	08 17 09	−41 39 39	8.50–9.40 V	39237.985	1.126411
V Pup	07 56 48	−49 06 30	07 57 48	−49 12 14	4.74–5.25 p	28648.3048	1.4544867
V525 Sgr	19 04 02	−30 14 24	19 06 16	−30 11 07	7.9–8.8 p	29662.4593	0.70512200
V453 Sco	17 53 00	−32 28 08	17 55 17	−32 28 26	6.36–6.73 V	41762.58	12.0061
V499 Sco	17 25 45	−32 57 54	17 28 03	−32 59 35	8.8–9.36 p	28340.405	2.3332977
AC Vel	10 44 18	−56 33 59	10 45 43	−56 45 03	8.5–9.0 p	23936.285	4.5622426
AY Vel	08 18 37	−43 43 27	08 19 48	−43 50 07	9.1–9.8 p	26308.903	1.617653

TABLE 10.8. W UMa- Type Binary System

Name	RA (1950.0)	Dec. (1950.0)	RA (1985.0)	Dec. (1985.0)	Mag.	Epoch	Period (Days)
AB And	$23^h09^m08^s$	$36°37'22''$	$23^h10^m48^s$	$36°48'47''$	10.4–11.27 p	2440128.7945	0.33189305
BX And	02 05 58	40 33 30	02 08 07	40 43 26	8.9–9.57 v	43809.8873	0.61011508
S Ant	09 30 07	–28 24 25	09 31 39	–28 33 43	6.7–7.22 B	35139.929	0.648345
OO Aql	19 45 47	09 11 03	19 47 28	09 16 18	9.2–10.0 v	40522.294	0.5067887
i Boo	15 02 08	47 50 51	15 03 19	47 42 41	6.5–7.10 v	39370.4222	0.2678160
XY Boo	13 46 48	20 26 02	13 48 28	20 15 37	10.0–10.36 p	39953.9621	0.37054663
TX Cnc	08 37 11	19 10 37	08 39 11	19 03 10	10.45–10.78 p	38011.3909	0.38288153?
RR Cen	14 13 25	–57 37 19	14 15 54	–57 47 03	7.46–8.1 B	29036.0321	0.60569121
VW Cep	20 38 03	75 24 57	20 37 31	75 32 23	7.8–8.21 p	41880.8027	0.2783161
RZ Com	12 32 35	23 36 51	12 34 20	23 25 17	10.96–11.66 B	34837.4198	0.33850604
ε CrA	18 55 21	–37 10 25	18 57 43	–37 07 34	4.96–5.22 p	39707.6619	0.5914264
YY Eri	04 09 46	–10 35 41	04 11 26	–10 30 19	8.8–9.50 B	33617.5198	0.321496212
AK Her	17 11 43	16 24 32	17 13 17	16 22 08	8.83–9.32 B	38176.5092	0.42152368
SW Lac	22 51 22	37 40 20	22 52 59	37 51 31	10.2–11.23 p	43459.7476	0.3207216
AM Leo	11 59 35	10 09 59	11 01 25	09 58 42	8.2–8.65 v	39936.8337	0.36579720
UZ Leo	10 37 53	13 49 42	10 39 45	13 38 44	9.58–10.15 V	40673.6666	0.6180429
XY Leo	09 58 55	17 39 07	10 00 50	17 29 00	10.43–10.93 B	41005.5351	0.28411
V502 Oph	16 38 48	00 36 07	16 40 35	00 32 06	8.34–8.84 V	41174.2288	0.45339345
V566 Oph	17 54 26	04 59 29	17 56 09	04 59 15	7.60–8.09 p	41835.8618	0.40964399
V839 Oph	18 06 59	09 08 26	18 08 39	09 08 50	9.4–9.99 p	36361.7317	0.4089946
U Peg	23 55 25	15 40 30	23 57 12	15 52 11	9.23–9.80 V	36511.6688	0.3747819
AW UMa	11 27 26	30 14 35	11 29 17	30 03 00	6.84–7.10 V	38044.7815	0.4387318
W UMa	09 40 16	56 01 52	09 42 43	56 01 15	7.9–8.63 V	41004.3977	0.33363696
AG Vir	11 58 29	13 17 12	12 00 17	13 05 31	8.4–8.98 V	39946.7472	0.64264787
AH Vir	12 11 48	12 06 00	12 13 35	11 54 20	9.6–10.31 p	35245.6522	0.40752189

TABLE 10.9. Eclipsing Binaries of the RS CVn Type

Name	RA (1950.0)	Dec. (1950.0)	RA (1985.0)	Dec. (1985.0)	Mag.	Epoch	Period (Days)
CQ Aur	06h00m39s	31°19′52″	06h02m55s	31°19′37″	9.6–10.6 p	2429558.78	10.62148
SS Boo	15 11 39	38 45 15	15 12 59	38 37 26	10.2–11.2 p	20707.375	7.606215
SS Cam	07 10 19	73 25 16	07 14 36	73 21 38	10.1–10.7 v	35223.28	4.8241
RU Cnc	08 34 34	23 44 15	08 36 38	23 36 55	9.9–11.5 v	22650.720	10.172988
RS CVn	13 08 18	36 12 00	13 09 55	36 00 50	8.4–9.92 p	38889.3300	4.797855
AD Cap	21 37 03	−16 14 00	21 38 58	−16 04 29	9.3–9.9 p	30603.55	6.11826
UX Com	12 59 07	28 53 50	13 00 48	28 42 32	10.91–11.8 B	25798.370	3.642386
RT CrB	15 35 59	29 39 01	15 37 25	29 32 10	10.2–10.7 v	28273.280	5.11712
WW Dra	16 38 21	60 47 45	16 38 50	60 43 41	8.29–9.49 V	28020.3693	4.629583
Z Her	17 55 52	15 08 31	17 57 27	15 08 21	7.3–8.1 p	13086.348	3.9928012
AW Her	18 23 27	18 15 50	18 24 59	18 17 04	9.5–10.9 v	27717.2151	8.80086
MM Her	17 56 32	22 08 58	17 58 01	22 08 50	9.8–10.8 p	31302.451	7.96037
PW Her	18 08 35	33 22 34	18 09 52	33 23 02	10.7–11.8 p	28248.564	2.8810016
GK Hya	08 28 13	02 26 55	08 30 02	02 19 50	9.1–9.8 p	26411.460	3.587035
RT Lac	21 59 29	43 38 55	22 00 54	43 49 03	8.84–9.89 V	39073.8020	5.074012
AR Lac	22 06 39	45 29 46	22 08 04	45 40 04	6.11–6.77 V	39376.4955	1.9831987
RV Lib	14 33 02	−17 49 05	14 34 59	−17 58 14	9.8–10.4 p	30887.236	10.722164
VV Mon	07 00 51	−05 39 42	07 02 34	−05 42 49	9.6–10.4 v	26037.529	6.05079
LX Per	03 09 52	47 55 05	03 12 18	48 02 56	8.4–9.4 p	27033.120	8.038044
SZ Psc	23 10 50	02 24 06	23 12 37	02 35 32	8.02–8.69 B	36114.565	3.96637
TY Pyx	8 57 36	−27 37 14	08 59 06	−27 45 26	6.87–7.47 V	27154.325	1.599292
RW UMa	11 38 05	52 16 29	11 39 58	52 04 50	10.3–11.9 p	33006.308	7.328251

relate any optical changes that occur simultaneously with radio and x-ray outbursts.

Finally, there is a group of binary systems that display dramatically the results of stellar evolution. The *cataclysmic systems* contain a faint red dwarf star and a white dwarf companion. The red dwarf star is slowly expanding, as stars are known to do when their evolution carries them from the main sequence to the red giant stage. In this case, however, as the star expands, its atmosphere is drawn by the gravitational attraction of the white dwarf. This transferred matter forms a disk around the white dwarf as it spirals into the star. This process of mass transfer occurs in many binary systems, including Algols, but nowhere are the effects as visibly dramatic as in cataclysmic systems. For these systems, the mass transfer is so rapid that a thick, extensive accretion disk forms. The continuous emission from the disk becomes so bright that it can outshine the stars. Matter streams in, striking the disk and producing a flickering hot spot. High-speed photometry (with time resolution of a fraction of a second) shows this flickering clearly. It disappears when the spot is eclipsed. There are several subclasses within the cataclysmic group, one of which is *novae*. It is believed that the explosive events in these systems are a direct result of the mass accretion process. A very interesting review of cataclysmic systems has been given by Trimble.[46] A review of high-speed photometry of these systems has been published by Warner and Nather.[12] Unfortunately, nearly all these systems are faint and beyond the grasp of small telescopes. For this reason, we have not included a list of these objects. However, one can be found in a well-written review article by Robinson.[47]

In general, a photometric study of an eclipsing system can have one or both of the following goals. The first is a set of observations to determine the time of mideclipse, which is often called the *time of minimum light*. This is done in order to determine the orbital period. In many eclipsing systems, the orbital periods are not constant. Mass transfer or mass ejected from the system may cause a slight shift in the position of the center of mass of the system. As a result, there is a small change in the orbital period. With observations of the time of minimum light, we can check to see if the eclipses are occurring "on time." Eclipses that occur earlier or later than predicted indicate a period change. Variations in the orbital period give us indirect information about the mass transfer or mass loss processes. A second reason for observing the times of minimum light is so that spectroscopists can determine accurately the orbital phases at which their spectrograms are taken. This is essential for the proper interpretation of the spectroscopic data.

The second observational goal mentioned above is the observation of the entire light curve. For short-period systems like the W UMa binaries, it is possible to observe a sizable portion of the light curve in one night. The data from several nights, covering portions of different orbital periods, can be combined into a composite light curve similar to Figure 1.1. Then you can calculate the orbital phase of each observation. This is explained below.

The analysis of a light curve can yield many physical parameters of the stars. However, analysis is a rather advanced topic and the reader is referred to Irwin,[48] Binnendijk,[49] and Tsesevich[50] for an introduction. A review of the most modern synthetic light curve analysis can be found in Binnendijk.[51] By no means should an observer be discouraged from obtaining a complete light curve even if the analysis appears too difficult. There are many professional astronomers who specialize in these studies and would welcome the data. Many binary systems show distorted light curves whose shapes can change because of circumstellar matter or starspots. Such systems should have their entire light curves monitored frequently. A comparison of light curves over many years can often provide clues as to the location and nature of the matter responsible for the distortions. See Bookmyer and Kaitchuck,[52] for example.

The determination of the time of minimum light is a good program for the newcomer to photometry. This involves simple differential photometry without the need to transform to the standard system. It provides valuable practice in observing techniques and data reduction. In addition, it provides valuable information. After the binary has been chosen and a comparison star has been found, the next step is to calculate the predicted time of minimum light. Tables 10.2 through 10.5 contain two numbers for each binary labeled *epoch* and *period*. These numbers are referred to as the *light elements*. They allow for calculation of binary orbital phase for any given date and time. The epoch is a heliocentric Julian date of a primary eclipse. The second number is the orbital period in days. You may be surprised by the accuracy to which the orbital periods are quoted. Although the time of minimum light can be determined only to an accuracy of a minute or so, we can measure the orbital period to an accuracy of a few seconds or even less. This is because for short-period systems a small error in the orbital period can accumulate in a few years to a large discrepancy between the predicted and observed times of eclipse.

Orbital phase is defined to be zero at mideclipse and 0.50 at one-half an orbit later, etc., as seen in Figures 1.1 and 10.2. The first step in predicting the time of minimum light is to compute the heliocentric Julian date (HJD) at 0 hour UT, as explained in Sections 5.3d and 5.3e for the day and time in question. We then compute a quantity called H, by

$$H = \text{fractional part of} \left(\frac{\text{HJD} - \text{epoch}}{\text{period}} \right)$$

if HJD is greater than the epoch, or

$$H = 1 - \text{fractional part of} \left(\frac{\text{HJD} - \text{epoch}}{\text{period}} \right)$$

if HJD is less than the epoch.

The heliocentric phase is then given by

$$\text{Phase (Hel.)} = H + \frac{\text{UT(in hours)}/24 \text{ hours}}{\text{period}}.$$

To predict the time of minimum light, we need only to find a day and time for which phase 0.0 or equivalently 1.0, occurs after dark and with our star above the horizon. Because the orbital periods change, the light elements in Tables 10.2 through 10.5 give only an approximate time of minimum light. This is especially true for light elements with small epoch values where the error could be as much as an hour. You should begin observing early to avoid the possibility of missing part of the eclipse. The most recent references to published light elements can be obtained from the astronomy department at the University of Florida.[53] They maintain a file of index cards containing literature references to research papers, times of minimum light, and the most recent light elements for hundreds of eclipsing-binary systems. A xerox of these cards can be obtained by writing to them and specifying which binary systems you are studying.

The actual observing procedure to follow is outlined in Section 9.3c and the data reduction is explained in Section 4.8. For short-period systems, you can observe most or all of the eclipse in a few hours. For long-period systems, the eclipse can last days and you have to spread your observing over several nights. It is important to observe as much of the descending and ascending branches of the curve as possible. This improves the accuracy of your determination of the time of minimum light greatly. Because of observational error, the time of minimum light does not simply refer to the time of your lowest measurement. There are several methods for determining the time of minimum light from the observations.

The *bisection of chords technique* can be applied if the eclipse curve is symmetric about mideclipse. First, a plot is made of Δm versus HJD. A smooth freehand curve is drawn through the data points and several horizontal lines are drawn connecting points of equal brightness on the ascending and descending sides of the curve. The midpoint of each line is measured and the average is taken to give an estimate of the time of mideclipse. The standard deviation from the mean gives an error estimate.

The *tracing-paper method* is still simpler. On the data plot, a vertical line is drawn to indicate your best estimate of the time of minimum light. The data points, the vertical line, and the horizontal axis are then transferred to a piece of tracing paper. The tracing paper is then turned over, so the time axis is reversed, and laid back on the original plot. The tracing paper is then moved horizontally so that data points and their tracing-paper counterparts give the best fit with a minimum of scatter. In general, the vertical lines on the plot and tracing paper do not align. The time of minimum light is midway between these two lines and can be read directly from the horizontal axis. Shifting the tracing paper to the right and left until the fit obviously worsens gives an error estimate.

A more analytical technique has been devised by Hertzsprung.[54] Because this reference is difficult to find in many libraries, we illustrate it in detail. This method is more time-consuming but has the advantage of eliminating most individual bias. It can also be implemented easily on a computer or programmable calculator. A plot of the observations, Δm versus HJD, is made on a large sheet of graph paper, say 50 $\times$ 100 centimeters. The plotting scales should be chosen so that the eclipse curve has a slope of about 45°. Each data point is then connected by straight line segments. Figure 10.3 shows this plot (greatly reduced in size) for an eclipse of V566 Oph observed by Bookmyer.[55] First, esti-

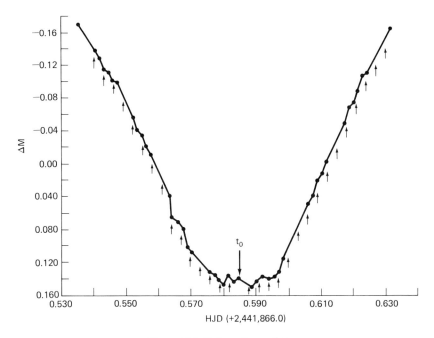

Figure 10.3. Eclipse observations.

mate the time of minimum light, called t_0, by simply looking at the plot. In this case, t_0 was estimated to be 0.5850, the Julian decimal. Place an arrow on the plot at 10 or more time intervals Δt to either side of t_0. In this case, Δt was chosen as 0.003 day. Read Δm from the plot at each arrow, using the straight line segments. These values are recorded in pairs that correspond to the same time interval on either side of t_0. The absolute value of the differences between the numbers in each pair is then found. The middle three columns of Table 10.10 shows these measurements and their differences. Ideally, if we had guessed t_0 exactly right, and the light was symmetric with no observational error, the column of differences would only contain zeros.

This process is then repeated, taking the first arrow on the right to be a new estimate of t_0 (the rightmost three columns of Table 10.10). This really does not involve much work; it requires moving some numbers in the table to different rows. The process is repeated a third time using the first arrow on the left as the new t_0 (the leftmost three columns in Table 10.10). We now let Y^-, Y^0, and Y^+ be the sum of the squares

TABLE 10.10. Time of Minimum Light Data

| $t_0 - \Delta t = 0.5820$ | | | $t_0 = 0.5850$ | | | $t_0 + \Delta t = 0.5880$ | | |
| Δm | | | Δm | | | Δm | | |
Descending	Ascending	|Diff.|	Descending	Ascending	|Diff.|	Descending	Ascending	|Diff.|
−0.141	−0.108	0.033	−0.117	−0.130	0.013	−0.101	−0.152	0.051
−0.117	−0.083	0.034	−0.101	−0.108	0.007	−0.083	−0.130	0.047
−0.101	−0.057	0.044	−0.083	−0.083	0.000	−0.058	−0.108	0.050
−0.083	−0.028	0.055	−0.053	−0.057	0.001	−0.033	−0.083	0.050
−0.058	−0.004	0.054	−0.033	−0.028	0.005	−0.007	−0.057	0.050
−0.033	+0.022	0.055	−0.007	−0.004	0.003	+0.018	−0.028	0.046
−0.007	+0.053	0.060	+0.018	+0.022	0.004	+0.048	−0.004	0.052
+0.018	+0.078	0.060	+0.048	+0.053	0.005	+0.076	+0.022	0.054
+0.048	+0.104	0.056	+0.076	+0.078	0.002	+0.107	+0.053	0.054
+0.076	+0.132	0.056	+0.107	+0.104	0.003	+0.119	+0.078	0.041
+0.107	+0.139	0.032	+0.119	+0.132	0.013	+0.132	+0.104	0.028
+0.119	+0.140	0.021	+0.132	+0.139	0.007	+0.142	+0.132	0.010
+0.132	+0.148	0.016	+0.142	+0.140	0.002	+0.137	+0.139	0.002
+0.142	+0.141	0.001	+0.137	+0.148	0.011	+0.141	+0.140	0.001
	$Y^- = \Sigma\|\text{Diff.}\|^2$			$Y^0 = \Sigma\|\text{Diff.}\|^2$			$Y^+ = \Sigma\|\text{Diff.}\|^2$	
	= 0.02834			= 0.00065			= 0.02553	

of the values in columns 3, 6, and 9, respectively. The time of minimum light, t_{min}, is then computed by

$$t_{min} = t_0 + \tfrac{1}{2}\left(\frac{Y^- - Y^+}{Y^- - 2Y^0 + Y^+}\right)\Delta t.$$

In this particular example,

$$t_{min} = 0.5850 + \tfrac{1}{2}\left(\frac{0.02834 - 0.02553}{0.02834 - 2(0.000650) + 0.02553}\right)0.003$$
$$= 0.5858$$

For even higher accuracy, the entire procedure can be iterated using the above value as a new estimate of t_0. However, this is probably not justified unless the observational data is of very high quality. An estimate of the error in the time of minimum light can be obtained by changing Δt and recomputing t_{min}.

When reporting a time of minimum light, it is customary to also include a second number called the $(O - C)$ value. This is the difference in decimal day between the *observed* and *computed* time of minimum light. Because this computed time of minimum light depends on the set of light elements used, they should also be reported. The computed time of minimum light can be calculated by

$$T_{min} = T_0 + P \cdot E,$$

where T_0 and P are the initial epoch and period, respectively, and E is an integer. With a little trial and error, a value of E can be found, which gives a value of T_{min} as close as possible to the observed time of minimum light. In our example above, the observed time of minimum light is calculated from a set of light elements published by Bookmyer[56] many years before, namely

$$T_{min} = 2{,}436{,}744.4200 + 0.40964091 E.$$

If E is chosen as 12,504, then the calculated time of minimum light is 2,441,866.5699. The *observed* time of minimum, t_{min}, minus the *computed* time of minimum light, T_{min}, is

$$(O - C) = 0.5858 - 0.5699$$
$$= 0.0159 \text{ day}$$

or about 23 minutes. This large difference is certainly easy to measure.

When $(O - C)$ values over intervals of a few years are collected and plotted versus Julian date, the story of period variations is told. For instance, if the $(O - C)$ values scatter about zero, resulting in a horizontal line on the plot, the period is constant. If the $(O - C)$ points trace out a curved path on the plot, this indicates that the period is continuously changing. The $(O - C)$ plot for V566 Oph shows a long interval of a constant period and then abruptly at about Julian date 2,440,000, or the year 1968, the $(O - C)$ values begin to rise following a straight line.[57] Such a straight, sloping line on an $(O - C)$ plot indicates that a single abrupt period change occurred. From this date (2,440,000) until the time of minimum light measured in our example above, the binary system completed a number of orbits given by

$$\frac{2{,}441{,}866.6 - 2{,}440{,}000}{\text{period}}$$

or

$$\frac{1866.6}{0.409644} = 4556.$$

If it took 4556 orbits to accumulate an $(O - C)$ value as large as 23 minutes, then the change in orbital period must have been

$$\frac{23 \text{ minutes}}{4556} = 0.005 \text{ minute}$$

or 0.3 second! This example demonstrates that times of minimum light determined to an accuracy of only a few minutes can often detect period changes of just a few tenths of a second.

10.6 SOLAR SYSTEM OBJECTS

In the past, less photoelectric photometry has been done on solar system objects than on stellar or galactic objects. There are many reasons for this, among them being, that there are fewer objects to study and an incorrect impression that little new information could be learned. There are three areas that amateurs can contribute most readily: comets, asteroids, and satellites.

Very little comet photometry has been performed. Most astronomers concentrate on spectra when comets appear, trying to decipher the composition of comets. However, recent research at GSFC by Niedner[58] has shown that magnetic sector boundaries emanating from the sun cause tail disconnections and brightness flareups. This latter phenomenon may be related to the brightness variations shown by periodic comets such as Schwassmann-Wachmann and therefore observing these comets photoelectrically may help in our understanding of the interplanetary magnetic field. Observing comets as they approach from beyond Jupiter can provide a probe at great distances from the sun. One problem with comets is that most of the light from the coma/nucleus region is from emission lines, making interpretation of wide-band colors as used in the *UBV* system difficult. We have the following suggestions for comet observations:

1. Use *UBV* photometry on new comets as soon as they are bright enough to be detected by your telescope. Be careful in measuring

a comet with a tail and coma, as your diaphragm will not contain the entire comet. Center on the nucleus and keep accurate records of what diaphragms you use. Sky determinations are also difficult because of contamination by the comet.

2. On large, bright comets, you can obtain intensity *isophotes* (contours of equal intensity) by centering on the comet nucleus and measuring its brightness in diaphragms from as small to as large as possible.
3. On bright comets, use narrow-band or interference filters to isolate particular emission features. Borra and Wehlau[59] used 75 Å wide bandpasses centered at 3878 Å (CN bands), 4870 Å (reflected sunlight continuum), and 5117 Å (Swan C_2 bands) for their photographic photometry.

In 1986, be sure to observe Halley's comet! Positions of comets are given monthly in *Sky and Telescope.*

There are several hundred asteroids brighter than eleventh magnitude. All are within range of amateur telescopes but only a few have been observed photoelectrically. Not only can photoelectric measurements provide information about their albedo and size, but most asteroids are nonspherical and show brightness variations during their rotations. Obtaining light curves can give clues as to their shape and the orientation of their spin axis. A study of 43 Ariadne by Burchi and Milano[60] gives examples of parameters obtainable from asteroid photometry. The current hypothesis that many asteroids have satellites also means that eclipsing asteroids should be visible, but have not yet been detected. Because asteroids are constantly moving with respect to stars, you need to obtain ephemerides from the Almanac[24] or another similar source. Comparison stars are difficult to obtain because of the asteroids' movement. Stars from the *UBV* catalog[61] should probably be used.

Planetary satellites are very difficult to measure. They are always much fainter than their parent body and lie very close in angular distance. You must be very careful to avoid contamination from the brighter planet. Eclipses behind the shadow of the planet can yield three main pieces of information: the diameter of the satellite, its orbit, and the structure of the planetary upper atmosphere. The Galilean satellites also have intervals of mutual eclipses twice per Jovian year. These eclipses of one satellite by another can give better information about their diameter and orbits, and can occur at a greater angular separation from the planet. In all eclipse observations, record the eclipse time to the nearest tenth of a second, preferably from inspection of the strip chart or microcomputer time series. Williamon[62] gives examples of the photometric appearance of this eclipse.

Observations of the planets have been carried out at major observatories. Uranus and Neptune have been observed for many years at Lowell Observatory to check the constancy of their solar luminosity. These two planets are faint enough that nearby comparison stars can be found. The rotation period of Pluto was only recently obtained photoelectrically[63] because of the low amplitude of its light variations, about 0.2 magnitude in V. The variation may be partially a result of the newly discovered satellite. You too can participate in such determinations if you have access to a meter-sized telescope.

10.7 EXTRAGALACTIC PHOTOMETRY

Extragalactic photometry is a very active area of research for many professional astronomers. This type of photometry can take many different forms. For example, drift scan photometry is a way of obtaining brightness and color profiles by drifting the photometer aperture across the face of a galaxy. Such data yield information about the distribution of dust and stellar populations within the galaxy. Aperture photometry is another technique that yields much the same information. Measurements of a galaxy are made through photometer diaphragms of various sizes centered on the galactic nucleus. Quasars, BL Lac objects, compact galaxies, and many radio galaxies appear starlike in a telescope. Photometry of these objects is done much as it is for ordinary stars.

The problem with all these projects is that we are dealing with very distant objects that appear very faint. Because much of this book has been aimed at small telescope users, it would be out of place to discuss in detail projects that exceed the capabilities of their equipment. We instead mention one particular project which is within the reach of a 30- to 40-centimeter (12- to 16-inch) telescope and is of great scientific value. Many quasars, BL Lac objects, and radio galaxies are variable in the optical region of the spectrum. For many such objects, there is a serious shortage of observations over the long term, necessary for an understanding of the nature of these variations. The reason for the shortage of data is simply that these objects greatly outnumber the number of astronomers working in this field.

There are very few quasars brighter than thirteenth visual magnitude. The quasar 3C273 is about visual magnitude 12.5 and its variations have been well documented. The AAVSO[3] can supply a finder chart for this object at a modest cost. There are several radio galaxies and BL Lac objects brighter than thirteenth visual magnitude which could be monitored profitably. Burbidge and Crowne[64] have compiled an optical catalog of radio galaxies. A similar catalog for quasistellar objects has been published by Hewitt and Burbidge.[65] Both references represent a good starting place for finding potential objects to study and for further references to published papers and finding charts. It should be emphasized that these objects are faint and are difficult to measure accurately with a small telescope. However, even observations with an error of 0.1 magnitude are valuable, particularly if they are made on a routine basis.

10.8 PUBLICATION OF DATA

The acquisition of photoelectric data is enjoyable in itself. As you gain experience and observational data, you may reach the conclusion that you would like to publish your data. Publication serves three purposes: it provides a means of rapid dissemination of important data or results, it gives permanence to results so that they may be used decades in the future, and it gets your name in print.

There are several methods to publish your data. These are listed below.

1. The American Association of Variable Star Observers (AAVSO) publishes newsletters and its own journal. They are the central depository for all visual observations and also most amateur photoelectric data. Check with them at their headquarters.[3] They are often consulted by professional observatories for information on long-period variables.
2. The International Astronomical Union (IAU) has a depository of photoelectric data.[66] This is a good alternative if you do not wish to publish your results in meeting abstracts or a journal. It is used heavily for extensive sets of data on a particular star such as RW Tau or SS 433.
3. Societies such as the American Astronomical Society, the Astronomical Society of the Pacific, and the International Astronomical

Union sponsor meetings and publish the proceedings. Generally, only the abstract of your paper or talk is published, but interested parties then know who to contact for further details.

4. Among the small journals are the *Information Bulletin of Variable Stars,* the *Journal of the AAVSO,* the *Journal of the Royal Astronomical Society of Canada,* and the *Monthly Notices of the Astronomical Society of South Africa.* In many cases, these journals do not have page charges and may not have your paper refereed formally.

5. Major astronomical journals include the *Publications of the Astronomical Society of the Pacific, Astronomy and Astrophysics,* the *Astronomical Journal,* the *Astrophysical Journal,* and the *Monthly Notices of the Royal Astronomical Society.* These journals send submitted papers to independent referees who assess the quality of the paper and who can ask for changes in addition to recommending acceptance or rejection. In addition, most major journals have page charges because of the technical nature of the papers, the cost of typesetting, and the large volume of papers. The cost may range from around $40 to $100 per page or more.

The large number of astronomers and the use of modern techniques yield more data than in the past, and the emphasis on professional publication forces more papers to be written, contributing to the necessity of these page charges. At the same time, avenues have been opened to amateur observers, as few professionals can devote time to long-term projects or projects that contain classical astronomy that do not yield immediate results.

When you decide that you would like to publish results, we suggest that you find a professional astronomer with whom to collaborate. The astronomer probably knows more about photometry theory and publishing procedures than you do, and in addition may be able to give you guidance and access to a library or computer center. Other reasons for collaboration are that the astronomer may have an ongoing project to which you can contribute, the astronomer may be a good reference if you are contemplating a career in astronomy, and the astronomer's institution may be able to pay page charges on a joint publication. Check with local universities or read journals to find someone who appears to be interested in the same field that you are.

There is not enough room in this text to give complete details of the techniques of technical writing. The IAU Style Book[67] and the AIP Style Manual[68] give the grammatical aspects for astronomical papers. There are several texts on scientific writing, and each journal mentions the format that they wish submitted papers to obey. The best method is to read the journals and see how others write and learn by example.

REFERENCES

1. Landis, H. J. 1977. *J. AAVSO* **6**, 4.
2. Stanton, R. H. 1978. *J. AAVSO* **7**, 14.
3. American Association of Variable Star Observers, 187 Concord Avenue, Cambridge, MA 02138.
4. Argue, A. N., Bok, B. J., and Miller, P. W. 1973. *A Catalog of Photometric Sequences.* Tucson: Univ. of Arizona Press.
5. Hertzsprung, E. 1924, *Bul. Ast. Inst. Neth.* **2**, 87.
6. Luyton, W. J. 1949. *Ap. J.* **109**, 532.
7. Gordon, K., and Kron, G. 1949. *Pub. A. S. P.* **61**, 210.
8. Moffett, T. J., 1974. *Sky and Tel.* **48**, 94.
9. Lovell, Sir Bernard 1971. *Quart. J.R.A.S.* **12**, 98.
10. Gershberg, R. E. 1971. *Flares and Red Dwarf Stars.* D. J. Mullan, trans. Armagh, Ireland: Armagh Observatory.
11. Gurzadyan, G. A. 1980. *Flare Stars.* Elmsford, NY: Pergamon.
12. Warner, B., and Nather, R. E. 1972. *Sky and Tel.* **43**, 82.
13. Nather, R. E. 1973. *Vistas in Astr.* **15**, 91.
14. McGraw, J. T., Wells, D. C., and Wiant, J. R. 1973. *Rev. Sci. Inst.* **44**, 748.
15. Povenmire, H. R. 1979. *Graze Observer's Handbook.* Indian Harbor Beach, Florida: JSB Enterprises.
16. Elliot, J. L. 1978. *A.J.* **83**, 90.
17. Walker, A. R. 1980. *I.A.U. Cir.* 3466.
18. Evans, D. S. 1977. *Sky and Tel.* **56**, 164.
19. Evans, D. S. 1977. *Sky and Tel.* **56**, 289.
20. Nather, R. E., and Evans, D. S. 1970. *A.J.* **75**, 575.
21. Nather, R. E. 1970. *A.J.* **75**, 583.
22. Evans, D. S. 1970. *A.J.* **75**, 589.
23. Nather, R. E., and McCants, M. M. 1970. *A.J.* **57**, 963.
24. *The Astronomical Almanac.* Washington, D.C.: Government Printing Office. Issued annually.
25. J. R. Percy, ed. *The Observer's Handbook.* Royal Astronomical Society of Canada. Toronto: Univ. of Toronto Press. Issued annually.
26. Smart, W. M. 1971. *Text-Book on Spherical Astronomy.* New York: Cambridge Univ. Press.
27. International Occultation Timing Association, P.O. Box 596, Tinley Park, IL 60477.
28. Glasby, J. S. 1969. *Variable Stars.* Cambridge: Harvard Univ. Press.

29. Strohmeier, W. 1972. *Variable Stars*. Edited by A. J. Meadows. Elmsford, New York: Pergamon.
30. Kukarkin, B. V. 1975. *Pulsating Stars*. New York: Halsted (translation).
31. Kukarkin, B. V., Kholopur, P. N., Efremuv, Yu. N., and Kurochkin, N. E. 1965. *The Second Catalogue of Suspected Variable Stars*. Moscow: U.S.S.R. Academy of Sciences.
32. Code 680, Goddard Space Flight Center, Greenbelt, MD 20771.
33. Becvar, A. 1964. *Atlas of the Heavens—II Catalogue*. Cambridge: Sky Publishing Co.
34. Eggen, O. J. 1979. *Ap. J. Supl. Ser.* **41**, 413.
35. Tsesevich, V. P. 1969. *RR Lyrae Stars*. Springfield, Virginia: National Technical Information Service, U.S. Department of Commerce.
36. Percy, J. R., Dick, R., Meier, R., and Welch, D. 1978. *J. AAVSO* **7**, 19.
37. Payne-Gaposchkin, C. H. 1971. *Smithsonian Contributions to Astrophysics*. No. 13 (LMC).
38. Payne-Gaposchkin, C. H., and Gaposchkin, S. 1966. *Smithsonian Contributions to Astrophysics*. No. 9 (SMC).
39. Sturch, C. 1966. *Ap. J.* **143**, 774.
40. Schaltenbrand, R., and Tammann, G. A. 1971. *Ast. and Ap. Supl.* **4**, 265.
41. Henden, A. A. 1979. *MNRAS* **189**, 149.
42. Henden, A. A. 1980. *MNRAS* **192**, 621.
43. Eggen, O. J. 1977. *Ap. J. Supl. Ser.* **34**, 1.
44. Zeilik, M., Hall, D. S., Feldman, P. A., and Walter, F. 1979. *Sky and Tel.* **57**, 132.
45. Dr. Douglas S. Hall, Dyer Observatory, Vanderbilt University, Nashville, TN 37235.
46. Trimble, V. 1980. *Mercury* **9**, 8.
47. Robinson, E. L. 1976. *Ann. Rev. Astr. and Ap.* **14**, 119.
48. Irwin, J. B. 1962. *Astronomical Techniques*. Edited by W. A. Hiltner. Chicago: Univ. of Chicago Press, p. 584.
49. Binnendijk, L. 1960. *Properties of Double Stars*. Philadelphia: Univ. of Pennsylvania Press, p. 258.
50. Tsesevich, V. P., ed. *Eclipsing Variable Stars*. New York: Halsted Press.
51. Binnendijk, L. 1977. *Vistas in Astr.* **21**, 359.
52. Bookmyer, B. B., and Kaitchuck, R. H. 1979. *Pub. A. S. P.* **91**, 234.
53. Curator, Card Catalog of Eclipsing Variables, Department of Physics and Astronomy, University of Florida, Gainesville, FL 32611.
54. Hertzsprung, E. 1928. *Bul. Astr. Inst. Neth.* **4**, 179.
55. Bookmyer, B. B. 1976. *Pub. A. S. P.* **88**, 473.
56. Bookmyer, B. B. 1969. *A. J.* **74**, 1197.
57. Kaitchuck, R. H. 1974. *J. AAVSO* **3**, 1.
58. Niedner, M. B. 1980. *Ap. J.* **241**, 820.
59. Borra, E. F., and Wehlau, W. H. 1971. *Pub. A. S. P.* **83**, 184.
60. Burchi, R., and Milano, L. 1974. *Ast. and Ap. Supl.* **15**, 173.
61. Blanco, V. M., Demers, S., Douglass, G. G., and Fitzgerald, M. P. 1968. *Photoelectric Catalog*. Washington, D.C.: U.S. Naval Observatory. **XXI**, second series.

62. Williamon, R. M., 1976. *Pub. A. S. P.* **88**, 73.
63. Neff, J. S., Lane, W. A., and Fix, J. D. 1974. *Pub. A. S. P.* **86**, 225.
64. Burbidge, G., and Crowne, A. H. 1979. *Ap. J. Supl. Ser.* **40**, 583.
65. Hewitt, A., and Burbidge, G. 1980. *Ap. J. Supl. Ser.* **43**, 57.
66. Breger, M. 1979. *Information Bulletin on Variable Stars,* No. 1659.
67. Anon., 1971. *International Astronomical Union Style Book.* Transactions IAU, **XIVB**, 261.
68. Hathwell, D., and Metzner, A. W. K., for the AIP Publications Board, *American Institute of Physics Style Manual.* New York: AIP. Revised periodically.

APPENDIX A
FIRST-ORDER EXTINCTION STARS

This appendix lists those stars that are particularly useful in determining the first-order extinction coefficients. Several sources were consulted with the following criteria used for star selection:

1. The star must be fainter than $4^m.0$ (V).
2. $-0.15 \leq B - V \leq +0.15$.
3. $-0.15 \leq U - B \leq +0.15$.
4. No star is a common close visual or spectroscopic binary.
5. There are no peculiar spectral types.
6. No star is a known variable.

The following tables indicate the SAO 1950.0 coordinates along with precessed 1985 coordinates (with no proper motion corrections). Those stars with no HR number or Flamsteed number are designated by their HD or BD number in an obvious manner.

Table A.1 lists the northern declination stars, while Table A.2 lists the southern declination stars. The observer is indicated in the rightmost column of each table.

AZ-TNT. = Iriarte et al.[1]
COUSINS = Cousins[2]
JOHNSON = Johnson[3]

For those who need fainter standards, the list of equatorial standards ($10 - 13^m$) by Landolt[4] is extremely useful.

REFERENCES

1. Iriarte, B., Johnson, H. L., Mitchell, R. I., and Wisniewski, W. K. 1965. *Sky and Tel.* **30**, 25.
2. Cousins, A. W. J. 1971. *Roy. Obs. Annals,* No. 7.
3. Johnson, H. L. 1963. In *Basic Astronomical Data.* Edited by K. Aa Strand. Chicago: Univ. of Chicago Press, p. 208.
4. Landolt, A. U. 1973. *A. J.* **78**, 959.

TABLE A.1. Northern First-Order Extinction Stars

HR	R.A. (1950.0) DEC.	R.A. (1985.0) DEC.	V	B-V	U-B	SPEC.	NAME	OBS.
0053	0 14 28.3 +38 24 14.9	0 16 18 38 35 54	4.61	0.06	0.07	A2V	THE AND	JOHNSON
0068	0 15 42.5 +36 30 29.7	0 17 32 36 42 9	4.50	0.06	0.07	A2V	SIG AND	AZ-TNT.
0343	1 8 2.6 +54 53 4.3	1 10 9 55 4 14	4.36	0.17	0.13	A7V	THE CAS	JOHNSON
0378	1 15 13.0 +21 6.2	1 17 11 1 32 0	5.15	0.07	0.10	A3V	89 PSC	COUSINS
0383	1 16 42.7 +27 6.8	1 18 38 27 11 8	4.76	0.03	0.10	A3V	UPS PSC	AZ-TNT.
0530	1 59 7.2 +72 10 50.8	2 2 8 72 20 57	3.98	-0.01	0.03	A1V	50 CAS	AZ-TNT.
0620	2 5 27.7 +37 22.8	2 7 34 37 47 20	4.83	0.12	0.14	A4V	58 AND	AZ-TNT.
0664	2 14 20.0 +33 37 1.5	2 16 24 33 46 44	4.01	0.00	0.00	A0V	GAM TRI	JOHNSON
0718	2 25 29.8 +8 14 13.1	2 27 47 8 23 36	4.28	-0.06	-0.13	B9III	CHI2 CET	AZ-TNT.
0879	2 55 33.3 +39 27 50.7	2 57 47 39 36 13	4.70	-0.06	-0.12	A2V	PI PER	AZ-TNT.
0932	3 5 27.7 +22 22.1	3 10 16 22 39 19	4.87	0.03	0.04	A0V		AZ-TNT.
0972	3 12 1.3 +20 51 37.0	3 14 22 20 59 25	4.89	-0.02	-0.01	A0IV	ZET ARI	AZ-TNT.
1002	3 18 5.0 +43 2.0	3 20 26 43 16 34	4.94	0.06	0.05	A3V	32 PER	AZ-TNT.
1148	3 45 2.9 +71 10 51.6	3 48 45 71 17 16	4.66	0.03	0.05	A3IV	GAM CAM	AZ-TNT.
1261	4 2 50.9 +50 13 3.3	4 5 27 50 18 43	4.29	0.02	-0.04	B9V	LAM PER	AZ-TNT.
1324	4 14 28.4 +50 10 28.9	4 17 6 50 15 37	4.65	0.05	0.05	A2	B PER	AZ-TNT.
1387	4 22 23.1 +22 10 51.9	4 24 28 22 15 38	4.23	0.12	0.15	A7V	KAP TAU	AZ-TNT.
1389	4 22 35.6 +17 48 55.2	4 24 36 17 53 41	4.29	0.04	0.08	A3V	68 TAU	AZ-TNT.
1448	4 31 28.5 +27 54.9	4 33 20 5 32 16	5.68	0.05	0.10	A3		AZ-TNT.
1544	4 47 53.0 +8 48 57.6	4 49 47 8 52 31	4.35	0.01	0.03	A0V	PI2 ORI	COUSINS
1570	4 52 8.4 +10 4 22.5	4 54 3 10 7 44	4.66	0.08	0.08	A0V	PI1 ORI	AZ-TNT.
1724	5 14 5.2 +1 53 36.5	5 15 50 1 55 53	6.42	-0.02	-0.02	B8V		AZ-TNT.
1807	5 23 57.1 +6 49 38.7	5 25 50 6 51 25	6.42	-0.04	-0.05	B9		COUSINS
1872	5 31 38.8 +3 44 3.5	5 33 29 3 45 27	5.35	0.04	0.06	A2	38 ORI	COUSINS
2034	5 50 11.0 +27 36 8.5	5 52 22 27 36 35	4.61	0.02	0.01	B9.5V	136 TAU	COUSINS
2103	5 56 15.2 +2 32 59.7	5 58 3 2 33 8	5.21	0.00	0.01	A1	60 ORI	AZ-TNT.
2174	6 6 21.2 +20 30 22.8	6 8 10 2 30 10	5.73	0.06	0.04	A0.5I		COUSINS
2209	6 13 20.4 +69 20 27.1	6 17 11 69 19 40	4.79	0.03	-0.01	A0V	22 CAM	AZ-TNT.
2238	6 15 12.7 +59 1 54.2	6 18 17 59 1 2	4.47	-0.01	-0.03	A2V	2 LYN	AZ-TNT.
2404	6 32 3.8 +7 36 46.8	6 33 57 7 35 6	6.44	-0.01	-0.03	A0	14 MON	COUSINS
2584	6 52 51.4 +8 23 23.1	6 54 45 8 20 39	6.38	0.04	0.06	A0		COUSINS
2535	6 53 58.3 +45 9 40.7	6 56 31 45 6 53	4.90	0.04	0.03	A2V	16 LYN	COUSINS
2629	7 5 22.2 +1 44 53 26.2	7 3 53 45 50 24	6.63	0.06	0.09	A0		AZ-TNT.
2654	7 9 44.8 +33 50.6	7 11 33 1 30 41	6.56	-0.01	-0.07	B9		COUSINS
2710	7 11 3.0 +5 44 22.7	7 13 3 5 40 49	6.08	-0.02	-0.05	A0		COUSINS
2751	7 14 44.3 +49 33 22.1	7 17 23 49 29 33	5.04	0.08	0.09	A3III		AZ-TNT.

281

NORTHERN FIRST ORDER EXTINCTION STARS

HR	R.A. (1950.0) DEC.	R.A. (1985.0) DEC.	V	B-V	U-B	SPEC.	NAME	OBS.
2818	7 22 56.8 +49 18 46.7	7 25 35 49 14 34	4.63	-0.02	-0.02	A1IV	21 LYN	AZ-TNT.
2946	7 38 47.2 +58 49 47.0	7 41 44 58 44 49	4.99	0.08	0.09	A3III	24 LYN	AZ-TNT.
3067	7 50 26.4 +26 53 48.7	7 52 34 26 48 20	4.99	0.09	0.12	A4V	PHI GEM	AZ-TNT.
3136	7 58 35.2 + 1 7.3	8 0 26 4 55 17	5.64	0.00	0.01	A0		COUSINS
3173	8 3 42.3 +51 39 9.9	8 7 20 51 33 3	4.84	0.05	0.00	A2V	27 LYN	AZ-TNT.
3410	8 35 9.0 + 5 50 45.6	8 36 51 5 45 24	4.17	0.02	0.04	A0V	DEL HYA	AZ-TNT.
3412	8 35 22.7 + 9 45 2.4	8 37 16 9 37 40	6.52	-0.02	-0.04	A0	37 CNC	COUSINS
3449	8 40 23.7 +21 38 58.0	8 42 25 21 31 24	4.66	0.02	0.01	A1V	GAM CNC	AZ-TNT.
3492	8 45 47.1 + 6 1 25.1	8 47 38 5 53 38	4.37	-0.05	-0.05	A0V	RHO HYA	AZ-TNT.
3573	8 55 33.2 + 4 44 8.5	8 57 21 4 36 0	6.59	0.06	0.08	A0		COUSINS
3651	9 9 36.2 + 4 23.2	9 11 25 3 55 44	6.14	0.00	0.00	A0		COUSINS
3799	9 31 24.6 +52 16 30.2	9 33 48 52 7 8	4.50	0.00	0.04	A2V	26 UMA	AZ-TNT.
3906	9 49 37.4 + 2 41 17.3	9 51 26 2 31 24	6.02	-0.04	-0.08	A1V	7 SEX	COUSINS
3974	9 59 29.2 +35 32 21.4	10 1 32 35 16 5	4.49	0.19	0.07	A7V	21 LMI	JOHNSON
4248	10 51 6.5 +43 27 24.0	10 53 26 43 16 12	4.71	-0.05	-0.06	A1V	OMI UMA	AZ-TNT.
4300	10 59 39.8 +20 26 54.2	11 1 31 20 15 36	4.41	-0.04	-0.04	A1V	60 LEO	AZ-TNT.
4356	11 11 11.9 + 9 12 9.9	11 12 59 9 0 43	5.42	-0.02	-0.05	A1V	69 LEO	COUSINS
4386	11 16 24.7 +38 27 36.4	11 18 19 38 16 42	4.06	-0.07	-0.12	B9V	SIG LEO	AZ-TNT.
4388	11 18 33.5 + 9 18 13.1	11 20 21 6 6 6	4.79	0.12	0.03	A2V	55 UMA	AZ-TNT.
4528	11 45 20.8 + 8 31 25.3	11 47 8 8 19 44	5.31	0.02	0.04	A1	4 VIR	COUSINS
4505	11 57 23.2 + 3 56 1.0	11 59 10 3 44 19	5.36	0.00	0.00	A0V	7 VIR	COUSINS
4589	11 58 18.6 + 6 53 35.1	12 0 6 6 41 53	4.67	-0.13	-0.10	A4V	PI VIR	AZ-TNT.
4789	12 32 21.6 +22 54 15.4	12 34 18 22 42 41	4.81	0.00	-0.01	A3III	23 COM	COUSINS
4805	12 35 31.4 + 3 33 26.5	12 37 18 3 21 53	6.32	-0.01	-0.01	A0		COUSINS
4828	12 39 21.2 +10 30 39.2	12 41 7 10 19 8	4.88	0.09	0.06	A1V	RHO VIR	AZ-TNT.
5021	13 16 19.0 + 3 57 1.6	13 18 5 3 45 59	6.62	0.05	0.02	A1IV		COUSINS
5037	13 19 0.2 + 2 57 57.8	13 20 55 2 45 55	5.70	0.02	0.02	A0		COUSINS
5062	13 23 16.5 +55 14 53.0	13 24 39 55 3 9	4.03	0.15	0.09	A5V	80 UMA	JOHNSON
5112	13 32 24.8 +49 16 15.8	13 33 50 49 5 31	4.70	0.12	0.11	A4V	24 CVN	AZ-TNT.
5264	13 59 5.9 + 1 47 8.5	14 0 52 1 37 0	4.27	0.09	0.14	A3III	TAU VIR	AZ-TNT.
5859	15 42 55.0 +22 56 31.0	15 44 38 22 50 42	5.58	0.07	0.05	A3V	PI SER	AZ-TNT.
5972	16 0 8.4 +68 52 34.8	16 1 38 68 48 0	5.01	-0.06	-0.10	B9IV	15 DRA	AZ-TNT.
6161	16 28 4.2 +42 32 21.1	16 28 33 42 28 28	4.20	-0.02	-0.10	B9V	SIG HER	AZ-TNT.
6168	16 32 29.3 +12 48 29.1	16 33 37 12 45 38	4.91	0.06	0.03	A3IV	60 HER	AZ-TNT.
6355	17 3 3.4 +37 20 34.1	17 4 40 37 18 21	4.66	0.05	0.03	A2V	69 HER	AZ-TNT.
6436	17 15 56.7	17 17						

282

HD	RA			Dec			V	B-V	U-B	Sp	Name	Source
6723	17 59	12.9		+ 1 18	17.3		4.42	0.04	0.02	A1V	68 OPH	AZ-TNT.
6789	17 48	18.3		+86 36	34.8		4.35	0.01	0.04	A1V	DEL UMI	AZ-TNT.
6923	18 23	10.7		+58 46	16.5		4.98	0.08	0.05	A1V	39 DRA	AZ-TNT.
7085	18 44	48.7		+18 7	28.0		4.36	0.13	0.07	A3V	111 HER	AZ-TNT.
7313	18 47	4.2		+ 0 46	40.6		6.24	0.03	0.01	A0		COUSINS
7371	19 15	16.6		+ 1 56	26.7		6.18	0.01	0.01	A0		COUSINS
7546	19 20	24.9		+65 37	5.2		4.59	0.02	0.06	A2IV	PI DRA	AZ-TNT.
7592	19 46	45.4		+19 20	55.5		5.00	0.10	-0.13	A3V	ZET SGE	AZ-TNT.
7724	19 51	20.1		+23 56	52.8		4.58	-0.06	0.01	B9.5I	13 VUL	AZ-TNT.
7736	20 11	57.7		+15 13	38.4		4.95	0.09	0.14	A2V	RHO AQL	AZ-TNT.
7740	20 12	43.6		+46 34	48.9		4.82	0.10	0.00	A3III	30 CYG	AZ-TNT.
7744	20 12	39.6		+36 49	7.8		4.99	0.12	0.08	A2III	29 CYG	AZ-TNT.
7748	20 14	14.1		+56 58	50.9		4.38	0.11	0.05	A3IV	33 CYG	AZ-TNT.
7857	20 31	29.1		+ 9 53	15.0		6.56	0.08	0.11	A0		COUSINS
7891	20 32	58.2		+14 30	2.2		4.69	0.11	-0.07	A3V	ZET DEL	AZ-TNT.
7891	20 36	17.2		+21 37	1 28.7		4.82	-0.02	-0.04	B9.5V	29 VUL	AZ-TNT.
8098	21 8	5.4		+ 9 59	38.4		6.07	0.02	0.01	A2V	6 EQU	COUSINS
8265	21 35	14.1		+ 6 33	33.9		6.20	0.02	0.01	A1V	3 PEG	COUSINS
8328	21 44	41.8		+ 2 36	14.6		5.64	0.00	0.01	A0	11 PEG	COUSINS
8491	22 13	30.3		+ 8 28	1.5		6.20	-0.04	-0.04	A0		COUSINS
8641	22 39	24.3		+29 13	46.1		4.79	-0.01	-0.00	A1V	OM PEG	AZ-TNT.
8717	22 52	42.5		+ 8 44	55.9		4.91	0.00	-0.01	A1V	RHO PEG	AZ-TNT.
8738	22 56	11.2		+ 7 15	21.7		6.33	0.06	0.01	A0		COUSINS
9042	23 50	31.0		+ 1 48	45.3		6.28	0.00	-0.01	A2	25 PSC	COUSINS

TABLE A.2. Southern First-Order Extinction Stars

HR	R.A. (1950.0) DEC.	R.A. (1985.0) DEC.	V	B-V	U-B	SPEC.	NAME	OBS.
0125	0 29 6.6 -49 4 47.3	0 30 41 -48 53 11	4.77	0.01	0.04	A0V	LAM1PHE	COUSINS
0191	0 41 6.8 -57 44 13.2	0 42 40 -57 32 43	4.36	-0.04	-0.03	A0V	ETA PHE	COUSINS
0444	1 30 33.9 -9 16 16.3	1 32 18 -9 5 29	6.59	-0.04	-0.10	A0		COUSINS
0558	1 52 17.6 -42 44 30.2	1 53 44 -42 34 12	5.11	-0.05	-0.14	B9	PHI PHE	COUSINS
0607	2 0 37.8 -6 42 42.3	2 2 25 0 32 22	5.42	-0.14	-0.13	A5	60 CET	COUSINS
0653	2 13 0.8 -9 41 52.7	2 14 43 -9 32 2	6.54	-0.01	-0.07	A0		COUSINS
0634	2 17 9.9 -4 34 28.9	2 18 55 -4 24 50	6.50	-0.07	-0.07	A2		COUSINS
0705	2 20 51.2 -68 53 11.8	2 21 29 -58 43 39	4.08	0.03	0.05	A2V	DEL HYI	COUSINS
0708	2 23 32.0 -12 60 54.3	2 25 13 -12 51 27	4.89	-0.02	-0.04	B9V	RHO CET	AZ-TNT.
0789	2 37 53.6 -43 19.7	2 39 13 -42 57 19	4.74	-0.06	0.06	A2V		COUSINS
0806	2 38 48.8 -68 28 50.9	2 39 20 -68 19 51	4.11	-0.06	-0.13	B9III	EPS HYI	COUSINS
0637	2 44 45.9 -67 49 35.4	2 45 18 -67 40 48	4.83	-0.05	0.08	A2	ZET HYI	COUSINS
0875	2 54 6.9 -3 54 45.0	2 55 52 -3 46 18	5.16	0.08	0.04	A1V		JOHNSON
0892	2 56 10.7 -2 58 51.2	2 57 56 -2 50 28	5.23	0.06	0.02	A2		COUSINS
1272	4 3 31.9 -8 59 23.9	4 5 13 -8 53 44	6.26	0.06	0.08	A3V		COUSINS
1383	4 21 11.3 -3 51 34.8	4 22 55 -3 46 44	5.18	0.07	0.08	A1V	XSI ERI	COUSINS
1522	4 43 53.7 -2 35.1	4 45 38 -2 58 48	6.33	0.04	0.06	A2		COUSINS
1596	4 55 39.6 -3 17 18.2	4 57 25 -2 14 6	6.34	0.09	0.11	A0		COUSINS
1621	4 59 16.3 -20 17 24.0	5 0 46 -20 8 22	4.91	-0.05	-0.15	B9		AZ-TNT.
1762	5 18 18.7 -21 17 19.2	5 19 48 -21 15 14	4.70	-0.05	-0.11	A0V		COUSINS
1826	5 26 26.9 -3 20 47.5	5 28 11 -3 19 7	6.38	-0.01	-0.06	B9		COUSINS
1937	5 36 27.8 -7 14 21.4	5 38 9 -7 13 12	4.81	0.14	0.09	A4IV	49 ORI	AZ-TNT.
2039	5 49 44.5 -1 9 41.5	5 51 24 -1 2 42	5.95	0.10	0.08	A0		COUSINS
2071	5 53 7.3 -4 47 45.5	5 54 50 -4 47 23	6.28	0.05	0.06	A0		COUSINS
2155	6 3 53.5 -14 55 45.0	6 5 28 -14 55 59	4.67	0.05	0.00	A1V	THE LEP	AZ-TNT.
2195	6 8 35.4 -6 44 31.6	6 10 17 -6 43 0	6.14	-0.01	0.03	A0		COUSINS
2210	6 9 13.8 -2 29 26.5	6 11 59 -2 30 0	6.62	-0.04	0.08	A0		COUSINS
2295	6 20 54.4 -4 39 37.4	6 22 38 -4 40 43	6.66	0.06	0.00	B9		COUSINS
2328	6 24 19.3 -7 28 49.6	6 26 1 -7 30 6	6.26	-0.03	-0.10	A0V	CHI2CMA	AZ-TNT.
2414	6 32 57.6 -22 55 26.1	6 34 25 -22 57 8	4.54	-0.04	-0.02	A0IV	DEL MON	AZ-TNT.
2714	7 9 18.6 -24 30 34.2	7 11 5 -28 21 25	4.15	0.03	0.03	A3V		AZ-TNT.
3131	7 57 37.5 -18 15 39.2	7 59 11 -18 2 59	4.61	0.07	0.08	A0		COUSINS
3383	8 31 29.9 -1 58 46.8	8 33 16 -2 56 53	5.80	0.10	0.00	A9II		COUSINS
3426	8 35 53.1 -42 30 47.9	8 37 40 -42 47 15	4.14	-0.02	0.12	A0		COUSINS
3437	8 38 36.1 -8 52 23.8	8 40 17 -8 55 47	6.62	-0.02	0.02	A5II		COUSINS
3452	8 39 34.6 -47 8 16.2	8 40 43 -47 15 21	4.77	0.12	0.12	A5II	DEL PYX	COUSINS
3556	8 53 22.8 -27 29 18.7	8 54 52 -27 37 21	4.88	0.12	0.15	A3V		COUSINS

3615	9	1	39.8	-66	11	46.2	9	2	12			4.00	0.15	0.14	A5V	ALF VOL	COUSINS
3787	9	29	26.0	0	57	48.0	9	31	13	-66	20	4.56	0.11	0.09	A3III	TAU2HYA	AZ-TNT.
3832	9	35	24.3	9	11	48.0	9	37	7	1	24	6.40	-0.04	-0.07	A0	34 HYA	COUSINS
3981	10	5	22.7	0	7	56.3	10	8	21	9	52	6.49	-0.04	-0.06	A0III	ALP SEX	AZ-TNT.
3989	10	7	38.4	8	35	42.0	10	9	20	8	2	5.90	0.02	-0.05	A0	17 SEX	COUSINS
4109	10	14	4.4	-1	29	9.7	10	16	20	-3	54	6.04	-0.04	-0.07	A0		COUSINS
4138	10	29	11.9	-3	44	38.2	10	27	39	-71	54	4.73	-0.03	0.05	A2V		COUSINS
4293	10	29	4.4	-71	57	7.0	10	29	54	-42	43	4.38	0.11	0.13	A2IV		COUSINS
4343	10	57	51.3	-41	33	26.4	10	59	44	-22	33	4.48	-0.03	0.07	A2III	BET CRT	COUSINS
5163	13	9	11.7	-22	2	9.0	11	10	25	5	25	6.52	0.00	0.02	A0		COUSINS
5342	14	13	17.9	5	57	52.0	13	43	30	-37	36	6.14	0.01	-0.01	A3		COUSINS
5367	14	17	54.5	-37	39	52.6	14	15	49	5	0	4.04	-0.03	-0.11	A0IV	PSI CEN	COUSINS
5489	14	41	30.4	5	2	23.1	14	19	38	-35	44	4.92	-0.02	-0.04	A1		COUSINS
5670	15	13	54.7	-34	37	52.6	15	16	3	-37	41	4.06	0.00	-0.08	A3V	BET CIR	COUSINS
5724	15	22	34.9	5	58	28.6	15	24	21	-58	51	4.60	0.09	-0.04	A0IV		COUSINS
5959	15	58	5.4	-38	33	58.6	15	59	22	-38	10	5.55	0.04	-0.04	A1	50 LIB	COUSINS
6031	16	9	15.8	8	16	17.1	16	11	10	-8	1	4.92	0.09	0.10	A2V	PSI SCO	AZ-TNT.
6070	16	18	12.7	-28	56	10.2	16	19	22	-10	33	4.77	0.02	-0.01	A0V	D SCO	AZ-TNT.
6446	17	21	7.7	-12	47	29.3	17	22	58	-28	36	4.31	0.03	-0.03	A1V	NU SER	AZ-TNT.
6519	17	28	21.7	-23	55	52.1	17	30	10	-12	56	4.81	0.00	-0.07	A1	51 OPH	COUSINS
6581	17	38	36.1	-12	51	33.0	17	40	29	-23	6	4.24	0.08	-0.10	A2V	OMI SER	AZ-TNT.
6930	18	26	20.8	-12	46	1.0	18	28	34	-12	3	4.71	0.07	-0.05	A3V	GAM SCT	AZ-TNT.
6963	18	30	42.2	-5	35	58.9	19	2	7	-5	34	6.36	0.02	-0.07	A0		COUSINS
7029	19	0	16.5	-3	46	40.0	19	1	47	-3	40	5.42	0.04	0.07	A2V	14 AQL	COUSINS
7254	19	6	4.3	-37	59	3.5	19	35	55	-37	55	4.12	-0.06	-0.12	A2V	ALF CRA	AZ-TNT.
7440	19	33	40.5	-24	44	4.2	19	19	48	-24	2	4.59	-0.04	-0.11	B9	52 SGR	AZ-TNT.
7773	19	17	53.5	-12	55	4.5	20	19	17	-17	32	4.76	-0.01	-0.01	B9V	NU CAP	AZ-TNT.
8075	20	28	8.3	-17	25	57.8	21	5	48	-33	43	4.07	-0.05	-0.05	A0V	THE CAP	AZ-TNT.
8431	22	5	28.3	-33	14	8.3	22	9	17	-33	17	4.50	0.05	-0.07	A2V	MU PSA	COUSINS
8451	22	28	45.1	0	8	23.9	22	29	58	0	17	6.27	-0.07	-0.12	A0		AZ-TNT.
8573	22	28	4.1	-10	56	9.2	22	29	51	-10	41	4.82	-0.06	0.05	A0IV	THE AQR	COUSINS
8840	23	12	9.7	-3	46	9.1	23	14	47	-3	31	5.55	0.06	0.08	A2		AZ-TNT.
8959	23	35	7.8	-45	46	17.4	23	37	34	-45	38	4.73	0.08	-0.13	B9.5V	OM2 AQR	COUSINS
9016	23	40	9.9	-14	24	24.7	23	48	37	-14	44	4.51	0.00	0.00	A0V	DEL SCL	AZ-TNT.
9078	23	46	19.5	-28	24	17.4	0	2	9	-28	12	4.57	0.04	-0.04	A2V		AZ-TNT.
0	0	1	10.8	-17	36	51.4	0	2	58	-17	25	4.56	-0.12		B9IV	2 CET	AZ-TNT.

APPENDIX B
SECOND-ORDER EXTINCTION PAIRS

As mentioned in Section 4.4, second-order extinction can be determined by observing a close optical pair of widely differing colors as they pass through varying air mass. A list of bright pairs in the equatorial plane was published by Crawford, Golson, and Landolt.[1] Barnes and Moffett[2] extended this list to include R and I magnitudes and one additional star. Their paper is extremely useful as it gives finding charts for all 37 stars of their extinction network.

An examination of the Johnson, Tonantzintla, and Cousins lists of standard stars yielded 23 more stars useful in the extinction determination. Table B.1 lists all 60 stars.

Extinction examples using stars from this list are presented in Appendix G.

REFERENCES

1. Crawford, D. L., Golson, J. C., and Landolt, A. U. 1971. *Pub. A. S. P.* **83**, 652.
2. Barnes, T. G., III, and Moffett, T. J. 1979. *Pub. A. S. P.* **91**, 289.

TABLE B.1. Second-Order Extinction Pairs

HR	R.A. (1950.0)	DEC.	R.A. (1985.0)	DEC.	V	B-V	U-B	SPEC.	NAME	OBS.
0011	0 5 10.3	-2 49 37.3	0 6 57	-2 37 55	6.45	-0.14	-0.46	B8		KITT PK.
0014	0 5 38.4	-2 43 33.6	0 7 25	-2 31 52	6.13	1.34	1.11	K0		KITT PK.
0607	2 0 37.5	-0 6 42.3	2 2 25	0 3 22	5.42	0.14	0.13	A5	60 CET	COUSINS
0610	2 1 14.3	-0 34 45.1	2 3 1	-0 24 40	5.92	0.88	0.51	G5II	61 CET	COUSINS
0718	2 25 29.8	8 14 13.1	2 27 21	8 23 36	4.28	-0.06	-0.13	B9III	CHI2 CET	JOHNSON
0725	2 26 54.9	9 20 36.9	2 28 47	9 29 57	6.07	1.02	0.86	K2III		COUSINS
	2 37 0.0	1 9 13.8	2 38 48	1 18 15	8.19	-0.06	-0.27	B9	HD16581	KITT PK.
	2 37 11.0	1 54 57.1	2 38 59	2 3 58	8.31	1.51	1.79	K4	HD16608	KITT PK.
1373	4 20 2.8	17 25 36.8	4 22 3	17 30 30	3.76	0.99	0.83	K1III	DEL TAU	AZ-TNT.
1389	4 22 35.6	17 48 55.2	4 24 36	17 53 41	4.29	0.04	0.08	A3V	68 TAU	AZ-TNT.
1534	4 46 1.6	3 33 44.7	4 47 51	3 37 24	7.32	-0.06	-0.31	B9	HD30544	KITT PK.
	4 46 6.9	3 30 7.0	4 47 57	3 33 46	6.01	1.21	1.15	K0	HD30545	KITT PK.
1826	5 26 26.9	-3 20 47.5	5 28 11	-3 19 7	6.38	-0.01	-0.06	B9		COUSINS
1830	5 26 54.0	-3 29 5.3	5 28 38	-3 27 26	5.80	1.15	1.07	G8		COUSINS
2071	5 53 7.3	-4 47 41.5	5 54 50	-4 47 23	6.28	0.05	0.06	A0		COUSINS
2070	5 53 1.8	-4 37 23.6	5 54 45	-4 37 4	5.87	1.17	1.22	K2		COUSINS
	5 59 47.5	1 5 25.9	6 1 35	1 5 23	8.56	0.00	-0.04	B9	HD40983	KITT PK.
	6 0 5.5	1 6 54.7	6 1 53	1 6 51	8.17	0.98	0.76	K0	HD41029	KITT PK.

SECOND ORDER EXTINCTION PAIRS

HR	R.A. (1950.0)	DEC.	R.A. (1985.0)	DEC.	V	B-V	U-B	SPEC.	NAME	OBS.
	6 50 0.8	1 9 5.3	6 51 49	1 6 30	8.17	-0.06	-0.28	B8	HD50279	KITT PK.
	6 49 29.0	1 18 45.5	6 51 17	1 16 12	7.85	1.55	1.72	K5	HD50167	KITT PK.
2710	7 9 11.3	5 44 20.7	7 11 3	5 40 49	6.08	-0.02	-0.05	A0		COUSINS
2713	7 9 27.6	5 33 33.3	7 11 19	5 30 1	6.15	1.15	0.97	K0		COUSINS
	7 46 9.6	-0 8 0.5	7 47 57	-0 13 16	8.74	0.05	0.05	B9	HD63390	KITT PK.
	7 46 3.8	-0 37 7.4	7 47 50	-0 42 23	8.43	0.95	0.57	K0	HD63368	KITT PK.
	8 45 0.9	0 15 44.1	8 46 48	0 7 59	7.83	0.08	0.09	B9	HD75012	KITT PK.
	8 45 44.0	0 44 25.3	8 47 31	0 36 38	7.24	1.48	1.78	K2	HD75138	KITT PK.
	9 45 12.4	-2 28 49.7	9 47 58	-2 38 36	8.65	-0.17	-0.76	B5	HD84971	KITT PK.
	9 45 50.7	-4 10 24.8	9 47 36	-4 20 11	8.66	1.16	1.12	K5	HD84916	KITT PK.
3989	10 7 38.4	-8 9 42.8	10 9 22	-8 20 2	5.90	0.02	-0.06	A0	17 SEX	COUSINS
3996	10 8 26.6	-8 10 15.8	10 10 10	-8 20 37	5.64	1.31	1.41	K2	18 SEX	COUSINS
	11 13 38.6	-3 11 56.9	11 15 25	-3 23 24	7.41	-0.24	-0.93	B3	HD98981	KITT PK.
	11 13 44.4	-3 29 18.7	11 15 31	-3 40 46	8.94	0.71	0.32	K0	HD98007	KITT PK.
4854	12 44 29.9	6 13 27.0	12 46 16	6 1 59	6.36	-0.06	-0.05	B9	HD111133	KITT PK.
	12 44 42.8	7 16 50.7	12 46 29	7 5 22	8.46	1.16	1.21	K0	HD111165	KITT PK.
	13 33 6.7	-5 54 3.6	13 34 56	-6 4 46	8.07	-0.16	-0.63	B8	HD118246	KITT PK.
	13 32 27.4	-6 42 44.3	13 34 17	-6 53 28	8.18	1.07	0.95	K2	HD118129	KITT PK.

```
5501  14 42 57.2   0 55 38.4   14 44  44 48   5.70 -0.03 -0.06  B9    HD129956  KITT PK.
      14 43  3.7 - 0  6 17.2   14 44  44  6   8.37  1.50  1.86  K5    HD129975  KITT PK.

      15 43 30.2 - 1 38 56.3   15 45  45 27   5.40 -0.03 -0.42  B8    HD140873  KITT PK.
      15 43 22.6 - 1 17 25.8   15 45  45 57   8.80  1.66  2.02  K5    HD140850  KITT PK.

      17 41 48.6   5 44  6.7   17 43  43 13   8.31  0.05 -0.14  B9    HD161261  KITT PK.
      17 41 45.4   5 16 16.6   17 43  43 23   7.80  1.28  1.10  K2    HD161242  KITT PK.

      18 33 55.8   3  7 13.0   18 35  35 26   9.12  0.28 -0.10  B9    HD171732  KITT PK.
      18 33 56.0   2 31 36.0   18 35  35 49   9.06  1.13  1.05  K2    HD171731  KITT PK.

7313  19 15 16.6 -    2 46.7   19 17   0 15   5.18  0.01  0.01  A0    23 AQL    COUSINS
7319  19 15 59.8 -    1 34.1   19 17   3 25   5.09  1.15  1.01  K2II            COUSINS

      20 34 44.6   0  4 41.2   20 36  36  2 39  8.12  0.17 -0.32  B8  HD184790  KITT PK.
      20 34 33.2   0 41 33.6   20 36  36 34 13  8.16  1.20  0.93  K5  HD184914  KITT PK.

7878  21 33 25.7   5 15  7.2   21 35  35 24 31  6.23 -0.09 -0.39  B8  HD196426  KITT PK.
      21 33 44.4   5 54 46.0   21 35  35  4 11  8.72  1.66  2.04  K5  HD196395  KITT PK.

      22 33 25.7   5 15  7.2   22 35  35 26  0  8.32 -0.06 -0.35  B9  HD205556  KITT PK.
      22 33 44.4   5 54 46.0   22 35  35  5 39  7.72  1.26  1.32  K2  HD205584  KITT PK.

      22  4  6.3   2 11 43.1   22  5   5 21 58  6.52 -0.06 -0.23  B9  HD209905  KITT PK.
      22  3 29.4   1  0 48.2   22  5   5 11  2  8.94  1.21  1.17  K2  HD209796  KITT PK.

8451  22  7 45.1 - 4  8 23.9   22  9   9 58  3  6.27  0.00 -0.07  A0            COUSINS
8451  22  7 57.4 - 4 30 48.9   22  9   9 20 28  6.06  0.98  0.84  K0            COUSINS

8988  23 40  7.8 -14 49 17.4   23 41 56 -14 37 38  4.51 -0.04 -0.13  B9.5V  OM2 AQR  AZ-TNT.
```

APPENDIX C
UBV STANDARD FIELD STARS

Having a photometric system without standard stars is like measuring the distance from New York to Paris in meters without defining the length of the meter. The standard stars are an integral part of a photometric system. They are as important as the filter responses themselves.

After the definition of the *UBV* system by Johnson and Morgan,[1] Johnson and Harris[2] published a list of 108 stars intended for use as photometric standards for the system. There were 10 primary standards, stars that were measured every possible night, and 98 additional stars that were measured from two to 17 times, averaging 7.3 measurements each. Because of the few measurements of some of these stars, internal accuracy of the system is on the order of $0^m.03$ in V. That is, if you select a large subset of these stars and measure them, your mean probable error of a measurement should be about this size.

The Johnson standard list has no stars south of $-20°$ declination. Traditionally, this has made Southern Hemisphere transformations difficult. For this reason, Cousins[3] has proposed a second list of standards between $+10°$ and $-10°$ declination, and secondary standards in the E and F regions of the southern sky.[4] The equatorial standards are presented by Cousins with their HR numbers but without coordinates, making them difficult for the amateur to use. The E and F region stars have the coordinates listed, but the amateur may have difficulty finding the reference in the local library. For this reason, the E and F region stars are included in this appendix and the list is separated into stars north and south of the celestial equator. Table C.1 lists the northern standard field stars, while Table C.2 lists the southern standard field stars.

Another problem with the standard stars listed in references 2 through 4 is the preponderance of bright stars. Transformations using bright stars and large telescopes are very difficult because of dead-time corrections and photomultiplier tube fatigue. For instance, with the Goethe Link 40-centimeter (16-inch) telescope, you cannot easily look at stars brighter than $4^m.0$ in V, and seldom use stars brighter than $6^m.0$ as standards. Fainter stars that can be used as secondary standards can be found in the list by Landolt.[5] These equatorial stars range from seventh to fourteenth magnitude, with most about twelfth magnitude in V. Another list of brighter UBV secondary standards is readily available from Sky Publishing[6] and is recommended for purchase.

To use the stars from this appendix for transforming to the standard system, pick 20 or more from the list, distributed over the sky but greater than 30° above the horizon. This removes effects caused by a small number of standards and minimizes systematic regional errors.

REFERENCES

1. Johnson, H. L. and Morgan, W. W. 1953. *Ap. J.* **117**, 313.
2. Johnson, H. L. and Harris, D. L., III 1954. *Ap. J.* **120**, 196.
3. Cousins, A. W. J. 1971. *Roy. Obs. Annals.* No. 7.
4. Cousins, A. W. J. 1973. *Mem. Roy. Astr. Soc.* **77**, 223.
5. Landolt, A. U. 1973. *A. J.* **78**, 959.
6. Iriarte, B., Johnson, H. L., Mitchell, R. I., and Wisniewski, W. K. 1965. *Sky and Tel.* **30**, 25. (Available as a reprint.)

TABLE C.1. Northern *UBV* Standard Field Stars

HR	R.A. (1950.0)	DEC.	R.A. (1985.0)	DEC.	V	B-V	U-B	SPEC.	NAME	OBS.
0039	0 10 39.4	14 54 20.6	0 12 27	15 6 1	2.83	-0.23	-0.87	B2IV	GAM PEG	JOHNSON
0063	0 14 28.3	38 24 14.9	0 16 18	38 35 54	4.61	0.06	0.04	A2V	THE AND	JOHNSON
0226	0 47 2.8	40 48 25.2	0 48 58	40 59 51	4.53	-0.15	-0.58	B5V	NU AND	JOHNSON
0343	1 8 2.6	54 53 4.3	1 10 9	55 4 14	4.33	0.17	0.11	A7V	THE CAS	JOHNSON
0403	1 22 31.5	57 58 34.4	1 24 48	60 9 29	2.68	0.13	0.12	A5V	DEL CAS	JOHNSON
0437	1 28 49.0	15 19.4	1 30 40	15 16 7	3.62	0.97	0.76	GBIII	ETA PSC	JOHNSON
0493	1 39 46.6	20 20 34.3	1 41 41	20 12 9	5.23	0.83	0.50	K1V	107 PSC	JOHNSON
0553	1 51 52.3	20 33 52.0	1 53 48	20 44 10	2.65	0.13	0.10	A5V	BET ARI	JOHNSON
0617	2 4 20.9	23 13 37.0	2 6 18	23 23 36	2.00	1.15	1.12	K2III	ALF ARI	PRIMARY
	2 5.3	57 54.0	2 7	3 7 52	10.03	1.44	1.08		2 348	JOHNSON
0718	2 25 29.8	8 14 13.1	2 27 21	8 23 30	4.28	-0.06	-0.13	B9III	CHI2 CET	JOHNSON
0753	2 33 20.1	6 38 57.8	2 35 11	6 48 8	5.82	0.97	0.79	K3V	A	JOHNSON
0753	2 33 20.1	6 38 57.8	2 35 40	6 48 8	11.65	1.61	1.12		B	JOHNSON
0996	3 16 44.1	3 11 15.2	3 18 33	3 18 53	4.82	0.68	0.18	G5V	KAP CET	JOHNSON
1030	3 22 7.1	8 51 50.8	3 24 9	8 58 38	3.59	0.89	0.62	G8III	OMI TAU	JOHNSON
1046	3 26 10.5	55 16 30.6	3 28 51	55 24 5	5.08	0.05	0.03	A1V		JOHNSON
1346	4 16 56.7	15 30 36.8	4 18 55	15 35 32	3.65	0.99	0.82	K0III	GAM TAU	JOHNSON
1373	4 20 41.6	17 25 36.8	4 22 3	17 30 30	3.76	0.98	0.82	K0III	DEL TAU	JOHNSON
1409	4 25 41.6	19 5 4.4	4 27 44	19 8 54	3.54	1.02	0.88	K0III	EPS TAU	JOHNSON
1543	4 47 7.4	6 52 32.3	4 49 0	6 56 8	3.19	0.45	-0.01	F8V	PI3 ORI	JOHNSON
1552	4 48 32.4	5 31 16.3	4 50 24	5 34 48	3.69	-0.17	-0.80	B2III	PI4 ORI	JOHNSON
1641	5 0 8.2	41 10 8.4	5 5 27	41 12 57	3.17	-0.18	-0.67	B3V	ETA AUR	JOHNSON
1790	5 22 26.8	6 18 21.6	5 24 19	6 20 13	1.64	-0.23	-0.87	B2III	GAM ORI	JOHNSON
1791	5 23 7.7	28 34 1.7	5 25 20	28 35 50	1.65	-0.13	-0.49	B7III	BET TAU	JOHNSON
2010	5 46 44.3	12 38 13.6	5 48 42	12 38 51	4.90	-0.07	-0.18	B9IV	134 TAU	JOHNSON
	5 28 51.7	17 40 12.0	6 30 54	17 38 40	9.63	1.50	1.18		17 1320	JOHNSON
2421	6 34 49.4	16 26 37.3	6 36 50	16 24 48	1.93	0.00	0.03	A0IV	GAM GEM	JOHNSON
2763	7 15 13.2	16 37 56.1	7 17 14	16 34 6	3.58	0.11	0.10	A3V	LAM GEM	JOHNSON
	7 19 35.0	5 39 42.0	7 21 27	5 35 39	9.82	1.56	1.12		5 1668	JOHNSON
2852	7 25 53.8	31 53 8.3	7 28 30	31 48 48	4.16	0.32	0.03	F0V	RHO GEM	JOHNSON
2985	7 41 25.9	24 31 10.5	7 43 32	24 26 7	3.57	0.93	0.68	G8III	KAP GEM	JOHNSON
3249	8 13 48.3	9 20 27.7	8 15 42	9 13 58	3.52	1.48	1.78	K4III	BET CNC	PRIMARY
3454	8 40 36.7	3 34 45.7	8 42 26	3 27 11	4.30	-0.19	-0.74	B3V	ETA HYA	PRIMARY
3565	8 55 47.6	48 14 21.8	8 58 12	48 6 12	3.15	0.18	0.07	A7V	IOT UMA	JOHNSON
3665	9 11 45.8	2 31 34.6	9 13 34	2 22 51	3.88	0.86	0.13	A0P	THE HYA	JQHNSON
3815	9 32 40.9	36 2 14.7	9 34 47	35 52 51	5.41	0.77	0.45	G8IV-	11 LMI	JOHNSON
3974	10 4 29.2	35 29 21.4	10 6 32	35 19 5	4.48	0.18	0.08	A7V	21 LMI	JOHNSON

HR	RA (h m s)	Dec (° ′ ″)	RA (h m s)	Dec (° ′)	V	B-V	U-B	Sp	Name	Source
3982	10 5 42.6	12 12 44.5	10 7 34	2 26	1.36	-0.11	-0.36	B8V	ALF LEO	JOHNSON
4033	10 14 5.4	43 10.8 53.5	10 16 12	42 24	3.06	0.03	-0.06	A2IV	LAM UMA	JOHNSON
4133	10 21 33.4	37 54 54.2	10 23 21	59 15	9.63	1.52	1.19		1 2447	JOHNSON
4456	10 30 10.8	6 33 52.2	10 32 1	23 2	3.85	-0.14	-0.95	B1IB	RHO LEO	JOHNSON
4534	11 32 6.4	17 4 24.6	11 33 55	16 52 47	5.95	-0.16	-0.64	B3V	90AB LEO	JOHNSON
4540	11 46 30.6	14 2 5.8	11 48 18	13 39 25	2.14	-0.09	-0.07	A3V	BET LEO	JOHNSON
4550	11 48 5.4	2 38 47.6	11 49 53	51 4	3.61	0.10	0.17	F8V	BET VIR	JOHNSON
4554	11 51 12.6	51 57 39.2	11 53 2	37 52 58	6.45	0.55	0.75	G8VP		JOHNSON
4660	11 52 57.0	58 18 22.0	11 55 40	57 46 49	2.44	0.08	0.01	A0V	GAM UMA	JOHNSON
4931	12 53 35.4	38 36.9	12 53 49	56 26 56	3.31	0.40	0.01	A3V	DEL UMA	JOHNSON
4983	13 9 32.4	7 52.0	13 11 4	27 49 43	4.93	0.36	0.07	F2V	78 UMA	JOHNSON
5052	13 21 55.8	25 55.8	13 23 19	54 59 57	4.28	0.57	0.08	G0V	BET COM	JOHNSON
5072	13 25 59.0	52 18.2	13 27 42	51 50 50	4.01	0.16	-0.26	A5V	70 VIR	JOHNSON
5235	13 52 43.1	18 43.1	13 45 45	13 28 33	4.98	0.71	0.19	G5V	80 UMA	JOHNSON
5511	14 43 6.0	2 9.0	14 45 29	18 57 20	2.69	0.58	-0.03	G0IV	ETA BOO	JOHNSON
5854	15 41 48.2	15 34 53.9	15 43 31	15 43 20	3.74	0.00	-0.03	A0V	109 VIR	JOHNSON
5867	15 43 52.7	15 45 30.7	15 45 29	26 16 6	2.65	1.17	1.24	K2III	ALF SER	PRIMARY
5868	15 44 0.0	15 49 24.8	15 45 44	28 53 55	3.67	0.06	0.06	A2IV	BET SER	JOHNSON
5933	15 54 8.5	27 1 17.4	15 55 55	7 15 44	4.43	0.48	0.10	G0V	LAM SER	JOHNSON
5947	15 55 30.9	46 25 53.6	15 57 36	26 43 20	3.85	0.23	-0.03	F6V	GAM SER	JOHNSON
5092	16 18 14.2	23 21 10 36.7	17 25 19	55 18 16	4.15	1.23	1.28	K3III	EPS CRB	PRIMARY
6556	17 23 36.7	12 35 41.8	17 34 13	46 20 53	3.89	1.36	-0.56	B5IV	TAU HER	PRIMARY
6603	17 32 41.0	35 11.8	17 34 42	42 8 24	7.54	1.26	0.10	K7V	157881	JOHNSON
6629	17 41 0.0	12 35 11.8	17 42 43	43 4 20	2.08	0.15	0.10	A5III	ALF OPH	JOHNSON
7001	17 50 23.0	2 43 28.3	17 52 26	42 34 16	2.77	1.16	1.24	K2III	BET OPH	JOHNSON
7178	18 35 14.7	38 44 9.6	18 36 25	4 46 46	3.75	0.04	0.04	A0V	GAM OPH	JOHNSON
7235	18 57 4.3	32 37 11.8	18 58 22	2 42 28	9.54	1.74	1.29	M5V	4 3561	JOHNSON
7557	19 31 7.2	3 44 54.0	19 32 43	38 45 6	0.04	0.00	-0.01	A0V	ALF LYR	JOHNSON
7602	19 48 20.6	6 39 7.8	19 50 1	32 40 29	3.25	-0.05	-0.09	B9III	GAM LYR	JOHNSON
7906	20 37 51.4	51 44 5.8	20 38 54	13 59 17	2.99	1.08	1.00	A0V	ZET AQL	JOHNSON
8522	22 37 18.9	16 44 49.8	22 38 35	5 43 42	9.13	1.16	1.49	M3.5V	4 4048	JOHNSON
8781	23 2 16.2	10 47 22.4	23 4 2	6 49 27	6.82	0.02	0.83	B5	184279	JOHNSON
8832	23 10 51.9	56 53 31.3	23 12 12	18 22 23	0.77	0.22	0.08	A7IV-	BET AQL	JOHNSON
8832	23 10 51.9	56 53 31.3	23 12 12	18 51 19	3.71	0.86	0.48	G8IV	ALF DEL	JOHNSON
				38 58 29	3.77	-0.06	-0.22	B9V	10 LAC	JOHNSON
				15 7 57	4.88	-0.28	-1.04	O9V	ALF PEG	PRIMARY
					2.49	-0.05	-0.06	B9V		JOHNSON
					5.57	1.01	-0.89	K3V		PRIMARY
8969	23 37 22.6	5 21 18.6	23 39 9	5 32 57	4.13	0.51	0.00	F7V	IOT PSC	JOHNSON
	23 46 35.6	2 8 11.7	23 48 23	2 19 52	8.98	1.48	1.09	M2V	1 4774	JOHNSON

TABLE C.2. Southern UBV Standard Field Stars

HR	R.A. (1950.0)	DEC.	R.A. (1985.0)	DEC.	V	B-V	U-B	SPEC.	NAME	OBS.
0331	1 5 31.0	-41 45 14.2	1 7 6	-41 34 1	5.21	0.16	0.08	A3	UPS PHE	COUSINS
0370	1 12 55.9	-45 47 53.1	1 14 28	-45 36 47	4.96	0.57	0.10	G0	NU PHE	COUSINS
	1 14 46.8	-42 47 43.9	1 16 20	-42 36 40	7.86	-0.08	-0.37	B9	HD7795	COUSINS
0411	1 22 31.7	-44 47 18.1	1 24 2	-44 36 22	6.27	1.14	1.08	K0		COUSINS
	1 32 13.7	-45 56 52.2	1 33 42	-45 46 7	6.92	0.90	0.48	G5	HD9733	JOHNSON
0509	1 41 44.7	-16 12 0.5	1 43 26	-16 1 20	3.50	0.72	0.20	G5	TAU CET	JOHNSON
	1 58 0.2	-18 18 30.0	1 59 40	-18 8 20	10.18	1.53	1.16	GBVP	-18 0359	JOHNSON
0875	2 54 6.9	-3 54 45.0	2 55 52	-3 46 18	5.17	0.08	0.05	A1V		PRIMARY
	3 25 47.0	-76 55 9.9	3 24 58	-76 47 51	6.80	0.20	0.10	A0	HD21940	COUSINS
1084	3 30 34.4	-9 37 34.8	3 32 15	-9 30 31	3.73	0.89	0.57	K2V	EPS ERI	JOHNSON
	3 36 47.6	-74 8 17.8	3 36 22	-74 1 26	7.61	1.14	1.01	K0	HD23128	COUSINS
	3 48 26.1	-45 32 8.6	3 49 33	-45 25 49	6.94	0.94	0.70	G5	HD24291	COUSINS
	3 49 3.4	-74 50 43.9	3 48 26	-44 44 23	7.12	0.39	-0.05	F2	HD24636	COUSINS
	4 0 40.2	-48 48 3.9	4 1 47	-42 42 16	8.20	0.13	0.15	A2	HD25653	COUSINS
	4 7 0.7	-45 59 46.3	4 8 5	-45 54 15	6.58	0.38	0.00	F0		COUSINS
1291	4 10 56.0	-44 29 42.2	4 12 2	-44 24 22	6.71	1.48	1.80	K0	HD27728	COUSINS
1316	4 15 55.2	-75 55 50.9	4 14 55	-75 50 41	7.23	1.64	1.97	K0	HD27471	COUSINS
	4 16 27.9	-41 33 20.4	4 17 32	-41 15 15	7.54	0.64	0.17	G0	HD29751	COUSINS
	4 35 9.1	-73 18 39.5	4 34 31	-73 14 24	6.81	0.96	0.66	K0		COUSINS
1666	5 5 23.4	-8 58 58.5	5 7 6	-8 50 15	2.80	0.13	0.18	A3III	BET ERI	JOHNSON
1781	5 21 8.8	-12 18.7	5 22 56	-0 10 23	5.70	-0.22	-0.87	B5V	35299	JOHNSON
	5 28 55.3	-3 41 4.1	5 30 39	-3 39 31	7.97	1.47	1.21	K2	36395	JOHNSON
1855	5 29 30.6	-7 20 12.9	5 31 11	-7 18 42	4.63	-0.26	-1.07	B0V	UPS ORI	JOHNSON
1861	5 30 9.5	-1 37 35.7	5 31 55	-1 36 7	5.35	-0.20	-0.94	B1V	36591	JOHNSON
1899	5 32 59.1	-5 56 28.1	5 34 41	-5 55 8	2.77	-0.25	-1.08	O9III	EPS ORI	JOHNSON
1903	5 33 40.5	-1 13 56.1	5 35 27	-1 12 38	1.70	-0.19	-1.04	B0IA	EPS ORI	JOHNSON
1998	5 44 41.3	-14 50 21.2	5 46 16	-14 49 36	3.55	0.10	0.06	A3V	ZET LEP	JOHNSON
	6 29 34.2	-43 40 51.4	6 30 37	-43 42 23	6.69	1.00	0.74	K0	HD46415	COUSINS
2462	6 30 17.6	-48 10 28.8	6 31 13	-48 12 2	4.94	0.88	0.61	K0		COUSINS
	6 43 47.5	-47 10 8.9	6 44 45	-47 12 23	7.22	-0.16	-0.74	B3		COUSINS
	6 46 47.2	-43 44 37.9	6 47 50	-43 47 1	7.42	-0.16	0.12	A3		COUSINS
2546	6 48 30.2	-45 23 24.0	6 49 31	-45 25 52	6.54	1.51	1.82	K0	HD49260	COUSINS
3314	8 23 9.7	-3 44 31.5	8 24 54	-3 51 23	3.90	-0.02	-0.02	A0V	HD49850	COUSINS
3654	9 9 15.4	-44 39 45.2	9 10 31	-44 48 22	4.98	0.23	-0.57	B5		JOHNSON
	9 17 1.1	-44 47 46.9	9 18 18	-44 56 39	7.20	1.11	0.94	K0	HD80527	COUSINS
3730	9 20 34.3	-45 50 0.5	9 21 51	-45 59 0	5.74	0.92	0.65	G5		COUSINS

HR	RA (1950) h m s	Dec (1950) ° ′ ″	RA (2000) h m	Dec (2000) ° ′	V	B-V	U-B	Sp	Name	System
4519	9 21 39.7	-48 4 19.1	9 22 53	-48 13 20	6.27	-0.14	-0.61	B5	HD81347	COUSINS
	9 24 15.7	-12 51 42.0	9 25 56	-13 0 49	10.06	1.53	1.15	F5	-12 2918	JOHNSON
4601	9 27 27.2	-42 58 13.0	9 28 47	-43 7 26	6.60	0.48	0.06	G0	HD82224	COUSINS
4624	11 40 1.3	-74 56 58.5	11 41 34	-75 8 37	6.47	0.52	0.05	B8	HD101805	COUSINS
	11 43 15.0	-45 24 44.4	11 44 52	-45 46 24	5.29	-0.12	-0.55	G5	HD103281	COUSINS
4662	11 51 4.2	-46 21 50.2	11 52 58	-46 39 31	7.22	1.04	0.83	F8	HD104138	COUSINS
	12 1 8.5	-73 56 16.2	12 2 37	-74 14 49	6.66	0.56	0.09	K0	HD106321	COUSINS
4804	12 6 18.1	-44 45 7.8	12 8 57	-44 32 31	6.45	1.22	0.98	A2	GAM CRV	COUSINS
5019	12 25 52.0	-17 45 9.3	12 13 15	-14 38 25	5.75	1.24	1.58	K0	HD107145	JOHNSON
5056	12 13 26.3	-76 26 3.4	12 15 18	-17 27 32	5.31	-0.11	-0.35	BBIII	HD107547	COUSINS
5401	12 51 1.3	-73 35 3.6	12 54 15	-43 25 14	2.60	-0.44	-0.01	F5	0 2989	COUSINS
	12 59 21.8	-42 22 35.6	13 0 27	-73 5 17	6.84	0.16	-0.14	A2	61 VIR	JOHNSON
5509	12 21 59.8	-75 58 42.5	13 17 40	-40 2 13	6.78	-0.09	-0.23	B9	ALF VIR	JOHNSON
5530	12 35 9.7	- 0 26 54.0	12 35 49	-11 8 51	6.48	1.41	1.26	M0.5V	ALF1 LIB	JOHNSON
5531	13 47 1.5	-18 29 9.7	13 48 56	-14 15 3	8.49	0.71	0.25	G6V	ALF2 LIB	COUSINS
5528	13 33 3.3	-10 1 29.4	13 54 24	-15 58 45	4.75	-0.23	-0.94	B1V	OMI LUP	COUSINS
5580	13 58 3.3	-45 22 26.3	13 54 26	-43 6 49	0.96	-0.31	-0.14	A3	BET LIB	JOHNSON
5685	13 21 3.9	-46 33 19.9	14 35 19	-46 10 1	5.82	1.49	1.71	K2	-7 4003	COUSINS
	14 16 4.6	-43 54 6.8	14 46 33	- 7 1 40	5.54	1.08	0.88	G5	-12 4523	COUSINS
6175	14 47 55.0	-15 54 25.0	14 49 29	-15 45 48	6.30	0.41	-0.04	F5IV	ZET OPH	PRIMARY
	14 48 6.4	- 6 41 11.1	14 50 17	-15 58 5	5.16	-0.15	-0.08	AM	154363	JOHNSON
6371	14 21 9.7	-43 42 39.6	14 50 30	-43 6 45	2.75	-0.16	-0.62	B5	-4 4226	JOHNSON
6427	15 9 1.7	-42 7 58.9	15 11 58	-44 41 49	4.32	-0.60	-0.28	F8	HD157487	JOHNSON
6460	15 52 1.2	- 7 8 4.4	16 4 16	- 7 3 1	6.10	-0.11	-0.37	B8V	HD159656	JOHNSON
	15 13 52.7	- 4 4 7.9	16 9 7	-12 22 53	2.61	1.61	1.20	O9.5V	HD159656	COUSINS
	16 22 27.0	-12 52 0.1	16 24 36	-32 7 16	10.56	1.60	1.18	K5V	-3 4233	COUSINS
	16 34 24.1	- 4 38 8.4	16 36 20	- 5 1 48	10.30	0.02	-0.86	M3.5V	KAP AQL	COUSINS
7446	16 57 39.6	- 4 4 58.6	17 2 59	- 4 44 5	2.56	1.16	1.05	G5	HD186502	COUSINS
7498	17 15 46.9	-44 44 41.0	17 18 30	-44 32 30	7.73	1.43	0.88	A0		
	17 20 35.3	-44 44 44.4	17 24 19	-44 48 19	10.07	0.02	0.56	B8		
	17 22 7.0	-44 48 42.7	17 24 36	-42 28 56	6.64	1.16	0.05	K0		
	17 34 18.3	-42 3 8.4	17 36 4	- 3 19 5	5.11	1.25	0.48	G5		
	18 2 28.3	- 7 7 1.7	18 32 33	-7 32 14	7.17	1.52	1.14	K5		
	19 34 12.1	-72 37 52.6	19 37 41	-72 30 42	9.38	0.01	0.65	B0.5I		
	19 41 13.3	-72 55 24.7	19 47 32	-72 3 41	4.96	-0.24	1.21	A3		
	19 45 27.3	-72 3 15.9	19 49 55	-72 49 59	5.39	0.24	-0.87	F8		
					7.30	0.46	-0.10			

SOUTHERN UBV FIELD STANDARD STARS

HR	R.A. (1950.0)	DEC.	R.A. (1985.0)	DEC.	V	B-V	U-B	SPEC.	NAME	OBS.
7590	19 54 50.8	-73 2 43.8	19 58 51	-72 57 1	3.95	-0.03	-0.06	A0	EPS PAV	COUSINS
	19 58 39.6	-44 36 21.4	20 1 7	-44 30 30	7.91	-0.04	-0.25	A0	HD189502	COUSINS
	19 59 4.0	-45 20 10.2	20 1 32	-45 14 18	6.57	1.22	1.22	K0	HD189563	COUSINS
	20 7 48.0	-43 48 40.3	20 10 13	-43 42 25	6.54	0.88	0.57	G5	HD191349	COUSINS
	20 12 51.2	-73 8 3.7	20 16 46	-73 1 34	6.56	1.41	1.67	K2	HD191937	COUSINS
	20 12 56.8	-72 57 50.3	20 16 50	-72 51 20	6.93	1.03	0.87	K0	HD191973	COUSINS
	20 14 50.3	-43 0 55.0	20 17 13	-42 54 22	7.01	0.33	0.02	F0	HD192758	COUSINS
	20 15 14.2	-42 46 30.8	20 17 37	-42 39 57	7.45	0.59	0.16	G0	HD192826	COUSINS
7950	20 44 58.2	-9 40 48.2	20 46 51	-9 33 3	3.77	0.01	0.04	A1V	EPS AQR	JOHNSON
	22 27 18.1	-42 32 51.5	22 29 22	-42 22 5	6.92	-0.05	-0.08	A0	HD213155	COUSINS
	22 29 13.8	-43 31 14.2	22 31 18	-43 20 25	6.91	0.98	0.77	G5	HD213457	COUSINS
8657	22 42 44.1	-46 48 38.1	22 44 47	-46 37 34	5.51	1.32	1.43	K0		COUSINS
8662	22 43 46.7	-47 12 12.3	22 45 50	-47 1 8	6.56	0.30	0.10	A5		COUSINS
	22 44 31.9	-45 13 44.8	22 46 34	-45 2 39	7.22	0.60	0.07	G0	HD215657	COUSINS
	22 45 27.5	-15 0 42.0	22 47 18	-14 49 36	10.17	1.60	1.15		-15 6290	JOHNSON
8704	22 50 50.8	-11 52 58.3	22 52 41	-11 41 47	5.81	-0.08	-0.32	B9	74 AQR	JOHNSON

APPENDIX D
JOHNSON UBV STANDARD CLUSTERS

When Johnson and Morgan created the *UBV* system, they included a list of standard stars. This list covered the entire sky visible from the United States and is used by some observers as the primary method of calibrating their instrumental magnitudes. However, at the same time, Johnson observed stars in or near three open clusters. These standard clusters were also to be used in determining the transforming coefficients.

The advantages of using star clusters to obtain coefficients include:

1. First-order extinction corrections are very small because all stars are within 1° of each other.
2. It is easy to observe many stars quickly because they can be located on a single finding chart and only small movements are required to change from one star to another.

The disadvantages of using one of these standard clusters include:

1. Red stars are almost always fainter than blue ones, because most stars are on the main sequence and are at the same distance.
2. A clear spot for obtaining a sky reading can be hard to find.
3. The density of stars makes the use of a large diaphragm difficult because "companions" become included.
4. Transformation difficulties can exist because each cluster is in an isolated region in the sky and was placed on the standard system differently than the whole-sky standard list.

Still, for many observers the cluster method of calculating color coefficients is highly useful. We recommend that you use both the whole-sky and the cluster methods sometime during your observing season.

In observing these clusters, Johnson picked from one to three regional standard stars, and measured all cluster stars with respect to these standards. He then tied the regional standards to the *UBV* system by using the whole-sky standards. Therefore, the clusters are placed on the *UBV* system in a two-step process, though with exactly the same equipment which was used to define the *UBV* system.

Each cluster is discussed in a similar manner. There is a brief description of the cluster and of the pertinent references. This is followed by a list of 26 stars for each cluster. These stars were picked from the much larger lists by Johnson to include stars over a wide magnitude and spectral range, and with as many Johnson observations as possible.

Each star's coordinates are taken from the SAO catalog, or from the catalog reference if the star is fainter than tenth magnitude. The stars are named on the right by a letter of the alphabet and the zone BD number where available. The last item for each cluster is a finding chart with 26 stars identified.

D.1 THE PLEIADES

This famous naked-eye cluster is often called the Seven Sisters. It is a very young cluster, containing many bright blue stars and nebulosity. There are about 600 stars brighter than fourteenth magnitude in the central 1° field.

Because of the youth of the cluster, only three red stars were observed that are brighter than tenth magnitude. This places a large weight on their measurement, especially star D, and this should be recognized by the observer. Table D.1 lists the information for this cluster.

Approximately 250 stars were observed with the 53- and 103-centimeter (21- and 42-inch) Lowell reflectors. All stars in the region were observed differentially with respect to the regional standard stars E, I, and Alcyone.

Catalog:
  Hertzsprung, E., 1947. *Ann. Leiden Obs.* **19**, 1A.
Photometry:
  Johnson, H. L., and Morgan, W. W. 1953. *Ap. J.* **117**, 313.
  Johnson, H. L., and Mitchell, R, I. 1958. *Ap. J.* **128**, 31.
Chart:
  Red Palomar chart MLP 357 (E-641).

D. 2 THE PRAESEPE

The Praesepe, M44, or the Beehive, is one of the largest and nearest open clusters. It is visible to the naked eye, though not as easily as the Pleiades, and has several hundred members.

TABLE D.1. Pleiades Cluster Standards

HR	R.A. (1950.0)	DEC.	R.A. (1985.0)	DEC.	V	B-V	U-B	SPEC.	NAME	OBS.
1165	3 44 30.4	23 57 7.6	3 46 35	24 3 35	2.87	-0.09	-0.34	B7II	A	JOHNSON
1145	3 42 13.6	24 18 42.9	3 44 18	24 25 17	4.31	-0.11	-0.46	B6V	B	JOHNSON
1172	3 45 22.9	23 16 8.8	3 47 27	23 22 34	5.45	-0.07	-0.32	B8V	C	JOHNSON
	3 45 7.0	24 50 9.1	3 47 12	24 56 35	6.46	1.70	2.07	K5	D	JOHNSON
	3 44 0.3	24 22 0.4	3 46 5	24 28 30	6.82	0.03	-0.07	B9V	E	JOHNSON
	3 45 31.1	24 11 36.4	3 47 36	24 18 2	6.95	0.12	0.09	A0	F	JOHNSON
	3 47 28.8	24 20 42.1	3 49 34	24 27 2	7.42	0.13	-0.12	A2	G	JOHNSON
	3 43 38.7	23 27 23.1	3 45 43	23 33 51	7.72	1.23	1.12	K0	H	JOHNSON
	3 43 36.2	23 28 12.2	3 45 40	23 34 42	8.11	0.35	0.29	A0	I	JOHNSON
	3 42 35.6	24 18 30.4	3 44 40	24 25 3	8.60	0.35	0.11	A2	K	JOHNSON
	3 44 44.2	23 23 26.6	3 46 48	23 29 54	8.79	1.15	0.81	K0	L	JOHNSON
	3 46 55.8	24 4 2.5	3 49 0	24 10 24	9.16	0.16	0.15	A0	M	JOHNSON
	3 44 11.1	24 7 26.0	3 46 16	24 13 55	9.46	0.47	0.02	F8	N	JOHNSON
	3 42 41.0	24 28 22.0	3 44 46	24 34 55	9.70	0.55	0.05	F8	O	JOHNSON
	3 45 27.5	23 53 48.0	3 47 32	24 0 13	10.02	0.56	0.09		P	JOHNSON
	3 43 55.7	23 25 50.0	3 46 0	23 32 19	10.52	0.64	0.16	F9	Q	JOHNSON
	3 46 25.5	23 41 19.0	3 48 30	23 47 42	11.35	0.78	0.38	G2	R	JOHNSON
	3 43 42.3	23 20 40.0	3 45 46	23 27 10	12.02	0.99	0.54	G8	S	JOHNSON
	3 43 8.3	24 42 47.0	3 45 13	24 49 18	12.05	1.01	0.84	G5	T	JOHNSON
	3 41 5.0	24 14 17.0	3 43 9	24 20 54	12.51	0.81	0.30	G1	U	JOHNSON
	3 44 19.0	24 14 18.0	3 46 24	24 20 46	12.61	1.18	1.00	G9V	V	JOHNSON
	3 43 7.7	25 0 51.0	3 45 13	25 7 22	14.36	1.01	0.47		X	JOHNSON
	3 46 26.1	23 49 58.0	3 48 30	23 56 21	15.72	1.15	0.98		Y	JOHNSON
	3 44 39.0	24 34 52.0	3 46 44	24 41 20	16.42	0.60	0.25		Z	JOHNSON

300 ASTRONOMICAL PHOTOMETRY

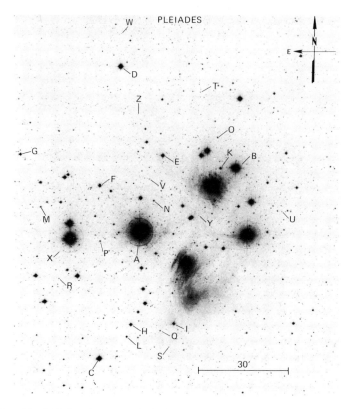

Figure D.1. The Pleiades cluster standards. Copyright by the National Geographic Society—Palomar Observatory Sky Survey. Reproduced by permission.

Johnson made observations of 150 stars in the region of the Praesepe cluster. The 26 stars listed in this table are therefore a small subset of the total. Two regional standards were used in this cluster, stars B and F. All stars in the cluster were then observed differentially with respect to these two standards. Table D.2 lists the information for this cluster.

Catalog:
  Klein-Wassink, W. J. 1927. *Pub. of Kapetyn Astronomical Laboratory at Gröningen,* No. 41, 5.
  Vanderlinden, H. L. 1933. *Étude de l'amas de Praesepe.* Gembloux: Joules Duculot.

Photometry:
  Johnson, H. L. 1952. *Ap. J.* **116**, 640.

Chart:
  Blue Palomar chart MLP 426 (O-1311).

TABLE D.2. Praesepe Cluster Standards

HR	R.A. (1950.0)	DEC.	R.A. (1985.0)	DEC.	V	B-V	U-B	SPEC.	NAME	OBS.
3429	8 37 35.1	19 43 23.1	8 39 35	19 55	6.30	0.17	0.16	A2	A,2171	JOHNSON
3428	8 37 14.0	20 11 8.0	8 39 14	20 3 41	6.39	0.98	0.83	K0II	B,2158	JOHNSON
3428	8 37 30.0	19 50 52.9	8 39 30	19 43 25	6.44	1.02	0.90	K0II	C,2166	JOHNSON
	8 32 27.1	19 45 48.3	8 34 27	19 38 32	6.53	0.67	0.25	G0II	D,2118	JOHNSON
	8 36 58.7	19 43 6.6	8 38 59	19 35 40	6.59	0.96	0.72	K0II	E,2150	JOHNSON
	8 37 19.1	20 8 56.6	8 39 19	20 1 29	6.61	0.01	0.02	A0	F,2159	JOHNSON
	8 38 4.4	19 45 32.3	8 40 4	19 38 3	6.78	0.17	0.14	A6V	G,2175	JOHNSON
	8 37 51.1	19 53 51.6	8 39 51	19 46 23	6.85	0.20	0.15	A9II	H,2172	JOHNSON
	8 38 57.9	20 3 13.1	8 40 58	19 55 42	6.90	0.96	0.74	K0II	J,2185	JOHNSON
	8 37 26.2	19 42 36.1	8 39 26	19 35 8	7.54	0.16	0.13	F0II	K,2163	JOHNSON
	8 36 17.1	19 46 10.0	8 38 17	19 38 45	8.50	0.25	0.07	A9V	L,2144	JOHNSON
	8 35 54.9	19 40 38.8	8 37 55	19 33 15	9.00	0.32	0.03	A5	M,2139	JOHNSON
	8 37 8.7	20 18 48.9	8 39 9	20 11 22	9.67	0.44	-0.02	F6V	N,2156	JOHNSON
	8 35 39.9	19 38 30.0	8 37 40	19 31 6	10.01	1.01	0.78	G5	P,2056	JOHNSON
	8 37 2.7	20 14 34.0	8 37 3	20 7 7	10.11	0.49	-0.00		Q	JOHNSON
	8 38 33.7	20 7 22.0	8 40 34	19 59 52	10.72	0.60	0.10		R,2181	JOHNSON
	8 37 48.2	19 50 53.0	8 39 48	19 43 24	10.87	0.68	0.19		S	JOHNSON
	8 36 22.1	20 23 17.0	8 38 23	20 15 52	11.31	0.70	0.24		T	JOHNSON
	8 38 42.0	20 8 54.0	8 40 42	20 1 23	11.71	0.78	0.38		U	JOHNSON
	8 36 13.8	20 28 55.0	8 38 21	20 21 30	12.37	0.46	-0.05		V	JOHNSON
	8 36 37.3	29 57 51.0	8 38 37	19 50 25	12.64	1.00	0.76		W	JOHNSON
	8 38 24.3	19 55 37.0	8 40 24	19 48 7	13.70	0.81	0.40		X	JOHNSON
	8 37 13.0	19 56 12.0	8 39 13	19 48 45	14.61	0.95	0.69		Y	JOHNSON
	8 36 23.0	20 14 51.0	8 38 23	20 7 26	14.94	1.48	1.10		Z	JOHNSON

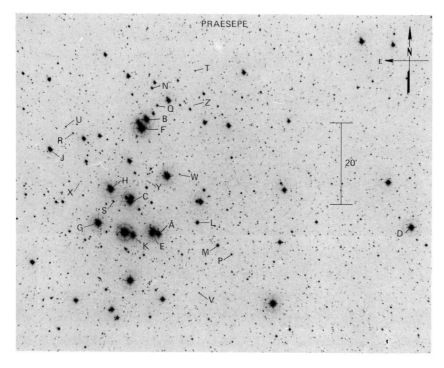

Figure D.2. The Praesepe cluster standards. Copyright by the National Geographic Society—Palomar Observatory Sky Survey. Reproduced by permission.

D.3 IC 4665

The standard region near the open cluster IC 4665 was observed by Johnson in 1954. The cluster itself contains a few blue stars, some of which were measured, but most of the standard stars in this region are field stars within 1° of the cluster center.

All stars in the region were measured differentially with respect to the regional standard, star A. Some of these stars were observed at McDonald and some at Mount Wilson observatories, but most were measured at Lowell observatory. Table D.3 lists the information for this cluster.

This region is interesting because several reasonably bright red stars are included, thereby reducing the error in the color transformation that can be troublesome on the other two standard regions.

Catalog:
Kopff. 1943. *Mitt. Hamburg Sternw.* **8**, 93.

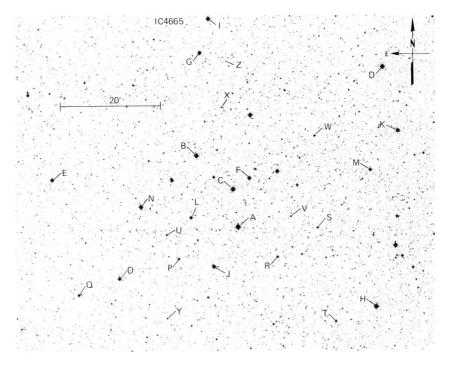

Figure D.3. The IC 4665 cluster standards. Copyright by the National Geographic Society—Palomar Observatory Sky Survey. Reproduced by permission.

Photometry:
Johnson, H. L. 1954. *Ap. J.* **119**, 181.

Chart:
Red Palomar chart MLP 569 (E-780).

TABLE D.3. IC 4665 Regional Standards

HR	R.A. (1950.0)	DEC.	R.A. (1985.0)	DEC.	V	B-V	U-B	SPEC.	NAME	OBS.
17	43 40.1	5 32 54.5	17 45 23	5 32 7	6.85	0.01	-0.54	B4V	A,3483	JOHNSON
17	44 14.0	5 47 31.0	17 45 56	5 46 45	7.12	0.02	-0.48	B8	B,3490	JOHNSON
17	43 43.7	5 40 35.2	17 45 26	5 39 48	7.34	0.02	-0.46	B9	C,3484	JOHNSON
17	41 36.9	6 4 51.4	17 43 19	6 3 57	7.43	0.33	0.15	A2	D,3514	JOHNSON
17	46 16.3	5 42 59.4	17 47 59	5 42 20	7.49	0.02	-0.41	B9	E,3504	JOHNSON
17	43 29.9	5 42 46.4	17 45 12	5 41 58	7.59	0.00	-0.49	B6V	F,3432	JOHNSON
17	44 9.8	6 8 18.1	17 45 52	6 7 32	7.74	1.28	0.55	B9	G,3525	JOHNSON
17	41 45.4	6 16 16.6	17 43 28	6 15 23	7.83	1.10		K2	H,3469	JOHNSON
17	44 2.1	6 15 14.2	17 45 43	6 14 27	7.89	1.03	0.77		I,3524	JOHNSON
17	44 0.8	5 24 54.8	17 45 43	5 24 8	7.94	0.45	0.01	KØ	J,3488	JOHNSON
17	41 24.8	5 51 57.2	17 43 7	5 51 12	8.05	0.07	-0.17	FØ	K,3466	JOHNSON
17	41 19.4	5 34 57.4	17 43 2	5 34 13	8.22	0.06	-0.30	AØ	L,3491	JOHNSON
17	41 48.6	5 44 6.7	17 43 31	5 43 13	8.31	1.73	2.08	B9	M,3471	JOHNSON
17	45 1.7	5 37 14.6	17 46 44	5 36 31	8.33	1.23	1.04	B9	N,3498	JOHNSON
17	45 21.0	5 22 43.3	17 47 1	5 22 1	8.40	0.11	-0.27	K5	O,3500	JOHNSON
17	44 30.3	5 26 35.1	17 46 13	5 25 50	8.89	1.07		K2	P,3493	JOHNSON
17	45 55.3	5 19 32.0	17 47 38	5 18 51	8.96	1.25		B9V	Q,3503	JOHNSON
17	43 7.4	5 26 44.1	17 44 50	5 25 55	9.10	0.26	0.10		R,3479	JOHNSON
17	42 33.6	5 32 33.6	17 44 16	5 31 42	9.39	0.31	-0.17	A2V	S,3473	JOHNSON
17	42 19.0	5 13 14.0	17 44 2	5 12 22	9.68	1.27	1.04		T,3472	JOHNSON
17	44 41.0	5 31 39.0	17 46 24	5 30 54	9.81	0.68	0.23		U,3496	JOHNSON
17	42 55.0	5 35 1.0	17 44 37	5 34 11	10.10	0.12	0.01	AØ	V,3477	JOHNSON
17	42 34.0	5 52 46.0	17 44 16	5 51 55	10.21	1.29	1.27		W,3474	JOHNSON
17	43 52.0	5 57 21.0	17 45 34	5 56 34	10.61	0.45	0.15	A2	X,3485	JOHNSON
17	44 41.0	5 13 15.0	17 46 24	5 12 30	10.75	0.37	0.22		Y,3495	JOHNSON
17	43 45.0	6 6 44.0	17 45 27	6 5 56	11.33	0.53	-0.05		Z,3523	JOHNSON

APPENDIX E
NORTH POLAR SEQUENCE STARS

The North Polar Sequence (NPS) was developed at Harvard College Observatory in 1906 to provide stars with standard photographic magnitudes that could be used to derive magnitudes for other stars. The sequence began with 10 and rapidly expanded to 96 stars. Three separate sequences evolved: red (r), blue (no suffix), and supplementary (s, mostly yellow). Mount Wilson collaborated in the latter stages, and the sequences formed the basis of the International System of magnitudes adopted by the IAU in 1922. See Pickering,[1] Leavitt,[2] and Stebbins et al.[3] for more detail.

The NPS has only historical significance now because it has been superseded by sequences in clusters and in the equatorial plane, accessible by major telescopes in both hemispheres. In addition, only photographic magnitudes and colors have been derived for the majority of the stars, and even these are in error for the fainter members of the NPS.

One modern application for which the NPS is well suited is to determine temporal consistency of the atmosphere. This is because none of the stars lies more than 3° from the north celestial pole, and therefore remain at the same air mass with time. It is a good idea to observe one of these stars several times during the night to be sure that the atmosphere is remaining constant.

Table E.1 lists the sequence stars that are brighter than eleventh magnitude on photovisual plates. The V magnitudes of those stars in the table with no $(U - B)$ color indicated are photovisual magnitudes, and the color index is photographic-photovisual. The accompanying finding chart identifies the table stars.

For amateurs in the Southern Hemisphere, a short sequence near the south celestial pole has been set up by Soonthornthum and Tritton.[4] Their photoelectric sequence includes nine stars ranging from $V = 6.53$ to $V = 12.66$. They include a finding chart in their article.

306 ASTRONOMICAL PHOTOMETRY

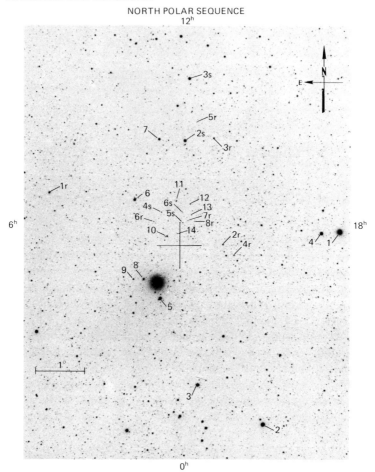

Figure E.1. The North Polar Region. Lick Observatory photograph.

REFERENCES

1. Pickering, E. C. 1912. *Harvard College Circular,* No. 170.
2. Leavitt, H. C. 1917. *Ann. Harvard College Obs.* **71**, 47.
3. Stebbins, J., Whitford, A. E., and Johnson, H. L. 1950. *Ap. J.* **112**, 469.
4. Soonthornthum, B., and Tritton, K. P. 1980. *Observatory* **100**, 4.

TABLE E.1. North Polar Sequence Stars

NP	(1950.0) R.A.	DEC.	(1985.0) R.A.	DEC.	V	B-V	U-B	SPEC.	NAME	OBS.
5	1 10 57.3	88 45 18.9	1 25 44	88 56 20	6.44	0.11	0.14	A2	BD88 4	JOHNSON
8	2 49 44.3	88 55 16.0	3 23 39	89 3 18	8.08	0.41	-0.06	F0	BD88 9	JOHNSON
9	3 18 47.5	88 45 56.0	3 51 12	88 52 51	8.89	0.20	0.00	A	BD88 13	NPS
10	7 12 31.2	89 44 52.0	9 16 26	89 38 9	9.05	0.12	0.00	A5	BD89 3	NPS
1R	7 17 50.8	87 7 34.5	7 33 54	87 13 17	5.04	1.57	0.00	M0	BD87 51	NPS
6R	8 22 39.7	89 21 30.6	9 11 39	89 13 40	9.24	1.23	0.00	G8	BD89 9	NPS
6	8 45 53.6	88 46 14.8	9 12 7	88 38 0	7.09	0.06	0.00	A0	BD89 13	NPS
4S	9 49 27.5	89 19 44.9	10 19 45	89 9 29	9.89	0.47	0.00	G0	BD89 12	NPS
7	11 13 59.6	87 54 43.3	11 19 36	87 43 14	7.55	-0.17	0.00	B8	BD88 64	NPS
14	11 23 15.7	89 51 54.0	11 46 14	89 40 44	10.57	0.45	0.00	F2	BD89 1	NPS
11	11 45 52.7	89 12 24.9	11 50 16	89 0 57	9.61	0.22	0.00	F0	BD89 18	NPS
2S	12 14 45.0	87 58 37.6	12 15 10	87 46 9	6.22	0.26	0.00	F2	BD88 71	NPS
3S	12 15 24.5	86 42 49.3	12 16 17	86 31 13	6.33	0.33	0.00	F0	BD87 107	NPS
6S	12 30 4.4	89 25 16.9	12 24 4	89 13 40	10.72	0.67	0.00	G8	BD89 26	NPS
5R	12 43 27.1	88 37 45.6	12 39 44	88 26 15	8.63	1.53	0.00	K0	BD89 22	NPS
3R	12 58 22.1	87 55 5.7	12 55 7	87 43 47	7.47	1.42	1.51	K2	BD88 76	JOHNSON
5S	13 1 40.6	89 38 11.9	12 41 40	89 26 47	10.06	1.08	0.00	G5	BD89 37	NPS
12	13 15 1.8	89 13 28.9	13 1 42	89 2 17	9.78	0.35	0.00	A3	BD89 25	NPS
7R	13 46 56.8	89 33 26.3	13 16 6	89 22 38	9.87	1.12	0.00	G8	BD89 35	NPS
13	13 52 20.5	89 25 1.7	13 26 12	89 14 23	10.27	0.24	0.00	A5	BD89 29	NPS
8R	14 47 18.5	89 29 12.7	14 4 12	89 19 45	10.41	1.02	0.00	G8	BD89 31	NPS
1	17 48 18.3	86 36 34.8	17 37 0	86 35 41	4.35	0.01	0.04	A1V	BD86 269	JOHNSON
4	17 49 12.2	86 59 31.9	17 36 14	86 58 39	5.74	0.14	0.00	A3	BD86 272	NPS
2R	18 21 21.8	89 3 3.5	17 36 2	89 2 59	6.34	1.56	0.00	M3	BD88 112	NPS
4R	18 58 29.5	89 46 55.6	17 36 38	88 49 1	8.22	1.02	0.00	K0	BD88 114	NPS
2	22 17 33.6	85 51 27.0	22 14 32	86 1 57	5.24	-0.02	-0.22	A0	BD85 383	JOHNSON
3	23 27 34.2	87 1 54.4	23 27 9	87 13 28	5.56	0.16	0.00	F0	BD86 344	NPS

APPENDIX F
DEAD-TIME EXAMPLE

This appendix gives an example of how to calculate the dead-time coefficient for a photomultiplier pulse-counting system. Section 4.2 presents the basis for this example and should be consulted for more detail.

The main requirement for determining the dead-time coefficient is a set of stars, some of which should have negligible correction and some with a large correction. For our system, this means stars with rates around 100,000 counts per second and 800,000 counts per second to bracket the changeover point. Our sample of eight stars should be considered the minimum necessary to determine the coefficient.

Several stars were measured with the neutral density filter used at Indiana University, which has the $v_1 - v_0$ relation:

$$v_1 - v_0 = -0.008(B - V) + 3.934 \qquad (F.1)$$

The data are listed in Table F.1. The steps required to obtain the coefficient follow.

1. Calculate $v_1 - v_0$ for each star from Equation F.1, using the known color index given in column 2.
2. Using Equation 4.7, calculate the intensity ratio $b \, (= I_0/I_1)$.
3. Using the observed count rate obtained with the filter in the light path (n_L) and the rate without the filter (n_H), calculate the true count rate without the filter using Equation 4.3a:

$$N_H = bn_L$$

TABLE F.1. Dead-Time Coefficient Data

	Collected Data				Calculated Results		
Star	$B - V$	n_L	n_H	$v_1 - v_0$	b	N_H	A
HR8334	0.51	10,670	374,200	3.930	37.32	398,200	0.0622
HR8465	1.55	25,170	824,700	3.922	37.04	932,200	0.1225
HR8469	0.23	5190	190,100	3.932	37.40	194,100	0.0208
HR8494	0.27	11,720	407,300	3.932	37.39	438,200	0.0731
HR8498	1.46	12,400	427,400	3.922	37.06	459,600	0.0726
HR8585	0.01	17,190	582,700	3.934	37.46	643,900	0.0999
HR8622	−0.20	6170	224,200	3.936	37.52	231,500	0.0320
HR8694	1.05	21,430	708,100	3.926	37.17	796,600	0.1178

4. Calculate the quantity

$$A = \ln\left(\frac{N_H}{n_H}\right)$$

for each star (where ln is the natural logarithm).

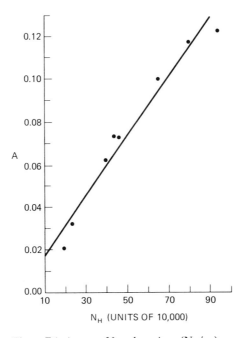

Figure F.1. A versus N_H, where $A = (N_H/n_H)$.

5. Now plot A versus N_H from Equation 4.4,

$$A = tN_H$$

The slope of the resultant line is t, the dead-time coefficient. The plot for this example is given in Figure F.1.

For our example using eight stars, the slope (using linear least squares) is $t = 1.40 \times 10^{-7}$ seconds. An example of the use of the dead-time coefficient can be seen in Appendix H.

APPENDIX G
EXTINCTION EXAMPLE

The approach used to determine the extinction coefficients depends, to some extent, on the type of observing program. In a program of differential photometry the atmospheric extinction corrections can usually be ignored. Occasionally, however, an extinction correction is advisable if the two stars are widely separated. In this case, the comparison star measurements themselves yield the extinction coefficients. When the observing program calls for measuring many different stars at various locations in the sky, it is necessary to follow another approach. In this case, a separate set of standard stars must be observed to determine the extinction coefficients. No matter which kind of observing program is conducted, another set of observations must be made to determine the second-order extinction coefficients.

We now take you through each of these procedures using actual photometric data.

G.1 EXTINCTION CORRECTION FOR DIFFERENTIAL PHOTOMETRY

Because the second-order extinction corrections applied to differential photometry are small enough to ignore, we concentrate on the principal extinction coefficients. As stated above, it is the comparison star measurements that allow us to find the principal extinction coefficients. The success of this method depends on the fact that the comparison star is usually observed through a large range of air mass. Second, the method assumes that the transmission of the atmosphere has remained constant and that the electronics remain free from gain drift throughout the night. These conditions are usually satisfied on a clear night and with well-designed electronics.

As the comparison star rises or sets, the amount of light detected changes as more or less light is absorbed by the earth's atmosphere. Using Equations 2.1 through 2.3 we can calculate the instrumental magnitude of the comparison

312 ASTRONOMICAL PHOTOMETRY

star through each filter at each air mass. These magnitudes can be corrected for atmospheric absorption by using Equation 4.20 for each filter, that is

$$v_0 = v - k'_v X \qquad \text{(G.1)}$$
$$b_0 = b - k'_b X \qquad \text{(G.2)}$$
$$u_0 = u - k'_u X \qquad \text{(G.3)}$$

where v, b, and u are the instrumental magnitudes, v_0, b_0, and u_0 are the instrumental magnitudes corrected for extinction and k'_v, k'_b, and k'_u are the principal extinction coefficients. These coefficients are, of course, unknown. However, a plot of v versus X yields a straight line slope k'_v. Similarly, plots of b versus X and u versus X yield the slopes k'_b and k'_u, respectively.

1. The comparison star used in this example has a right ascension of 2^h07^m and a declination of $40°23'$. The observatory has a latitude of $39°33'$ north. In Table G.1, column 1 contains the hour angle of each comparison star observation. Column 2 contains the air mass, X, calculated by Equations 4.17 and 4.18. Try calculating X to check a few entries in column 2. If you have difficulties, refer to the example in Section 4.4.
2. Columns 3, 4, and 5 of Table G.1 contain the counts per second recorded through each filter with the sky background subtracted. These have not been corrected for dead time because the count rates are rather low. If these observations had been made with a DC system, these three columns would contain the deflection (usually in percent of full scale) and the

TABLE G.1.   Comparison Star Data

Star: $\alpha = 2^h07^m$, $\delta = 40°23'$ Observatory: 39°33' north latitude

HA	X	Counts per Second			Instrumental Magnitudes		
		v Filter	b Filter	u Filter	v	b	u
2:42E	1.165	5660	7550	1413	−9.382	−9.695	−7.876
1:53	1.076	5854	7948	1530	−9.419	−9.751	−7.962
1:14	1.032	5878	8110	1596	−9.423	−9.772	−8.008
0:30	1.005	5887	8143	1638	−9.425	−9.777	−8.036
0:09W	1.001	5897	8088	1611	−9.427	−9.770	−8.018
0:14	1.001	5883	8083	1617	−9.424	−9.769	−8.022
1:00	1.021	5838	8006	1588	−9.416	−9.759	−8.002
1:43	1.062	5655	7821	1528	−9.381	−9.733	−7.961
2:24	1.127	5568	7511	1424	−9.364	−9.689	−7.884
3:15	1.252	5415	7195	1297	−9.334	−9.643	−7.782
4:01	1.423	5187	6819	1188	−9.287	−9.584	−7.687
4:31	1.579	5042	6450	1033	−9.256	−9.524	−7.535

amplifier gain settings. Columns 6, 7, and 8 are the instrumental magnitudes calculated by Equations 2.1, 2.2, and 2.3. Check some of these entries. If DC measurements had been made, we would substitute the deflection for the count rate and add the amplifier gain (in magnitudes). That is,

$$v = -2.5 \log(d_v) + G_v$$
$$b = -2.5 \log(d_b) + G_b$$
$$u = -2.5 \log(d_u) + G_u.$$

3. We now plot v versus X, b versus X, and u versus X. These are shown in Figure G.1. A linear least-squares analysis yields

$$k'_v = 0.306$$
$$k'_b = 0.441$$
$$k'_u = 0.840.$$

(See Section 3.5 for an explanation of linear least-squares analysis.)

4. The magnitude difference between the variable and the comparison star can now be corrected for extinction by using Equations 2.35 through 2.37 (or 2.38 through 2.40) and 2.43 through 2.45. The color indices are computed by Equations 2.41 and 2.42. These colors can be corrected for extinction by noting that

$$k'_{bv} = k'_b - k'_v$$
$$k'_{ub} = k'_u - k'_b$$

and applying Equations 2.46 and 2.47. The third term on the right side of Equation 2.46 can usually be dropped because k''_{bv} is small. Furthermore, if the variable and comparison star are nearly the same color, $\Delta(b - v)$ is nearly zero. However, if you wish, this term can be included by finding k''_{bv} as outlined in Section G.3.

G.2. EXTINCTION CORRECTION FOR "ALL-SKY" PHOTOMETRY

The procedure described in this section is applied to the situation in which stars have been measured at various positions in the sky. Each star is observed briefly through a limited range of air mass and hence the procedure of Section G.1 cannot be used. There are two approaches that can be followed, depending on whether or not the transformation coefficients to the standard system are known.

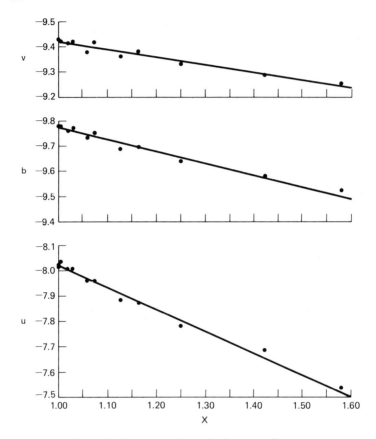

Figure G.1. Instrumental magnitudes versus air mass.

Unknown Transformation Coefficients

The observing procedure is quite simple. At various times during the night, one observes an early A-type standard star from the list in Appendix A or C. The reason for choosing this type of star becomes apparent in the following. If we substitute Equation 4.20 into 4.32, we obtain

$$V = v - k'_v X + \epsilon(B - V) + \zeta_v$$

or

$$V - v = -k'_v X + \epsilon(B - V) + \zeta_v. \tag{G.4}$$

For early A stars, $(B - V)$ is very small. Furthermore, ϵ is a small number, so their product is extremely small. Thus, to a good approximation the above equation becomes

$$V - v = \zeta_v - k'_v X \tag{G.5}$$

A plot of $V - v$ versus X for the early A stars yields a straight line with slope k'_v. Substituting equation 4.29 into 4.33 yields

$$(B - V) = \mu(b - v) - \mu(b - v)k''_{bv}X - \mu k'_{bv}X + \zeta_{bv} \tag{G.6}$$

Because μ is usually very nearly equal to one for most photometers and k''_{bv} is very small, we can make the following approximation.

$$(B - V) - (b - v) \simeq -k'_{bv}X + \zeta_{bv} \tag{G.7}$$

A plot of $(B - V) - (b - v)$ versus X yields a straight line with slope $-k'_{bv}$. A similar procedure with substitution of Equation 4.22 into 4.34 yields

$$(U - B) = \psi(u - b) - \psi k'_{ub}X + \zeta_{ub}. \tag{G.8}$$

We again note that $\psi \simeq 1$ for most photometers. We thus obtain our final equation

$$(U - B) - (u - b) \simeq -k'_{ub}X + \zeta_{ub}. \tag{G.9}$$

A plot of $(U - B) - (u - b)$ versus X yields a straight line with slope k'_{ub}.

1. Table G.2 contains the data for a number of extinction stars obtained during one night. Some stars appear more than once because they were observed later in the night at an appreciably different air mass. Most of the observing time was spent on program stars that do not appear in the table. Columns 2, 3, and 4 contain data taken from Appendices A and C. Column 5 contains the air mass calculated by Equations 4.17 and 4.18. Columns 6, 7, and 8 contain the count rates through each filter (after sky subtraction and a dead-time correction was made). Check some of the entries in column 9, using Equation 4.14 for v.
2. Check some of the entries in columns 10 and 11 using Equations 4.15 and 4.16.
3. Figure G.2 shows the plots of $(V - v)$ versus X, $(B - V) - (b - v)$ versus X and $(U - B) - (u - b)$ versus X. A linear regression analysis

316 ASTRONOMICAL PHOTOMETRY

TABLE G.2. Extinction Star Measurements

Star	V	B−V	U−B	X	v	b	u	V−v	(B−V)−(b−v)	(U−B)−(u−b)
80 UMa	4.03	0.15	0.09	1.042	399,726	799,296	140,792	18.014	0.912	−1.805
109 Vir	3.74	0.00	−0.03	1.269	524,038	1,152,529	201,823	18.038	0.856	−1.922
B Ser A	3.67	0.06	0.07	1.161	559,793	1,209,184	195,872	18.040	0.896	−1.906
τ Her	3.89	−0.15	−0.56	1.080	452,832	1,204,795	380,022	18.030	0.910	−1.813
γ Oph	3.75	0.04	0.04	1.868	454,939	887,150	122,538	17.895	0.765	−2.109
π Ser	4.83	0.07	0.05	1.097	194,252	407,561	71,059	18.051	0.875	−1.846
68 Oph	4.42	0.04	0.02	1.931	234,468	456,535	65,537	17.845	0.764	−2.088
68 Oph	4.42	0.04	0.02	1.340	260,411	559,110	94,773	17.959	0.870	−1.907
π Ser	4.83	0.07	0.05	1.082	192,633	402,426	71,980	18.042	0.870	−1.819
109 Vir	3.74	0.00	−0.03	1.673	458,829	974,150	148,048	17.894	0.817	−2.076
γ Oph	3.75	0.04	0.04	1.245	501,527	1,102,259	180,078	18.001	0.895	−1.927
57 Cyg	4.77	−0.14	−0.58	1.006	197,291	515,202	175,144	18.008	0.902	−1.752

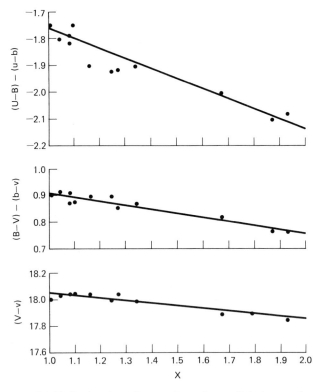

Figure G.2. Extinction plots when transformation coefficients are unknown.

yields

$$k'_v = 0.200$$
$$k'_{bv} = 0.153$$
$$k'_{ub} = 0.353.$$

Known Transformation Coefficients

When the transformation coefficients are known, the procedure of finding the extinction coefficients is simplified. Now all that is required are observations of five or more standard stars, of any color, at various air masses. These can be the same standards you would observe to determine your nightly zero-point constants (ζ_v, ζ_{bv}, ζ_{ub}). If Equation G.4 is rearranged, we obtain

$$V - v - \epsilon(B - V) = -k'_v X + \zeta_v. \qquad (G.10)$$

Because ϵ is known, we do not need to use A stars in order to approximate the $\epsilon(B - V)$ term as zero. A plot of $V - v - \epsilon(B - V)$ versus X yields a straight line with slope k'_v. Likewise, Equation G.6 becomes (assuming $k''_{bv} \sim 0$)

$$(B - V) - \mu(b - v) = -\mu k'_{bv} X + \zeta_{bv}. \qquad (G.11)$$

A plot of $(B - V) - \mu(b - v)$ versus X yields a slope $\mu k'_{bv}$. Rearranging Equation, G.8 becomes

$$(U - B) - \psi(u - b) = -\psi k'_{ub} X + \zeta_{ub}. \qquad (G.12)$$

Again, a plot of $(U - B) - \psi(u - b)$ versus X yields a slope $\psi k'_{ub}$.

1. Table G.3 contains data on 10 standard stars observed on one night. The first four columns contain values taken from Appendix C. Column 5 contains the air mass and columns 6, 7, and 8 contain the instrumental magnitudes and colors calculated by Equations 4.14, 4.15, and 4.16. The transformation coefficients were determined on a previous night to be

$$\epsilon = -0.084$$
$$\mu = 1.083$$
$$\psi = 1.006.$$

Compute the left-hand side of Equation G.10 for a few stars to check the entries in column 9. Likewise, compute the left-hand side of Equation G.11 and G.12 to check the entries in columns 10 and 11.

TABLE G.3. Observations of Standard Stars

Star	V	$(B-V)$	$(U-B)$	X	v	$(b-v)$	$(u-b)$	$V-v-$ $\epsilon(B-V)$	$(B-V)-$ $\mu(b-v)$	$(U-B)-$ $\psi(u-b)$
i Psc	4.13	0.51	0.00	2.183	−13.721	−0.001	2.183	17.851	0.511	−2.196
HR8832	5.57	1.01	0.89	1.262	−12.502	0.334	2.741	18.157	0.648	−1.867
10 Lac	4.88	−0.20	−1.04	1.183	−13.206	−0.794	0.709	18.069	0.660	−1.753
ε Aqr	3.77	0.01	0.04	1.562	−14.199	−0.564	2.033	17.970	0.621	−2.005
α Del	3.77	−0.06	−0.22	1.103	−14.312	−0.694	1.612	18.077	0.692	−1.842
B Aql	3.71	0.86	0.48	1.191	−14.344	0.190	2.269	18.126	0.654	−1.803
γ Oph	3.75	0.04	0.04	1.514	−14.293	−0.514	2.002	18.046	0.597	−1.974
τ Her	3.89	−0.15	−0.56	1.361	−14.179	−0.721	1.288	18.056	0.631	−1.856
ε Crb	4.15	1.23	1.28	1.754	−13.862	0.618	3.342	18.115	0.561	−2.082
β SerA	3.67	0.06	0.07	2.340	−14.177	−0.352	2.286	17.852	0.441	−2.230

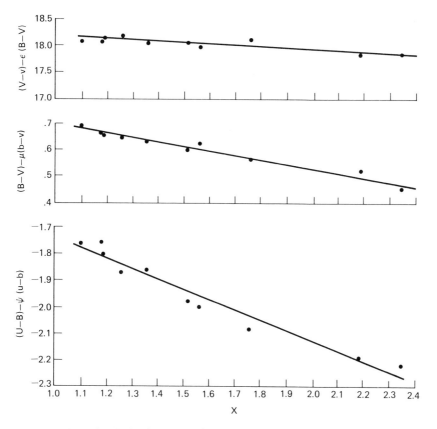

Figure G.3. Extinction plots when transformation coefficients are known.

2. Plots of $V - v - \epsilon(B - V)$ versus X, $(B - V) - \mu(b - v)$ versus X, and $(U - B) - \psi(u - b)$ versus X appears in Figure G.3. A linear regression analysis of each plot yields the following slopes.

$$k'_v = 0.212$$
$$k'_{bv} = 0.163$$
$$k'_{ub} = 0.373.$$

Note that this same linear regression analysis gives the intercepts that are the zero-point constants. These turn out to be

$$\zeta_v = 18.359$$
$$\zeta_{bv} = 0.875$$
$$\zeta_{ub} = -1.381.$$

G.3 SECOND-ORDER EXTINCTION COEFFICIENTS

Because k_v'' is very small and k_{ub}'' is defined as zero, we confine this example to k_{bv}''. Experience has shown that k_{bv}'' is both small and fairly stable. Therefore, k_{bv}'' need be determined only once or twice a year. The procedure is to observe a closely spaced pair of stars, of very different colors, at various air masses. If we let subscripts 1 and 2 refer to each star and use Equation 4.29 to form the differences in color indices, we obtain

$$(b - v)_{01} - (b - v)_{02} = (b - v)_1 - k_{bv}''X_1(b - v)_1 \\ - k_{bv}'X_1 - (b - v)_2 + k_{bv}''X_2(b - v)_2 + k_{bv}'X_2.$$

Because $X_1 \simeq X_2$, this reduces to

$$\Delta(b - v)_0 = \Delta(b - v) - k_{bv}''X\Delta(b - v). \tag{G.13}$$

Because $\Delta(b - v)_0$ is constant, a plot of $\Delta(b - v)$ versus $X\Delta(b - v)$ gives a straight line with slope k_{bv}''.

TABLE G.4. Observations of HD30544 and HD30545

HD30544 $(b - v)$	HD30545 $(b - v)$	$\Delta(b - v)$	X	$\Delta(b - v)X$
−0.628	0.458	−1.086	2.079	−2.258
−0.676	0.396	−1.072	1.786	−1.915
−0.718	0.376	−1.094	1.551	−1.697
−0.756	0.344	−1.100	1.367	−1.504
−0.778	0.339	−1.117	1.279	−1.429
−0.783	0.335	−1.118	1.231	−1.376

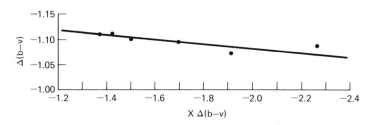

Figure G.4. Second order extinction plot.

1. The pair of stars used in this example is HD30544 and HD30545. These and other pairs can be found in Appendix B. Columns 1 and 2 of Table G.4 contain the color indices of each star calculated by Equation 4.15 and 4.16. Column 3 is just column 1 minus column 2. Column 4 is the mean air mass at the time of observation. Finally, column 5 is the product of columns 3 and 4.
2. Figure G.4 shows a plot of $\Delta(b-v)X$ versus $\Delta(b-v)$. A linear regression analysis yields a slope

$$k''_{bv} = -0.042.$$

APPENDIX H
TRANSFORMATION COEFFICIENTS EXAMPLE

H.1 DC EXAMPLE

The transformation coefficients are often determined from standard stars within a cluster that is near the zenith. This procedure reduces the impact of the extinction corrections greatly. Appendix D contains finder charts and lists of standard stars for three clusters. In this particular example, however, a cluster was not used. Standard stars near the meridian were selected from the list in Appendix C. While this technique works, it usually gives inferior results under variable sky conditions to the cluster method. A 60-centimeter (24-inch) telescope was used. A completely different telescope and photometer were used for the pulse-counting example to follow.

1. The first step is to measure each star through each filter. The sky background should be measured in each filter using the same amplifier gain and diaphragm as used for each star. In fact, the same diaphragm should be used for all measurements. Record the gain setting of *each* gain switch, *not their total*. This is necessary because there is more than one combination of gain switch positions which gives the same apparent total gain. However, each combination has a different set of gain corrections as determined by the gain correction table.

Table H.1 lists the coordinates, magnitudes, and colors of the stars observed. For now, ignore the three columns on the right. While in this particular example, only nine stars were used, it is recommended that you observe as many as is practically possible. Table H.2 lists the actual observations.

2. The next step is to subtract the sky background to produce the net deflection shown in column 5 of Table H.2. The settings of the two gain switches are shown in column 6, separated by a slash. For the particular amplifier used in this example, the coarse switch requires no gain corrections. The corrections

TABLE H.1. Standard Stars

Star	RA	Dec.	V	Standard B − V	U − B	V	Calculated B − V	U − B
11 LMi	9^h 34^m	35° 56′	5.41	0.77	0.45	5.43	0.77	0.38
21 LMi	10 06	35 22	4.48	0.18	0.08	4.52	0.16	0.15
λ UMa	10 16	43 03	3.45	0.03	0.06	3.47	0.02	0.11
α Leo	10 07	12 05	1.36	−0.11	−0.36	1.35	−0.05	−0.37
ε Leo	9 44	23 58	2.98	0.81	0.46	2.95	0.78	0.43
υ UMa	9 48	59 14	3.82	0.29	0.10	3.77	0.28	0.09
31 Leo	10 06	10 12	4.36	1.45	1.75	4.33	1.43	1.82
72 Leo	11 13	23 18	4.60	1.66	1.85	4.64	1.70	1.81
β Leo	11 48	14 43	2.14	0.09	0.07	2.13	0.08	0.05

TABLE H.2. Stellar Measurements

Star	Filter	Deflection	Sky	Net	Gain	v	b − v	u − b
11 LMi	v	41.0	10.9	30.1	5/2	3.306	0.189	2.433
	b	36.2	10.9	25.3	5/2			
	u	37.8	10.9	26.9	7.5/2			
21 LMi	v	36.3	7.6	28.7	5/1	2.378	−0.462	2.214
	b	51.5	7.6	43.9	5/1			
	u	36.6	14.1	22.2	5/2.5			
λ UMa	v	35.5	5.8	29.5	5/0	1.325	−0.613	2.184
	b	57.7	5.8	51.9	5/0			
	u	36.6	8.7	27.9	5/1.5			
α Leo	v	37.5	6.4	31.1	2.5/0.5	−0.738	−0.643	1.847
	b	67.8	6.4	61.4	2.5/0.5			
	u	55.7	10.7	45.0	2.5/2			
ε Leo	v	52.3	6.0	46.3	5/0	0.836	0.212	2.487
	b	44.1	6.0	38.1	5/0			
	u	35.9	11.5	24.4	5/2			
υ UMa	v	40.6	6.7	33.9	5/0.5	1.668	−0.312	2.220
	b	51.9	6.7	45.2	5/0.5			
	u	35.5	12.0	23.5	5/2			
31 Leo	v	46.0	13.7	32.3	5/1	2.250	0.925	3.727
	b	34.6	13.0	21.6	5/1.5			
	u	27.5	10.2	17.3	7.5/2.5			
72 Leo	v	52.7	13.0	39.7	5/1.5	2.513	1.167	3.650
	b	44.6	10.3	34.3	5/2.5			
	u	45.0	14.1	30.6	10/1			
β Leo	v	53.3	13.8	39.5	2.5/1.5	0.019	−0.527	2.178
	b	78.0	13.8	64.2	2.5/1.5			
	u	69.2	14.1	55.1	5/1			

for the fine gain switch were determined by the procedure outlined in Section 8.6. The table below gives the actual fine gain values. The amplifier gain is then the total of the coarse switch position and the actual fine gain read from the table.

Fine Gain Switch Position	Actual Gain in Magnitudes
0	0
0.5	0.494
1.0	1.023
1.5	1.510
2.0	2.003
2.5	2.496

The atmospheric, instrumental magnitudes and colors can be computed by Equations 4.11, 4.12, and 4.13. Check some of the results in the last three columns of Table H.2.

3. The magnitudes and colors determined above must now be corrected for atmospheric extinction. It is necessary to determine the air mass, X, for each star. Because the three filter measurements were made within a few minutes of each other, a single value of X is used for each star. Column 2 of Table H.3 lists the hour angle at the time of observation of each star. In this case, the telescope had an hour angle read-out device. If your telescope is not so equipped, you need to calculate hour angle as described in Section 5.3c. The east-west notation in the table is really unnecessary because it makes no dif-

TABLE H.3. Extinction Correction and Transformation Coefficient Determination

Star	HA	X	v_0	$(b-v)_0$	$(u-b)_0$	$V-v_0$	$(B-V) - (b-v)_0$	$(U-B) - (u-b)_0$
11 LMi	1:23W	1.044	2.909	−0.124	1.898	2.501	0.895	−1.449
21 LMi	0:47W	1.014	1.993	−0.766	1.695	2.487	0.946	−1.615
λ UMa	0:33W	1.014	0.940	−0.917	1.665	2.510	0.948	−1.605
α Leo	1:24W	1.160	−1.179	−0.991	1.253	2.539	0.881	−1.613
ε Leo	1:18W	1.069	0.430	−0.109	1.940	2.550	0.919	−1.479
υ UMa	1:18W	1.118	1.243	−0.647	1.648	2.577	0.938	−1.547
31 Leo	1:06W	1.152	1.812	0.579	3.137	2.551	0.871	−1.387
72 Leo	0:11W	1.026	2.123	0.859	3.125	2.476	0.801	−1.275
β Leo	0.58E	1.102	−0.400	−0.858	1.614	2.540	0.948	−1.544

ference for the calculation of X. As an example, consider 21 LMi.

$$\begin{aligned}
\text{hour angle} &= H = 0\text{:}47 \\
&= \frac{47 \text{ minutes}}{60 \text{ minutes/hour}} \times (15°/\text{hour}) \\
&= 11.75° \\
\delta &= 35.37° \text{ (declination from Table H.1)} \\
\phi &= 35.92° \text{ (latitude of observatory)} \\
X &= (\sin \delta \sin \phi + \cos \delta \cos \phi \cos H)^{-1} \\
X &= 1.014
\end{aligned} \qquad (4.18)$$

The more precise method described in Section 4.4a (Equation 4.19) is not needed because measurements were made near the zenith.

The extinction coefficients were determined by the method outlined in Appendix G. They were found to be

$$\begin{aligned}
k'_v &= 0.380 \\
k'_{bv} &= 0.300 \\
k'_{ub} &= 0.512 \\
k''_{bv} &\simeq 0
\end{aligned}$$

The extinction corrections were then made by calculating for each star, assuming that $k''_{ub} = 0$:

$$\begin{aligned}
v_0 &= v - k'_v X \\
(b - v)_0 &= (b - v) - k'_{bv} X \\
(u - b)_0 &= (u - b) - k'_{ub} X.
\end{aligned}$$

The results of these calculations appear in columns 4 through 6 of Table H.3. These are the instrumental magnitudes and colors. Check some of these results.

4. To determine ϵ and ζ_v, a plot of $V - v_0$ versus $(B - V)$ is needed. The values of V and $(B - V)$ are found in Table H.1. The computed values of $V - v_0$ appear in Table H.3. The graph appears in Figure H.1. A linear least-squares fit yields a slope of -0.012 and an intercept of 2.532. From Equation 4.38,

$$\begin{aligned}
\epsilon &= -0.012 \\
\zeta_v &= 2.532.
\end{aligned}$$

5. To determine μ and ζ_{bv} a plot of $[(B - V) - (b - v)_0]$ versus $(B - V)$ is required. Again, $(B - V)$ is found in Table H.1 and $[(B - V) -$

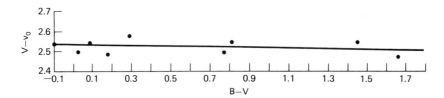

Figure H.1. Plot to determine ϵ and ζ_v.

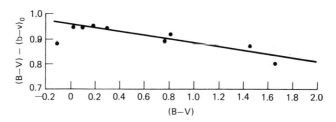

Figure H.2. Plot to determine μ and ζ_{bv}.

$(b - v)_0]$ is calculated and shown in Table H.3. The graph appears in Figure H.2. A linear least-squares fit yields a slope of -0.059 and an intercept of 0.940. From Equation 4.40,

$$1 - \frac{1}{\mu} = -0.059$$

and

$$\mu = 0.944$$
$$\frac{\zeta_{bv}}{\mu} = 0.940$$
$$\zeta_{bv} = 0.887.$$

6. The procedure in step 5 is repeated for $[(U - B) - (u - b)_0]$ versus $(U - B)$. The graph appears in Figure H.3. A linear least-squares fit yields a slope of 0.139 and an intercept of -1.570. From Equation 4.41,

$$1 - \frac{1}{\psi} = 0.139$$
$$\psi = 1.161$$

and

$$\frac{\zeta_{ub}}{\psi} = -1.570$$
$$\zeta_{ub} = -1.823.$$

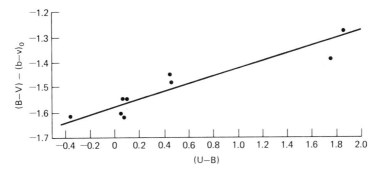

Figure H.3. Plot to determine ψ and ζ_{ub}.

Now the standardized magnitudes and colors of any star measured that same night can be found by the following expressions:

$$(B - V) = 0.944 (b - v)_0 + 0.887$$
$$V = v_0 - 0.012 (B - V) + 2.532$$
$$(U - B) = 1.161 (u - b)_0 - 1.823.$$

As a check on the quality of these transformation equations, the magnitudes and colors of the standard stars were computed and compared to the known values. The results appear in the rightmost three columns of Table H.1. Compare Figures H.1, H.2, and H.3 to Figures H.4, H.5, and H.6. This last set of figures, from the pulse-counting example, represents a very good set of observations and hence, a good set of transformation equations. The results of the DC example show that the observations are of marginal quality and the transformation equations are not very dependable. This is primarily because of poor sky conditions. The cluster method tends to work better when the sky conditions are less than excellent. Note that, because different telescopes and photometers were used, the color coefficients from the two examples are not the same.

As explained in Chapter 4, the coefficients ϵ, μ, and ψ are constant over fairly long time periods. However, ζ_v, ζ_{vb}, and ζ_{ub} *must be determined nightly.*

H.2 PULSE-COUNTING EXAMPLE

The transformation coefficients can be determined from standard stars within a cluster that is near the zenith. This procedure reduces the impact of the extinction corrections greatly. Appendix D contains finder charts and lists of standard stars for three clusters. In this case, the cluster IC 4665 was observed.

328   ASTRONOMICAL PHOTOMETRY

A 40-centimeter (16-inch) telescope was used. A completely different telescope and photometer were used for the DC example.

1. The first step is to measure each star through each filter. The sky background should be measured several times during this process. The same diaphragm should be used for all measurements. Table H.4 lists the coordinates, magnitudes, and colors of the stars observed. For now, ignore the three columns on the right. Table H.5 lists the actual observations. The third column contains the number of counts per second, which was found by averaging the counts obtained in several 10-second intervals and dividing by 10.

2. The next step is to correct the observations for dead time. The dead-time coefficient, t, was found by the technique outlined in Chapter 4 with a worked example in Appendix F. For the particular photomultiplier used in this example, $t = 1.54 \times 10^{-7}$ seconds per count. The observed count rate n is related to the true count rate N by

$$n = N e^{-Nt}.$$

A FORTRAN subroutine to solve this equation in an iterative fashion can be found in Section I.1. To illustrate the essential idea behind the solution, a numerical example is given using the first observation in Table H.5. The observed count rate is 28,925. As an initial guess, suppose that N is equal to this observed rate. Then

$$n = 28{,}925\ e^{-(28{,}925)1.54 \times 10^{-7}}$$
$$n = 28{,}796$$

TABLE H.4.   The Standard Stars

Star	RA	Dec.	V	Standard B − V	U − B	V	Calculated B − V	U − B
A	17^h 45^m	5° 32′	6.85	0.011	−0.54	6.84	0.018	−0.554
F	17 45	5 42	7.59	0.002	−0.49	7.58	0.003	−0.491
G	17 45	6 08	7.74	−0.009	−0.55	7.76	−0.007	−0.544
I	17 45	6 15	7.89	1.028	0.77	7.88	1.034	0.735
J	17 45	5 24	7.94	0.449	−0.01	7.95	0.433	−0.019
N	17 46	5 37	8.33	1.728	2.08	8.34	1.711	2.098
O	17 47	5 22	8.40	1.232	1.04	8.40	1.243	0.992
P	17 46	5 26	8.89	0.106	−0.27	8.88	0.099	−0.264
S	17 44	5 32	9.39	0.314	0.17	9.39	0.306	0.221
U	17 46	5 31	9.81	0.676	0.23	9.80	0.701	0.201
V	17 44	5 34	10.10	0.122	0.01	10.11	0.111	0.036
W	17 44	5 51	10.21	1.292	1.27	10.21	1.296	1.301

TABLE H.5. Stellar Measurements

Star	Filter	Counts per Second	Dead-time Corrected	Net	v	b − v	u − b
A	v	28,925	29,055	29,001	−11.156		
	b	61,718	62,313	62,220		−0.829	
	u	18,575	18,628	18,599			1.311
F	v	14,819	14,853	14,799	−10.426		
	b	32,152	32,312	32,219		−0.845	
	u	9136	9149	9120			1.370
Sky	v	54.1					
	b	93.2					
	u	28.6					
G	v	12,539	12,563	12,510	−10.243		
	b	27,547	27,664	27,575		−0.858	
	u	8287	8298	8270			1.308
I	v	10,708	10,726	10,673	−10.071		
	b	10,139	10,155	10,066		0.064	
	u	911	911	883			2.642
Sky	v	51.8					
	b	84.6					
	u	28.2					
J	v	10,387	10,404	10,352	−10.038		
	b	15,994	16,033	15,949		−0.469	
	u	2917	2918	2891			1.854
N	v	6810	6817	6765	−9.576		
	b	3756	3758	3674		0.663	
	u	115	115	87			4.062
O	v	6613	6620	6568	−9.544		
	b	5310	5314	5230		0.247	
	u	387	387	360			2.906
P	v	4524	4527	4475	−9.127		
	b	9165	9178	9094		−0.770	
	u	2129	2129	2102			1.590
Sky	v	51.4					
	b	82.8					
	u	26.6					
S	v	2826	2827	2776	−8.609		
	b	4856	4860	4781		−0.590	
	u	723	723	698			2.090
U	v	1926	1926	1875	−8.183		
	b	2416	2416	2337		−0.239	
	u	373	373	347			2.070
V	v	1508	1508	1457	−7.909		
	b	3028	3029	2950		−0.766	
	u	542	542	516			1.893
W	v	1296	1296	1245	−7.738		
	b	1034	1034	955		0.288	
	u	76	76	50			3.211
Sky	v	50.2					
	b	75.8					
	u	25.2					

Because this value is lower than the actual observed rate, we *increase* the next guess of N by the difference between these two numbers, i.e.

$$N \text{ (second guess)} = 28{,}925 + (28{,}925 - 28{,}796)$$
$$= 29{,}054$$

We can now recompute n by

$$n = 29{,}054 \, e^{-(29{,}054)1.54 \times 10^{-7}}$$
$$= 28{,}924.$$

This differs from the observed rate by only 1. Thus as a third guess,

$$N \text{ (third guess)} = 29{,}054 + 1$$

and

$$n = 29{,}055 \, e^{-(29{,}055)1.54 \times 10^{-7}}$$
$$n = 28{,}925.$$

Because this value agrees with the observed rate, we know that the true count rate is 29,055. This number appears in column 4 of Table H.5 along with the other corrected count rates.

3. The sky background is now subtracted from each value in column 4. When the stellar measurement is sandwiched between two sky measurements, an average sky measurement is subtracted.

The atmospheric instrumental magnitudes and colors can be computed by Equations 4.14 through 4.16. The results appear in columns 6, 7, and 8 of Table H.5. Check some of the results with your calculator.

4. The magnitudes and colors determined in step 3 must be corrected for atmospheric extinction. It is necessary to determine the air mass, X, for each star. Because the three filter measurements were made within a few minutes of each other, a single value of X is used for each star. Column 2 in Table H.6 lists the hour angle at the time of observation of each star. In this case, the telescope had an hour angle read-out device. If your telescope is not so equipped, you need to calculate the hour angle as described in Section 5.3c. The east-west notation in the table is really unnecessary because it makes no difference for the calculation of X. As an example, consider star A.

$$\text{hour angle} = H = 1{:}54$$
$$H = \left(1 + \frac{54}{60}\right) \times (15°/\text{hour})$$

$$H = 28.50°$$
$$\delta = 5.53° \text{ (declination from Table H.4)}$$
$$\phi = 39.17° \text{ (latitude of observatory)}$$
$$X = (\sin \delta \sin \phi + \cos \delta \cos \phi \cos H)^{-1}$$
$$X = 1.353$$

The more precise method of finding X described in Section 4.4 (Equation 4.19) is not needed since the measurements were made near the zenith.

The extinction coefficients were determined by the method outlined in Appendix G. They were found to be

$$k'_v = 0.209$$
$$k'_{bv} = 0.162$$
$$k'_{ub} = 0.337$$
$$k''_{bv} = -0.088$$

The extinction corrections were then made by calculating, for each star, assuming $k''_{ub} = 0$,

$$v_0 = v - k'_v X$$
$$(b - v)_0 = (b - v)(1 - k''_{bv} X) - k'_{bv} X$$
$$(u - b)_0 = (u - b) - k'_{ub} X.$$

The results of these calculations appear in columns 4 through 6 of Table H.6. These are instrumental magnitudes and colors. Check some of these results.

TABLE H.6. Extinction Correction and Transformation Coefficient Determination

Star	HA	X	v_0	$(b-v)_0$	$(u-b)_0$	$V-v_0$	$(B-V) - (b-v)_0$	$(U-B) - (u-b)_0$
A	1:54E	1.353	−11.439	−1.146	0.855	18.289	1.157	−1.395
F	1:50E	1.339	−10.705	−1.161	0.919	18.295	1.163	−1.409
G	1:46E	1.320	−10.519	−1.171	0.863	18.259	1.162	−1.413
I	1:43E	1.311	−10.345	−0.141	2.200	18.235	1.169	−1.430
J	1:38E	1.313	−10.312	−0.736	1.412	18.252	1.185	−1.422
N	1:34E	1.300	−9.847	0.528	3.624	18.177	1.200	−1.544
O	1:31E	1.297	−9.815	0.065	2.468	18.215	1.167	−1.428
P	1:27E	1.288	−9.396	−1.066	1.156	18.286	1.172	−1.426
S	1:19E	1.270	−8.874	−0.862	1.662	18.264	1.176	−1.492
U	1:18E	1.269	−8.448	−0.471	1.642	18.258	1.147	−1.412
V	1:13E	1.259	−8.172	−1.054	1.469	18.272	1.176	−1.459
W	1:08E	1.247	−7.999	0.118	2.791	18.209	1.174	−1.521

5. To determine ϵ and ζ_v, a plot of $V - v_0$ versus $(B - V)$ is needed. The values of V and $(B - V)$ are found in Table H.4. The computed values of $V - v_0$ appear in Table H.6. The graph appears in Figure H.4. A linear least-squares fit yields a slope of -0.057 and an intercept of 18.284. From Equation 4.38,

$$\epsilon = -0.057$$
$$\zeta_v = 18.284.$$

6. To determine μ and ζ_{bv}, a plot of $[(B - V) - (b - v)_0]$ versus $(B - V)$ is required. Again $(B - V)$ is found in Table H.4 and $[(B - V) - (b - v)_0]$ was calculated and appears in Table H.6. The graph appears in Figure H.5. A linear least-squares fit yields a slope of 0.011 and an intercept of 1.164. From equation 4.40,

$$1 - \frac{1}{\mu} = 0.011$$
$$\mu = 1.011$$

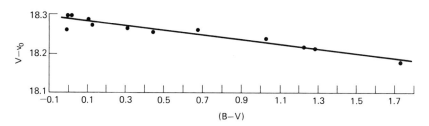

Figure H.4. Plot to determine ϵ and ζ_v.

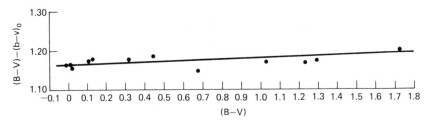

Figure H.5. Plot to determine μ and ζ_{bv}.

and

$$\frac{\zeta_{bv}}{\mu} = 1.164$$
$$\zeta_{bv} = 1.177$$

7. The procedure in step 6 is repeated for $[(U - B) - (u - b)_0]$ versus $(U - B)$. The graph appears in Figure H.6. A linear least-squares fit yields a slope of -0.045 and an intercept of -1.432. From Equation 4.41,

$$1 - \frac{1}{\psi} = -0.045$$
$$\psi = 0.957$$

and

$$\frac{\zeta_{ub}}{\psi} = -1.432$$
$$\zeta_{ub} = -1.370.$$

Now the standardized magnitudes and colors of any "unknown" star observed that same night can be found by the following expressions.

$$(B - V) = 1.011 \, (b - v)_0 + 1.177$$
$$V = v_0 - 0.057 \, (B - V) + 18.284$$
$$(U - B) = 0.957 \, (u - b)_0 - 1.370$$

As a check on the quality of these transformation equations, the magnitudes and colors of the standard stars were computed and compared to the known values. The results appear in the rightmost three columns of Table H.4. This

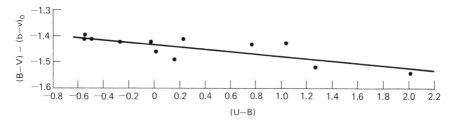

Figure H.6. Plot to determine ψ and ζ_{ub}.

comparison shows that the transformations are fairly good. Compare Figures H.4, H.5, and H.6 to H.1, H.2, and H.3. This last set of Figures, from the DC example, represents a rather poor set of observations and the resulting transformation equations are of questionable quality. Note that, because different telescopes and photometers were used, the color coefficients from the two examples are not the same.

As explained in Chapter 4, the coefficients ϵ, μ, and ψ are constant over fairly long time periods. However, ζ_v, ζ_{bv}, and ζ_{ub} *must be determined nightly.*

APPENDIX I
USEFUL FORTRAN SUBROUTINES

These routines were written by one of the authors (AAH) for the FORTRAN IV compiler of Control Data Corporation computers. The programming is functional but not elegant. Most of the routines are in constant use at Indiana University and have not been found in error.

Further information on programming and numerical analysis is available in many texts. Three books the authors have found most useful are:

- Bevington, A. R. 1969. *Data Reduction and Error Analysis for the Physical Sciences.* New York: McGraw-Hill.
- Carnahan, B., Luther, H. A., and Wilkes, J. O. 1969. *Applied Numerical Methods.* New York: Wiley and Sons.
- Murril, P. W., and Smith, C. L. 1969. *FORTRAN IV Programming for Engineers and Scientists.* Scranton: International Textbook Co.

I.1 DEAD-TIME CORRECTION FOR PULSE-COUNTING METHOD

```
      FUNCTION CNTCR (OBSD,TAU)
C
C     *** DEAD-TIME CORRECTION FOR PULSE COUNTING ***
C
C     INPUTS -
C             TAU    DEAD-TIME COEF IN SECONDS
C             OBSD   OBSERVED COUNT RATE IN INVERSE SECONDS
C     OUTPUTS -
C             CNTCR  CORRECTED COUNT RATE IN INVERSE SECONDS
C
C     *** COMMENTED VERSION BY ARNE A. HENDEN 1978 ***
C     *** FROM ORIGINAL FROM UBVLSQ AT I.U. ***
C
C     CALCULATE INITIAL COUNT VALUE FOR ITERATION
      A=OBSD*EXP(OBSD*TAU)
C     ITERATE FOR 20 TRIES.  IF NO CONVERGENCE,GET HELP
      DO 10 J=1,20
C     CALCULATE NEXT GUESS FOR RATE
      B=OBSD*EXP(A*TAU)
C     IS THIS GUESS WITHIN 0.001 COUNTS?
C     (ACTUALLY, PERCENTAGE ERROR WOULD BE BETTER)
      IF (ABS(A-B)-0.001) 20,20,5
C     NO...RESET OLD RATE VALUE TO NEW VALUE AND TRY AGAIN
5     A=B
10    CONTINUE
      WRITE (6,900)
900   FORMAT( *NO CONVERGENCE AFTER 20 TRIES*)
20    CNTCR=B
      RETURN
      END
```

I.2 CALCULATING JULIAN DATE FROM UT DATE

```
      SUBROUTINE JDAY (ID,IM,IY,UTHR,DATE)
C
C     *** CALCULATE JULIAN DATE FROM UT DATE ***
C
C     INPUTS -
C             ID     INTEGER UT DAY
C             IM     INTEGER UT MONTH
C             IY     INTEGER UT YEAR
C             UTHR   F.P. UT HOUR (24HR CLOCK)
C     OUTPUTS -
C             DATE   (JULIAN DATE - 2400000)
C
C     *** PROGRAMMED BY ARNE A. HENDEN 1977 ***
C
C     MONTH IS NUMBER OF ELAPSED DAYS IN NORMAL YEAR
      DIMENSION MONTH(12)
      DATA MONTH/0,31,59,90,120,151,181,212,243,273,304,334/
C     LEAP IS NUMBER OF LEAP DAYS SINCE 1900
      LEAP=IY/4
C     CHECK TO SEE IF THIS YEAR IS LEAP AND MONTH=JAN,FEB
      IF ((4*LEAP-IY).EQ.0.AND.IM.LT.3) LEAP=LEAP-1
C     CALCULATE INTEGRAL NUMBER OF JULIAN DAYS
      JD0=15020+IY*365+LEAP+MONTH(IM)+ID
C     ADD IN UTHR AND SUBTRACT A HALF-DAY
      DATE=FLOAT(JD0)+UTHR/24.-0.5
      RETURN
      END
```

I.3 GENERAL METHOD FOR COORDINATE PRECESSION

```fortran
      SUBROUTINE PRCS (R,DD,YR0,YR,IH,IM,IS,ID,IDM,IDS)
C
C     *** RIGOROUS COORDINATE PRECESSION W/O P.M. CORRECTIONS ***
C
C     INPUTS-
C              R       R.A. IN DECIMAL HOURS
C              DD      DEC. IN DECIMAL DEGREES
C              YR0     COORDINATE EPOCH
C              YR      YEAR TO BE PRECESSED TO
C     OUTPUTS-
C              IH,IM,IS    R.A. H,M,S IN INTEGER FORM
C              ID,IDM,IDS  PRECESSED DEC. IN INTEGER FORM
C
C     *** WRITTEN BY A. HENDEN 1978 ***
C
      DATA A,B,C,D,E/1.60017,5.25E-4,7.8E-5,1.60017,1.9E-3/
      DATA F,G,H,P/8.3E-5,1.39214,7.39E-4,1.8E-4/
      DATA PICON/0.0174532925/
C      CONVERT INPUTS TO RADIANS
      RA=R*PICON*15.
      DEC=DD*PICON
      TAU=(YR-YR0)/250.
      ZET=TAU*(A+TAU*(B+TAU*C))*PICON
      Z=TAU*(D+TAU*(E+TAU*F))*PICON
      THET=TAU*(G-TAU*(H+TAU*P))*PICON
      AMU=ZET+Z
      BET=RA+ZET
      Q=SIN(THET)*(TAN(DEC)+COS(BET)*TAN(THET*0.5))
      GAM=0.5*ATAN(Q*SIN(BET)/(1.-Q*COS(BET)))
      DEE=2.*ATAN(TAN(THET*0.5)*COS(BET+GAM)/COS(GAM))
      DR=GAM+GAM+AMU
      RA=(RA+DR)/(PICON*15.)
      DEC=(DEC+DEE)/PICON
C      WE NOW HAVE DECIMAL FORMS...CONVERT TO INTEGER FORM
      IH=INT(RA)
      X=(RA-FLOAT(IH))*60.
      IM=INT(X)
      IS=INT((X-FLOAT(IM))*60.)
      ID=INT(DEC)
      RM=(DEC-FLOAT(ID))*60.
      IF (RM.LT.0.) RM=-RM
      IDM=INT(RM)
      IDS=INT((RM-FLOAT(IDM))*60.)
      RETURN
      END
```

I.4 LINEAR REGRESSION (LEAST-SQUARES) METHOD

```
      SUBROUTINE SMOOTH (X,Y,M,A,B)
C
C     *** LINEAR LEAST SQUARES ROUTINE FROM NIELSON ***
C     SIMPLE SOLUTION WITH NO WEIGHTING FACTORS
C
C
C     INPUTS -
C             X,Y   DATA ARRAYS
C             M     NUMBER OF POINTS IN DATA ARRAYS
C     OUTPUTS -
C             A,B   WHERE  Y=AX+B
C
C     *** WRITTEN BY ARNE A. HENDEN 1973 ***
C
      DIMENSION X(M),Y(M)
C     INITIALIZE SUMMING PARAMETERS
      A2=A3=C1=C2=0.
      A1=M
C     LOOP TO SET UP MATRIX COEFFICIENTS
      DO 10 I=1,M
      A2=A2+X(I)
      A3=A3+X(I)*X(I)
      C1=C1+Y(I)
      C2=C2+Y(I)*X(I)
10    CONTINUE
C     SOLVE MATRIX - SIMPLE SINCE ONLY 2X2
      DET=1./(A1*A3-A2*A2)
      A=-(A2*C2-C1*A3)*DET
      B=(A1*C2-C1*A2)*DET
      RETURN
      END
```

I.5 LINEAR REGRESSION (LEAST-SQUARES) METHOD USING THE *UBV* TRANSFORMATION EQUATIONS

```
      SUBROUTINE SOLVE (N,X,SMAG,URMAG,SEXT,FEXT,COEF,ZERO)
C
C     LEAST SQUARES SOLUTION OF THE UBV TRANSFORMATION EQUATIONS
C  INPUT -
C     SMAG(N,3)   STANDARD MAGNITUDES AND COLORS IN THE FORM:
C                    SMAG(N,1) = V
C                    SMAG(N,2) = B-V
C                    SMAG(N,3) = U-B
C     URMAG(N,3)  YOUR CORRESPONDING INSTRUMENTAL MAGNITUDES & COLORS
C     X(N)        AIR MASS VALUES
C     N           NUMBER OF OBSERVATIONS TO REDUCE
C     SEXT(3)     SECOND ORDER EXTINCTION
C  OUTPUT -
C     FEXT(3)     FIRST ORDER EXTINCTION
C     COEF(3)     TRANSFORMATION COLOR COEFFICIENTS
C     ZERO(3)     ZERO POINTS
C
C     TO HAVE FEXT, COEF AND ZERO TO ALL BE VALID, YOU
C     MUST USE STANDARDS WITH BOTH A WIDE RANGE OF COLORS
C     AND A WIDE RANGE OF AIRMASSES.  WITH A CLUSTER, YOU
C     WILL GET VALID COEF PARAMETERS, BUT FEXT AND ZERO MAY
C     BE INVALID...SO BEWARE!
C
C     WRITTEN BY A. HENDEN 1980
C
      DIMENSION X(N),SMAG(N,3),URMAG(N,3),SEXT(3)
      DIMENSION FEXT(3),COEF(3),ZERO(3)
C     LOOP OVER COLORS
      DO 20 K=1,3
C     ZERO MATRIX ELEMENTS
      A1=0. $ A2=0. $ A3=0. $ A4=0. $ A5=0. $ A6=0.
      A7=0. $ A8=0. $ A9=0. $ A10=0. $ A11=0. $ A12=0.
C     LOOP OVER NUMBER OF STANDARDS
      DO 10 I=1,N
C     CALCULATE MATRIX ELEMENT SUMS
      TEMP=URMAG(I,K)*(1.-SEXT(K)*X(I))
      STD=SMAG(I,K)
      IF (K.NE.1) GO TO 5
C  NOTE:  THE V EQUATION HAS SLIGHTLY DIFFERENT FORM
      TEMP=SMAG(I,2)
      STD=STD-URMAG(I,K)
5     CONTINUE
C
C  THE MATRIX LOOKS LIKE:
C     ( A1   A2   A3  . A4  )
C     ( A5   A6   A7  . A8  )
C     ( A9   A10  A11 . A12 )
C  NOTE: A2=A5, A3=A9, AND A7=A10, BUT EXPLICIT HERE FOR CLARITY
C
      A1=A1+TEMP*TEMP
      A2=A2+TEMP*X(I)
      A3=A3+TEMP
      A4=A4+STD*TEMP
      A5=A5+TEMP*X(I)
      A6=A6+X(I)*X(I)
      A7=A7+X(I)
      A8=A8+STD*X(I)
      A9=A9+TEMP
      A10=A10+X(I)
      A11=A11+1
      A12=A12+STD
```

(Continued on p. 340)

I.5 LINEAR REGRESSION (LEAST-SQUARES) METHOD USING THE UBV TRANSFORMATION EQUATIONS (Continued)

```
              BB=A5*A11-A7*A9
              CC=A6*A9-A5*A10
              DD=A8*A11-A7*A12
              EE=A6*A12-A8*A10
              FF=A5*A12-A8*A9
C       CALCULATE DETERMINANT
              DET=A1*AA+A2*BB+A3*CC
C       SOLVE FOR WANTED VALUES
              COEF(K)=(A4*AA+A2*DD+A3*EE)/DET
              ZERO(K)=(A2*FF-A1*EE+A4*CC)/DET
              FEXT(K)=(A1*DD-A4*BB+A3*FF)/DET
              IF (K.NE.1) FEXT(K)=FEXT(K)/COEF(K)
20      CONTINUE
        RETURN
        END
10      CONTINUE
C       CALCULATE MINORS
              AA=A7*A10-A6*A11
```

I.6 CALCULATING SIDEREAL TIME

```
              SUBROUTINE STIME (DATE,UT,ALON,ST,IST)
C
C       *** CALCULATE SIDEREAL TIME ***
C
C
C       INPUTS -
C               DATE     JULIAN DATE - 2400000.
C               UT       UT IN DECIMAL HOURS
C               ALON     OBSERVER LONGITUDE IN HOURS
C                          ( LONGITUDE IN DEG / 15)
C       OUTPUTS -
C               ST       SIDEREAL TIME IN DECIMAL HOURS
C               IST      S.T. IN FORMAT HHMM  (INTEGER)
C
C       *** PROGRAMMED BY ARNE A. HENDEN 1977 ***
C
C               STHR IS # SIDEREAL HRS SINCE JAN 1, 1900
              STHR=6.6460556+2400.0512617*(DATE-15020.)/36525.+
     $        UT*1.00273-ALON
C               GET S.T. FROM STHR - # WHOLE S.T. DAYS SINCE JAN 1,1900
              ST=STHR-INT(STHR/24.)*24.
C               NOW COMBINE TO GET IST
              IST=INT(ST)
              IST=IST*100+INT((ST-FLOAT(IST))*60.)
              RETURN
              END
```

I.7 CALCULATING CARTESIAN COORDINATES FOR 1950.0

```fortran
      SUBROUTINE XYZ (DATE,X,Y,Z)
C
C     CALCULATE HELIOCENTRIC X,Y,Z COORDINATES FOR 1950.0
C
C     INPUTS-
C          DATE    JULIAN DAY - 2400000
C     OUTPUTS-
C          X,Y,Z   HELIOCENTRIC RECTANGULAR COORDINATES
C                  IN A.U.
C
C     EQUATIONS FROM ALMANAC FOR COMPUTERS, DOGGETT. ET. AL.
C        USNO 1978
C     *** PROGRAMMED BY ARNE A. HENDEN 1978 ***
C
      DATA PICON/0.01745329/
C     T IS A RELATIVE JULIAN CENTURY
      T=(DATE-15020.)/36525.
C     EL IS THE MEAN SOLAR LONGITUDE, PRECESSED BACK TO 1950.0
      EL=279.696678+36000.76892*T+0.000303*T*T-
     $  (1.396041+0.000308*(T+0.5))*(T-0.499998)
C     G IS THE MEAN SOLAR ANOMALY
      G=358.475833+35999.04975*T-0.00015*T*T
C     AJ IS THE MEAN JUPITER ANOMALY
      AJ=225.444651+2880.0*T+154.906654*T
C     CONVERT DEGREES TO RADIANS FOR TRIG FUNCTIONS
      EL=EL*PICON
      G=G*PICON
      AJ=AJ*PICON
C     CALCULATE X,Y,Z USING TRIGONOMETRIC SERIES
      X=0.99986*COS(EL)-0.025127*COS(G-EL)+0.0008374*COS(G+EL)+
     $  0.000105*COS(G+G+EL)+0.000063*T*COS(G-EL)+
     $  0.000035*COS(G+G-EL)-0.000026*SIN(G-EL-AJ)-
     $  0.000021*T*COS(G+EL)
      Y=0.917308*SIN(EL)+0.023053*SIN(G-EL)+0.007683*SIN(G+EL)+
     $  0.000097*SIN(G+G+EL)-0.000057*T*SIN(G-EL)-
     $  0.000032*SIN(G+G-EL)-0.000024*COS(G-EL-AJ)-
     $  0.000019*T*SIN(G+EL)
      Z=0.397825*SIN(EL)+0.009998*SIN(G-EL)+0.003332*SIN(G+EL)+
     $  0.000042*SIN(G+G+EL)-0.000025*T*SIN(G-EL)-
     $  0.000014*SIN(G+G-EL)-0.000010*COS(G-EL-AJ)
      RETURN
      END
```

APPENDIX J
THE LIGHT RADIATION FROM STARS

J.1 INTENSITY, FLUX, AND LUMINOSITY

The concepts of intensity, flux, and luminosity are very important and yet they are often confused. These terms cannot be used interchangeably as they have very different meanings. The following discussion is based on treatments by Aller[1] and Gray.[2] Figure J.1 shows a small area ΔA on the surface of a star. What we wish to consider is the amount of energy emitted by this small area in the direction θ from the normal (a line perpendicular to the surface). Throughout this discussion, we assume symmetry about the normal. That is to say, the results are the same if θ is drawn to the left of the normal in Figure J.1 instead of to the right. In order to measure the energy, some detector must be used to collect the energy falling on its surface. In practice, you cannot measure energy at a "point" because a point has no area. In Figure J.1, $\Delta A'$ represents the energy-collecting area, and $\Delta \omega$ is the solid angle subtended by this area as seen from the star. Strictly speaking, the only point on the surface that can radiate energy into the cone, $\Delta \omega$, is the point at the vertex of this cone. For the moment, assume that ΔA is very small compared to this cone, so that ΔA looks almost like a point.

The amount of energy emitted each second into the cone depends on $\Delta \omega$. Obviously, if the cone is made larger, it will contain a larger fraction of the total energy emitted by ΔA. A second factor is the range of wavelength or frequency to be measured. No detector can be made to measure light in an infinitely narrow wavelength interval. Likewise, no single detector can measure every portion of the electromagnetic spectrum. It is for this reason we build x-ray, γ-ray, radio, and infrared as well as optical telescopes. A large bandpass contains more energy than a narrow one. Finally, the projected size of ΔA affects the energy in the cone $\Delta \omega$. To understand this, imagine you are viewing ΔA by looking down the axis of the cone. If θ equals zero, you are viewing ΔA perpendicular to its surface and it appears its actual size. As θ increases, the

THE LIGHT RADIATION FROM STARS

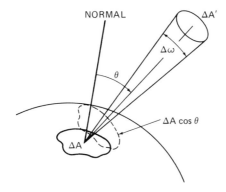

Figure J.1. Geometry used to define flux and intensity.

area appears smaller and smaller until at $\theta = 90°$ the apparent area is zero. Likewise, as θ increases, the apparent brightness decreases until at $\theta = 90°$ it reaches zero. The projected area is $\Delta A \cos \theta$.

Combining all of the above ideas, it can be said that the energy emitted (ΔE) in a time interval (Δt) into the cone is proportional to the size of the cone ($\Delta \omega$), the wavelength interval ($\Delta \lambda$), and the projected surface area ($\Delta A \cos \theta$). Symbolically,

$$\frac{\Delta E}{\Delta t} \propto (\Delta A \cos \theta) \Delta \lambda \Delta \omega. \tag{J.1}$$

The "constant of proportionality" must contain the information that describes the radiation emerging from the star through its surface. Its value is set by the physical conditions in the star's atmosphere such as temperature, pressure, and gravity. This quantity, called the *specific intensity*, I_λ, is defined by rearranging Equation J.1 and taking the limit as Δt, $\Delta \omega$, $\Delta \lambda$, and ΔA go to zero. Then

$$I_\lambda \equiv \lim_{\substack{\Delta t \to 0 \\ \Delta \omega \to 0 \\ \Delta \lambda \to 0 \\ \Delta A \to 0}} \frac{\Delta E_\lambda}{\Delta t \Delta A \cos \theta \, \Delta \omega \, \Delta \lambda}. \tag{J.2}$$

or

$$I_\lambda = \frac{dE_\lambda}{dt \, dA \cos \theta \, d\omega \, d\lambda}. \tag{J.3}$$

The subscript λ has been added to remind us that intensity is a function of wavelength. Specific intensity is not really a constant because it can be a function of direction, wavelength, and perhaps time. The units of specific intensity (usually just called intensity) are ergs per second, per Angstrom, per square centimeter, per square radian.

An important point to note is that intensity can be defined at any point in space, not just on the surface of the star. The area ΔA could be an imaginary surface at any distance from the star, and the definition of intensity could proceed exactly as before. A second point to note is that intensity is independent of the distance from the source. Intensity is a measure of the energy flowing in a solid angle and this does not depend on distance. As an example, suppose that a star radiates evenly in all directions, that is, *isotropically*. A solid angle of one *steradian* (one square radian) then contains $1/4\pi$ times the total energy flow from the star. One steradian always contains this much energy no matter how far one is from the star.

Flux is defined as the *net* energy flow across an element of area per second per wavelength interval. For any given surface of area ΔA, we must sum all the energy flowing in and out of this surface from every angle. Figure J.2 illustrates this idea. Because we want the net energy crossing ΔA, inward and outward energy flows have opposite algebraic signs. Flux is then defined as

$$F_\lambda \equiv \lim_{\substack{\Delta A \to 0 \\ \Delta t \to 0 \\ \Delta \lambda \to 0}} \frac{\Sigma \Delta E_\lambda}{\Delta A \, \Delta t \, \Delta \lambda} \qquad (J.4)$$

or

$$F_\lambda = \frac{\int dE_\lambda}{dA \, dt \, d\lambda}. \qquad (J.5)$$

Solving Equations J.3 for dE_λ and substituting into Equation J.5 results in

$$F_\lambda = \int_{\text{all angles}} I_\lambda(\theta) \cos\theta \, d\omega = 2\pi \int_0^\pi I_\lambda(\theta) \sin\theta \cos\theta \, d\theta. \qquad (J.6)$$

At the surface of a star, the integration can be broken into two parts, the flux leaving the star and the flux directed inward, or

$$F_\lambda = 2\pi \int_0^{\pi/2} I_\lambda(\theta) \sin\theta \cos\theta \, d\theta + 2\pi \int_{\pi/2}^{\pi} I_\lambda(\theta) \sin\theta \cos\theta \, d\theta. \qquad (J.7)$$

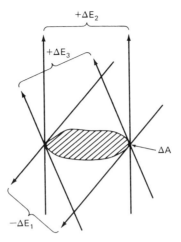

Figure J.2. Energy flow across ΔA.

In the second term, $I_\lambda(\theta)$ must be zero because at the surface of a star all of the energy is directed outward. Therefore,

$$F_\lambda = 2\pi \int_0^{\pi/2} I_\lambda(\theta) \sin\theta \cos\theta \, d\theta. \tag{J.8}$$

If I_λ is independent of direction, then

$$F_\lambda = 2\pi I_\lambda \int_0^{\pi/2} \sin\theta \cos\theta \, d\theta \tag{J.9}$$

or

$$F_\lambda = \pi I_\lambda. \tag{J.10}$$

Keep in mind that this result applies to a rather special case and in general flux and intensity are not related so simply.

To illustrate the difference between intensity and flux, consider the following experiment (based partly on Gray,[2] page 103). Imagine that a telescope lens focuses an image of the sun onto a projection screen. Assume that the solar disk is illuminated uniformly, that is, no limb darkening. The screen has a small hole of area A_D that allows light to reach a photomultiplier tube. Figure J.3 illustrates the experiment. A_s is the area on the sun's surface that corre-

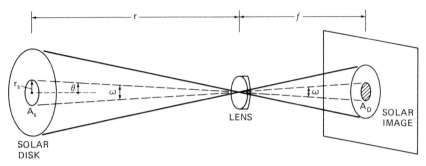

Figure J.3. Illustration to clarify difference between flux and intensity. See text.

sponds to A_D on the screen. In other words, the light entering the hole, A_D, was emitted from area A_s on the sun. Light emitted elsewhere on the sun is imaged to another portion of the screen. The only energy reaching the photomultiplier tube is that contained in the solid angle ω defined by

$$\omega = A_s/r^2 \tag{J.11}$$

where r is the distance between the lens and the sun. Because the output of the PM tube depends on this solid angle, this must be a measure of intensity. (Flux does not depend on solid angle.) If the distance to the sun could be slowly increased, the intensity measured by the photomultiplier tube would remain the same as long as the solar image is larger than A_D. To see this, note that

$$I \propto \frac{E_s}{A_s \omega}, \tag{J.12}$$

where E_s is the energy emitted each second by the surface element A_s into the lens. However, E_s is proportional to A_s for a uniformly emitting solar disk. Therefore,

$$I \propto \frac{1}{\omega} = \frac{r^2}{A_s}. \tag{J.13}$$

If we assume that the light rays pass through the center of the lens, and hence are undeviated, then

$$\frac{A_s}{r^2} = \frac{A_D}{f^2} = \omega \tag{J.14}$$

where f is the focal length of the lens. This last equation says that ω is determined by the size of our detector, A_D, and the focal length of our telescope. The projected area of the detector on the source is A_s. According to Equation J.14, ω is constant and Equation J.13 shows that intensity is constant.

Another way of obtaining the same result is to note that as the distance to the sun increases, A_s must increase because the angle at which the cone diverges from the lens is unchanged. If we assume, for the sake of simplicity, that A_s is circular, then

$$A_s = \pi r_s^2 \tag{J.15}$$

However, r_s increases with distance so that

$$r_s = r\theta \tag{J.16}$$

where θ is the angle subtended by r_s. Combining Equations J.15 and J.16, one finds

$$A_s \propto r^2 \tag{J.17}$$

Combining this with Equation J.13,

$$I \propto \text{constant.} \tag{J.18}$$

In other words, as the distance to the source increases, the area of the source imaged on the detector, A_s, increases in a compensating manner to keep the intensity constant. However, a transition occurs when the distance increases to the point where A_s equals the area of the disk of the sun. At this point, A_s has reached its maximum value and it no longer increases as the distance to the sun increases. The sun is now unresolved because the projected image is smaller than the PM tube opening. It is at this point that the photometer is no longer measuring intensity, but instead is measuring flux. The output of the PM tube is independent of ω, as long as ω contains the entire light source. As the distance between the telescope and the sun continues to increase, the output of the PM tube decreases because flux does depend on distance to the light source.

We can distinguish between two types of photometry. The surface photometry of galaxies is in fact a measurement of intensity. The procedure is to move the image of the galaxy across a small diaphragm opening. The data from this type of research specify a *magnitude per square arc second* at points across the visible surface of the galaxy. Stellar photometry, on the other hand, involves

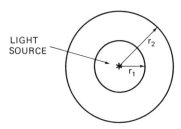

Figure J.4. Two concentric spheres.

unresolved point sources* and hence there is no specification of a solid angle, i.e., stars just have a "magnitude." We should note also that the use of a telescope or optical system can in no way increase the intensity reaching the detector. However, the total energy collected by the telescope depends on the area of the primary lens or mirror. This results in an increase in the flux passing through the focal point, which is a primary function of the telescope. When attempting to measure a faint star, the largest possible telescope should be used. When measuring an extended source (and therefore measuring intensity), the advantage of the large telescope is increased resolution. A small telescope can measure the intensity across the surface of a galaxy as well as a large telescope. However, in a small telescope, large regions may appear as an unresolved "blob," whereas in a large telescope, fine detail may be resolved and measured.

Flux obeys the familiar inverse square law of light. Consider Figure J.4, which shows a point light source surrounded by two transparent spheres of radii r_1 and r_2. If the light in the center is the only light source, then the same total energy that crosses the inner sphere per second must cross the outer sphere. The flux at each sphere is just the total energy radiated per second divided by the area of each sphere, that is,

$$F = \frac{E}{4\pi r^2} \qquad (J.19)$$

Thus the flux is inversely proportional to the square of the distance from the source:

$$\frac{F_1}{F_2} = \frac{r_2^2}{r_1^2}. \qquad (J.20)$$

*Distant galaxies can also appear as unresolved point sources.

The *luminosity* of a star is the total energy emitted per second at all wavelengths in all directions from the entire surface. We must therefore sum the energy contribution from each small element of stellar surface over all solid angles and all wavelengths. The geometry is the same as in Figure J.1. As in the previous discussion of intensity, the projected area, $dA \cos \theta$, must be used. The luminosity, L, is given by

$$L = \int \int \int I_\lambda (\cos \theta \, dA) \, d\omega \, d\lambda. \qquad (J.21)$$

If I_λ is constant over the surface of the star, then

$$L = 4\pi R_*^2 \int \int I_\lambda \cos \theta \, d\omega \, d\lambda \qquad (J.22)$$

where R_* is the radius of the star. By Equation J.6,

$$L = 4\pi R_*^2 \int F_\lambda^* \, d\lambda \qquad (J.23)$$

where F_λ^* is the flux at the surface of the star. We demonstrate later that, to a good approximation, the above integral can be replaced by Stefan's law.

J.2 BLACKBODY RADIATORS

The description of the radiation leaving a star is an enormously complex problem, and there is no simple mathematical expression that accurately describes I_λ for a star. It would be extremely helpful to have a simple expression that at least approximates I_λ for a star. The *blackbody radiator*, a highly idealized radiation source described by theoretical physics, fulfills this need. A blackbody is an object that absorbs all radiation falling upon it. This object also emits as much energy as it receives and is therefore in an equilibrium state at some temperature. Note that a blackbody radiates and therefore need not appear black. For a blackbody, the intensity of the radiation it emits, I_λ, depends only on its temperature T and in the wavelength interval $d\lambda$:

$$I_\lambda(T) \, d\lambda = \frac{2hc^2/\lambda^5}{e^{(hc/\lambda kT)} - 1} \, d\lambda. \qquad (J.24)$$

This expression is known as *Planck's law*. In this expression, T is the temperature in degrees Kelvin, h is Planck's constant (6.63×10^{-27} erg-seconds), c is the speed of light in a vacuum (3.00×10^{10} centimeters/second), k is the Boltzmann constant (1.38×10^{-16} erg/°K), and the wavelength is in centimeters. The units of I_λ are ergs per square centimeter per second per square radian per wavelength interval of one centimeter. Note that these are the units of intensity. By substituting all the constants, Equation J.24 takes on a more usable form of

$$I_\lambda(T)d\lambda = \frac{1}{\lambda^5}\left(\frac{1.19 \times 10^{-5}}{e^{(1.44/\lambda T)} - 1}\right)d\lambda \tag{J.25}$$

where λ is still in centimeters.

The continuum spectrum of most stars at least roughly approximates a black body spectrum. Figure J.5 shows the shape of the blackbody spectrum (a plot of Equation J.25) for two different temperatures. Note that the curves for different temperatures do not cross; a hot blackbody is brighter than a cool one at all wavelengths provided they are viewed from the same distance. The wavelength at which the curve reaches its maximum height, λ_{max}, is a function of temperature. To find this wavelength, it is necessary to solve

$$\frac{dI_\lambda}{d\lambda} = 0 \tag{J.26}$$

for λ, which then equals λ_{max}. The result, when T is in degrees Kelvin, is

$$\lambda_{max} = \frac{0.290}{T} \text{ centimeter.} \tag{J.27}$$

This result is referred to as *Wien's displacement law*. It is important to note that the constant in this equation applies to wavelength intervals only. That is, this equation cannot be used to find the frequency of maximum emission by substituting c divided by the frequency for λ_{max}. The constant must also be changed. That is, when T is in degrees Kelvin,

$$\nu_{max} = 5.88 \times 10^{10} \, T \text{ (Hertz).} \tag{J.28}$$

where ν_{max} is the frequency of maximum emission. Wien's displacement law is very useful for determining the temperature of a blackbody. It is simply a matter of measuring λ_{max} or ν_{max} to find the temperature. This technique can also be applied to stars because they radiate approximately as blackbodies.

Stefan's law is an expression that gives the total energy radiated per second per square centimeter per square radian *at all wavelengths*. Stefan's law is obtained by integrating Planck's law over all wavelengths. The result is

$$I(T) = \frac{\sigma}{\pi} T^4 \tag{J.29}$$

THE LIGHT RADIATION FROM STARS

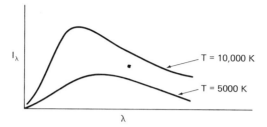

Figure J.5. Blackbody curves.

where $\sigma = 5.67 \times 10^{-5}$ ergs per square centimeter per Kelvin degree to the fourth power. $I(T)$ is simply the area under the curve in Figure J.5. For a blackbody radiator because I is independent of direction, Equation J.29 can be substituted into Equation J.10 to give the flux at its surface,

$$F = \sigma T^4. \tag{J.30}$$

This says that the total flux at all wavelengths depends solely on temperature. In this equation, F is related to F_λ by

$$F = \int_0^\infty F_\lambda \, d\lambda. \tag{J.31}$$

Equation J.30 can be substituted into Equation J.23 to yield an approximation for the luminosity of a star, that is

$$L = 4\pi\sigma R_*^2 \, T_*^4. \tag{J.32}$$

J.3 ATMOSPHERIC EXTINCTION CORRECTIONS

Figure J.6 shows the earth's atmosphere as a plane-parallel slab, i.e., the curvature of the earth is ignored. This is a valid approximation for stars that are more than 30° from the horizon. The relative loss of light flux dF_λ in traveling a distance ds in the earth's atmosphere is shown in Figure J.7. This loss must be proportional to F_λ (the more flux, the greater the absorption, in an absolute sense), to the absorption coefficient, α_λ (the fraction of flux lost per unit distance, in units of cm^{-1}), and the distance traveled through the atmosphere, ds. Stated mathematically,

$$dF_\lambda = -F_\lambda \, \alpha_\lambda \, ds. \tag{J.33}$$

352 ASTRONOMICAL PHOTOMETRY

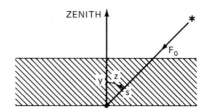

Figure J.6. Path Length is a function of Zenith angle.

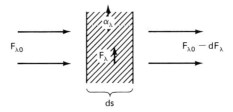

Figure J.7. Absorption of light.

The minus sign indicates that F_λ is decreasing, or being absorbed, with distance traveled. Equation J.33 can be rewritten as

$$\frac{dF_\lambda}{F_\lambda} = -\alpha_\lambda \, ds \tag{J.34}$$

and integrated over the path length, s, traveled in the atmosphere to yield

$$\ln(F_\lambda/F_{\lambda 0}) = -\int_0^s \alpha_\lambda \, ds \tag{J.35}$$

or

$$\frac{F_\lambda}{F_{\lambda 0}} = \exp\left(-\int_0^s \alpha_\lambda \, ds\right) \tag{J.36}$$

where $F_{\lambda 0}$ is the flux at the top of the atmosphere and F_λ is the flux reaching the ground. Astronomers often define the *optical depth*, τ, by

$$\tau_\lambda = \int_0^s \alpha_\lambda \, ds, \tag{J.37}$$

as this term is dependent only on the absorbing material and the geometric orientation, not on the source of radiation. Then we can rewrite Equation J.36 as

$$\frac{F_\lambda}{F_{\lambda 0}} = e^{-\tau_\lambda}. \tag{J.38}$$

Note that if $\tau_\lambda = 1$, then the flux reaching the ground is $1/e$ of the incident flux on the top of the atmosphere.

To convert this flux ratio to a magnitude difference, we apply Equation 1.3 to yield

$$\begin{aligned} m_\lambda - m_{\lambda 0} &= -2.5 \log (F_\lambda/F_{\lambda 0}) \\ m_\lambda - m_{\lambda 0} &= -2.5 \log (e^{-\tau_\lambda}) \end{aligned} \tag{J.39}$$

where m_λ and $m_{\lambda 0}$ are the apparent magnitude of the star at the earth's surface and above the atmosphere, respectively. Equation J.39 becomes

$$m_\lambda - m_{\lambda 0} = 2.5 \, (\log e) \tau_\lambda \tag{J.40}$$

or

$$m_{\lambda 0} = m_\lambda - 1.086 \, \tau_\lambda. \tag{J.41}$$

This equation can be placed in a more useful form if we show the variation of τ_λ with location in the sky. By Figure J.6

$$\cos z = y/s \tag{J.42}$$

or

$$s = y \sec z \tag{J.43}$$

and

$$ds = dy \sec z \tag{J.44}$$

where z is the zenith angle, y is the thickness of the atmosphere at the zenith, and s is the path length of the light. By Equations J.37 and J.44, the optical depth can be expressed as

$$\tau_\lambda = \sec z \int_0^y \alpha_\lambda \, dy. \tag{J.45}$$

The integral is simply the optical depth at the zenith, a constant factor. Thus, Equation J.41 becomes

$$m_{\lambda 0} = m_\lambda - 1.086 \sec z \int_0^y \alpha_y \, dy$$

or

$$m_{\lambda 0} = m_\lambda - k'_\lambda \sec z \qquad (J.46)$$

where k'_λ is called the principal extinction coefficient and $\sec z$ represents the air mass, or the relative amount of atmosphere traversed. Air mass is often designated as X.

There are two different kinds of particles in our atmosphere that cause extinction. Each has different wavelength dependences. The major constituent are the molecules, where k'_λ varies as λ^{-4}. These particles are roughly the same size as the wavelength of light. Larger particles, like dust, are called *aerosols* and cause k'_λ to vary as λ^{-1} or λ^0. The relative fraction of each type depends on the atmospheric conditions and the zenith distance.

We have taken a simple approach to extinction. Our hypothetical case accounts for the wavelength dependence of the absorption, but we assume a plane-parallel atmosphere and filters with infinitely sharp bandpasses in order to obtain a magnitude at a specific wavelength. The actual case of a spherical earth is covered in Chapter 4. The bandwidth effect is discussed here and gives rise to a correction to the extinction known as second-order extinction.

Within a filter's bandpass, some wavelengths suffer more extinction than others. In general, blue wavelengths are absorbed and scattered more readily. An average extinction in the bandpass could be used, but then stars with rising flux toward the blue end of the bandpass (in general, hot stars) would systematically be given less extinction than is the real case, and those rising toward the red (cool stars) would be given more extinction and a fainter magnitude. In other words, using an average extinction introduces a systematic error in the magnitude determination, an error that is dependent on the color of the star and the air mass through which it is observed. Correcting Equation J.46 for this color-dependent term, we obtain

$$m_{\lambda 0} = m_\lambda - k'_\lambda \sec z - k''_\lambda c \sec z \qquad (J.47)$$

or

$$m_{\lambda 0} = m_\lambda - (k'_\lambda + k''_\lambda c) \sec z \qquad (J.48)$$

where k_λ'' is called the second-order extinction coefficient and c is the instrumental color index of the star.

Because atmospheric extinction is wavelength-dependent, the apparent color index of a star is also affected. Equation J.47 can be written with a subscript 1 or 2 for the two wavelengths. That is,

$$m_{\lambda 01} = m_{\lambda 1} - k'_{\lambda 1} \sec z - k''_{\lambda 1} c \sec z \qquad (J.49)$$
$$m_{\lambda 02} = m_{\lambda 2} - k'_{\lambda 2} \sec z - k''_{\lambda 2} c \sec z. \qquad (J.50)$$

Subtracting, we then obtain the color index

$$(m_{\lambda 01} - m_{\lambda 02}) = (m_{\lambda 1} - m_{\lambda 2}) - (k'_{\lambda 1} - k'_{\lambda 2}) \sec z$$
$$- c(k''_{\lambda 1} - k''_{\lambda 2}) \sec z. \qquad (J.51)$$

If we let the color indices c and c_0 be defined as

$$c = (m_{\lambda 1} - m_{\lambda 2})$$
$$c_0 = (m_{\lambda 01} - m_{\lambda 02}) = (m_{\lambda 1} - m_{\lambda 2})_0$$

and let

$$k'_c = k'_{\lambda 1} - k'_{\lambda 2}$$
$$k''_c = k''_{\lambda 1} - k''_{\lambda 2}$$

then

$$c_0 = c - k'_c \sec z - k''_c (\sec z) \, c. \qquad (J.52)$$

In practice, the extinction coefficients can be determined by a few measurements of stars at various altitudes without any knowledge of α_λ or the actual physical processes of absorption. There are pitfalls in the determination of these coefficients because of their temporal and spatial variability. The practical details are presented in Chapter 4.

J.4 TRANSFORMING TO THE STANDARD SYSTEM

For observers at different observatories to be able to compare observations, observations must be transformed from the instrumental systems (which are all different) to the standard system. The main reason for this difference between photometer systems is that the equivalent wavelengths of observation are slightly different. The equivalent wavelength (λ_{eq}) is an average wavelength of observation weighted by the stellar flux and the response function of the equipment, and is defined by

$$\lambda_{eq} = \frac{\int_0^\infty \lambda \phi(\lambda) \, d\lambda}{\int_0^\infty \phi(\lambda) \, d\lambda} \tag{J.53}$$

where $\phi(\lambda) = \phi_A(\lambda)\phi_T(\lambda)\phi_F(\lambda)\phi_{PM}(\lambda)$, as defined in Chapter 1. Small changes in the spectral response of the measuring equipment change λ_{eq}. This means that a magnitude determined by one instrumental system differs slightly from that found by a second system. For example, suppose you are measuring a hot star whose flux is rising very rapidly toward the blue. If your equivalent wavelength is slightly blueward of the standard system, your measured magnitude will be brighter than the accepted value. Because the difference in λ_{eq} is usually small, the magnitude on the standard system, M, can be approximated by a Taylor expansion in λ_{eq} about the instrumental magnitude, m_0. That is,

$$M = m_0 + \left(\frac{\Delta m_0}{\Delta \lambda_{eq}}\right) \Delta \lambda_{eq} + \text{higher-order terms.} \tag{J.54}$$

Figure J.8 illustrates this situation. Note that in this discussion the magnitudes are assumed to have already been corrected for atmospheric extinction. The term in parentheses is the slope of the star's continuum, which is proportional to the star's color or color index in that region. Any instrumental magnitude, $m_{\lambda 0}$, can be converted to the standard magnitude, M, by an expression of the

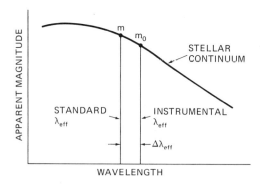

Figure J.8. Difference between the standard and instrumental equivalent wavelength.

form

$$M_\lambda = m_{\lambda 0} + \beta_\lambda C + \gamma_\lambda. \tag{J.55}$$

β_λ and γ_λ are constants which are unique to the photometer in use and C is the standard color index of the star, near the wavelength in question.

The transformation of a color index (that has already been corrected for extinction) is found by applying Equation J.55 to each spectral region and forming the difference. That is,

$$(M_{\lambda 1} - M_{\lambda 2}) = (m_{\lambda 1} - m_{\lambda 2}) + (\beta_{\lambda 1} - \beta_{\lambda 2})C + \gamma_{\lambda 1} - \gamma_{\lambda 2}$$

or

$$C = c_0 + \beta_c C + \gamma_c \tag{J.56}$$

where C is the color index on the standard system, c is the instrumental color index and β, γ, β_c, and γ_c are constants. If we define a constant, δ, by

$$\delta = \frac{1}{1 - \beta_c},$$

then

$$C = \delta c_0 + \gamma_c. \tag{J.57}$$

For both the magnitude and color index transformations, we have assumed that the higher-order terms beyond a linear relation to the color index are negligible. This is not always the case. An example is the use of a "neutral" density filter that adds additional color dependence to the instrumental response. These situations are apparent when the coefficient determination is attempted, as departures from a linear dependence are evident. For these situations, the next higher-order term (quadratic in color) must be added to the procedure. As these conditions rarely arise, further treatment is not presented in this text. but you should be aware of the possibility.

REFERENCES

1. Aller, L. H. 1963. *The Atmospheres of the Sun and Stars*. New York: Ronald Press.
2. Gray, D. F. 1976. *The Observation and Analysis of Stellar Photospheres*. New York: John Wiley and Sons.

APPENDIX K
ADVANCED STATISTICS

Appendix K explains some of the finer points in statistics and their application to astronomy. The subject matter treated here is not essential for basic photometry, but is not covered elsewhere in this text and is given here to stimulate the interested to read further.

Much of the presented material has been drawn from elementary statistics texts[1,2,3] or from the astronomical literature.[4,5,6] You should know the meaning of the terms *matrix, determinant, derivative,* and *partial derivative* before you can make full use of this appendix.

K.1 STATISTICAL DISTRUBUTIONS

To introduce the idea of a probability distribution, assume that we are executing a coin toss. It is relatively simple to calculate the probability of tossing three heads and seven tails out of 10 tosses, but what is the probability of tossing n heads and $(m - n)$ tails out of m tosses? These numbers can be thought of as describing the function $p(n)$, where p is the probability that n of the tosses will be heads. Such a function is called a *probability distribution*.

If the index n varies between 1 and m, then we have included all possible events and the sum of these probabilities must be unity. That is,

$$\sum_{n=1}^{m} p(n) = 1 . \tag{K.1}$$

The two types of probability distributions extensively used in experimental data in photometry are the Gaussian, for experimental values, and the Poisson, for photon arrivals. You need not know the functional forms of these distributions to use the results that mathmaticians have derived from them, such as the standard deviation; but it is important to at least look at the shapes of the distributions.

The Gaussian (or normal) distribution is defined by Equation 3.10, rewritten here as

$$p(x) = \frac{1}{\sigma\sqrt{2\pi}} \exp\left[-\frac{1}{2}\left(\frac{x-\bar{x}}{\sigma}\right)^2\right] \qquad (K.2)$$

This distribution is shown in Figure 3.1 and repeated in Figure K.1 for completeness. It has been shown to approximate errors of measurement very closely and therefore is the most widely known and used of all distributions.

Often, Equation K.2 is *standardized* when given in tables in the back of statistics texts. That is, the function is given with respect to the standardized normal variable, z, where

$$z = \frac{x - \bar{x}}{\sigma} \qquad (K.3)$$

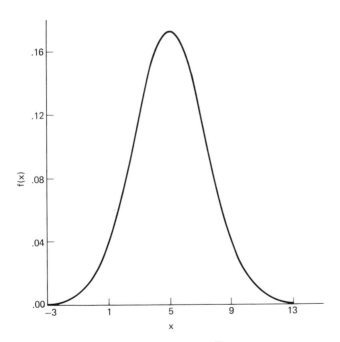

Figure K.1. Gaussian distribution for $\bar{x} = 5.0$, $\sigma = 2.3$.

then $p'(z)$ is tabulated where

$$p'(z) = \sigma p(z) = \frac{1}{\sqrt{2\pi}} e^{-z^2/2} \tag{K.4}$$

This allows tables of the function to be given without having to list function points for every combination of σ and $\bar{x}$.

For approximating arrival rates of photons, the Poisson distribution is the most commonly used function. It is defined by

$$p\begin{pmatrix} y \text{ occurrences in} \\ \text{a given time unit} \end{pmatrix} = \begin{cases} \dfrac{e^{-\lambda}\lambda^y}{y!}, \text{ for } \begin{cases} y = 0, 1, 2, \ldots \\ \lambda > 0 \end{cases} \\ 0, \text{ otherwise} \end{cases} \tag{K.5}$$

where p is the probability of occurrence and λ is the mean arrival rate, the number of photons per second. This function looks much like the Gaussian, except:

1. The variance is given by $\sigma^2 = \lambda$.
2. It is a discrete distribution, not continuous like the Gaussian, because a photon either arrives or it does not arrive.
3. It is skewed to the right because there are no negative occurrences, that is, y is never negative.
4. As $\lambda \to \infty$, the Poisson distribution approaches the Gaussian distribution in form.

An example of the Poisson distribution for $\lambda = 5.0$ is given in Figure K.2.

Just as the standard deviation was defined for a Gaussian distribution, a similar "standard deviation" can be derived for the Poisson distribution. If Equation K.5 is substituted in a generalized form of Equation 3.6 (a messy proposition!), then the standard deviation of Poisson-distributed data is given by

$$\sigma_p = \sqrt{N} \tag{K.6}$$

where N is the total number of observations. Note that this value is not the same as for the Gaussian distribution. In fact, no two distributions have the same value for the standard deviation. This means that if there is a bias in your sample so that it does not fit a Gaussian distribution curve, the standard deviation that you derive using Equation 3.6 is not the correct one for your sample! *You must be very careful to remove any chance of bias from your calculations.*

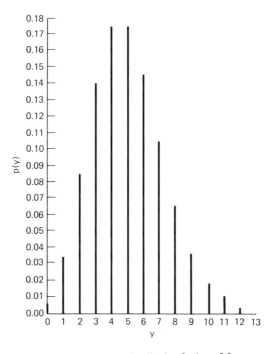

Figure K.2. Poisson distribution for $\lambda = 5.0$.

K.2 PROPAGATION OF ERRORS

If you use various experimental observations to calculate a result, and these observations each have uncertainties associated with them, then the error in the result will be a function of the errors of the individual observations. For example, if you want to calculate the density, or weight per unit volume, of a liquid in a container, you need to both find the volume of the container and its weight. It is not entirely obvious what the density error is if you know the volume to $\pm\ 0.1$ percent and the weight to $\pm\ 1.0$ percent. To find the resultant error, you need to propagate each individual error through the calculation and see how it affects the final result.

In a general form, assume that the function $f(x)$ is similar to that shown in Figure K.3. We can see that an error in the measurement of x, ϵ_x, causes a corresponding error, ϵ_f, in our evaluation of the function. Now

$$\frac{\epsilon_f}{\epsilon_x} = \frac{\Delta f(x)}{\Delta x} \cong \frac{df}{dx} = f'(x)$$

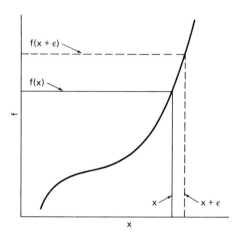

Figure K.3. Error in f(x) due to ϵ(x).

or

$$\epsilon_f = f'(x)\epsilon_x \tag{K.7}$$

where $f'(x)$ is a shorthand way of expressing the derivative of $f(x)$, i.e., how fast it varies with a change in x.

We can extend Equation K.7 to functions of more than one variable:

$$\epsilon_F = \frac{\partial f}{\partial x}\epsilon_x + \frac{\partial f}{\partial y}\epsilon_y + \frac{\partial f}{\partial z}\epsilon_z + \cdots \tag{K.8}$$

where $\frac{\partial f}{\partial x}$ is the partial derivative of f with respect to x, which means simply that we recognize f is a function of many variables, but we are evaluating the derivative with respect to x and hold the variations due to the other variables constant during the process.

Now

$$\sigma_F^2 = \frac{\sum_{i=1}^{N} \epsilon_f^2}{N} \tag{K.9}$$

$$\sigma_f^2 = \frac{1}{N}\sum_{i=1}^{N}\left[\frac{\partial f}{\partial x}\epsilon_{x_i} + \frac{\partial f}{\partial y}\epsilon_{y_i} + \cdots\right]^2 \tag{K.10}$$

which reduces upon expansion and removal of cross terms to:

$$\sigma_f = \left\{ \frac{1}{N} \left[\left(\frac{\partial f}{\partial x}\right)^2 \sigma_x^2 + \left(\frac{\partial f}{\partial y}\right)^2 \sigma_y^2 + \cdots \right] \right\}^{1/2} \quad (K.11)$$

that is, errors add in *quadrature*.

We seldom know the mathematical form of the function f. There are some examples in photometry, however, where Equation K.11 can be applied. The *UBV* transformation equations depend on the values of the zero points, the color terms, the extinction coefficients, and the air mass. By knowing the standard deviation in each of these parameters, you can calculate the expected deviation in the derived standard magnitudes and also see which terms are relatively unimportant.

K.3 MULTIVARIATE LEAST SQUARES

In Chapter 3, we considered the case of fitting a least-squares line to the function

$$f(x_i) = a + bx_i. \quad (K.12)$$

That is, by knowing the approximate $f(x_i)$ for several values of x_i, we could solve for the parameters a and b. This is simple linear regression in that the function $f(x)$ is dependent only in a linear fashion on one variable, x.

There are two ways that this result could be generalized: (1) $f(x)$ could have a nonlinear dependence on x, such as

$$f(x_i) = a + bx_i^2$$

or (2) the function f could be dependent on more than one variable, such as

$$f(x_i, y_i) = a + b_1 x_i + b_2 y_i. \quad (K.13)$$

Equation K.13 is known as *multiple linear regression* and is covered in this section. The nonlinear case can be found in Young,[1] Harnell,[2] and Bevington.[3]

Multiple linear regression is very important in photometry in solving the transformation equations. For example, substituting Equation 4.28 into 4.32 yields

$$V = \epsilon(B - V) + (v - k_v' X - k_v''(b - v)X) + \zeta_v, \quad (K.14)$$

which shows that V is a function of the star, the instrument, and the air mass through which the star is observed.

A mathematical representation of the regression equation is

$$\hat{y} = a + b_1 z_1 + b_2 z_2 + \cdots + b_m z_m \tag{K.15}$$

where $z_1, z_2, \ldots$ are *not* the individual values of one variable (z), but are independent variables. For example, a comparison with Equation K.13 gives $z_1 = x$, $z_2 = y$. Then we minimize the sum of the squares of the deviations, D:

$$\begin{aligned} D &= \sum_{i=1}^{n} (y_i - \hat{y}_i)^2 \\ &= \sum_{i=1}^{n} (y_i - a - b_1 z_{1i} - b_2 z_{2i} - \cdots - b_m z_{mi})^2 \end{aligned} \tag{K.16}$$

where n is the number of observations. The procedure for minimizing D is the same as in the simple linear case, except that partial derivatives must be taken for $m + 1$ variables instead of just two. Setting these $m + 1$ partial derivatives to zero and solving, we obtain the following set of normal equations:

$$\begin{aligned} an + b_1 \Sigma z_{1i} + b_2 \Sigma z_{2i} + \cdots + b_m \Sigma z_{mi} &= \Sigma y_i \\ a\Sigma z_{1i} + b_1 \Sigma z_{1i}^2 + b_2 \Sigma z_{1i} z_{2i} + \cdots + b_m \Sigma z_{1i} z_{mi} &= \Sigma z_{1i} y_i \\ a\Sigma z_{2i} + b_1 \Sigma z_{1i} z_{2i} + b_2 \Sigma z_{2i}^2 + \cdots + b_m \Sigma z_{2i} z_{mi} &= \Sigma z_{2i} y_i \\ \vdots \\ a\Sigma z_{mi} + b_1 \Sigma z_{1i} z_{mi} + b_2 \Sigma z_{2i} z_{mi} + \cdots + b_m \Sigma z_{mi}^2 &= \Sigma z_{mi} y_i \end{aligned} \tag{K.17}$$

In matrix form,

$$\begin{aligned} A_{jk} &= \Sigma z_j z_k \quad \text{where } j = 0,n \text{ and } k = 0,n \\ N_j &= \Sigma z_{ji} y_i \\ B_j &= b_j \quad \text{where } B(0) = a \end{aligned}$$

then

$$A_{jk} B_j = N_j.$$

Removing subscripts,

$$\mathbf{A}\mathbf{B} = \mathbf{N}$$
$$\mathbf{B} = \mathbf{A}^{-1}\mathbf{N} \qquad (K.18)$$

Equation K.18 gives the coefficient vector B, providing that matrix A can be inverted. There are more elegant methods of solving the normal equations, for instance, by assuming they are separable, solving for $b_1, b_2 \ldots, b_m$ and then back-substituting for the constant term b_0.

As an example, we solve Equation 2.10 for the zero-point, first-order extinction, and the transformation coefficient, that is a "full-blown" solution. If we rewrite Equation 2.10 by substituting Equation 2.7 to obtain

$$(B - V) = \zeta_{bv} + \mu(b - v)[1 - k''_{bv}X] - \mu k'_{bv}, \qquad (K.19)$$

We can see by correspondence that we have

$$y_i = a + b_1 z_1 + b_2 z_2$$
$$y_i = B - V$$
$$a = \zeta_{bv}$$
$$b_1 = \mu$$
$$z_1 = (b - v)[1 - k''_{bv}X]$$
$$b_2 = -\mu k'_b v$$
$$z_2 = X.$$

Then, writing the normal equations,

$$\begin{aligned} an + b_1 \Sigma z_{1i} + b_2 \Sigma z_{2i} &= \Sigma y_i \\ a\Sigma z_{1i} + b_1 \Sigma z_{1i}^2 + b_2 \Sigma z_{1i} z_{2i} &= \Sigma z_{1i} y_i \\ a\Sigma z_{2i} + b_1 \Sigma z_{1i} z_{2i} + b_2 \Sigma z_{2i}^2 &= \Sigma z_{2i} y_i, \end{aligned} \qquad (K.20)$$

we obtain

$$\begin{pmatrix} n & \Sigma z_{1i} & \Sigma z_{2i} \\ \Sigma z_{1i} & \Sigma z_{1i}^2 & \Sigma z_{1i} z_{2i} \\ \Sigma z_{2i} & \Sigma z_{1i} z_{2i} & \Sigma z_{2i}^2 \end{pmatrix} \begin{pmatrix} a \\ b_1 \\ b_2 \end{pmatrix} = \begin{pmatrix} \Sigma y_i \\ \Sigma z_{1i} y_i \\ \Sigma z_{2i} y_i \end{pmatrix}, \qquad (K.21)$$

a matrix equation, which can be solved by back-substitution or the use of minors. See Arfken[7] for more detail.

The transformation equations for V and $(U - B)$ can be solved in a similar manner. A subroutine to calculate the full treatment UBV least-squares solution can be found in Section I.5. You should be very wary of solving for both the color and extinction coefficients at once, as they interact with each other and bad data points are not obvious. We suggest using simple linear least squares for the color coefficients and a separate solution for extinction, as presented in Chapter 4.

K.4 SIGNAL-TO-NOISE RATIO

Each time a star is measured, there is an uncertainty in the value obtained. If systematic errors are ignored, the uncertainty can be defined as the *noise,* or the standard deviation of a single measurement from the mean of all the measures made on the star. The ratio of the value derived for the observed count rate to the uncertainty in that number is called the output *signal-to-noise ratio* (S/N).

If an idealized detector is considered, where no background is present and the only source of noise is from statistical fluctuations from the star, then the arriving signal is given by

$$S_{in} = \dot{C}_s t \tag{K.22}$$

where S_{in} is the number of input photons to the detector in time t, $\dot{C}_s$ is the rate of photon arrival in photons per second, and t is the integration time, or the time required for the measurement (seconds). For a weak beam of photons, Poisson statistics are a good approximation to the statistical fluctuations of the beam. These statistics state that the noise, or standard deviation of the signal from the mean, is given by Equation K.6, or in another form,

$$N_{in} = S_{in}^{1/2} = (\dot{C}_s t)^{1/2}. \tag{K.23}$$

The S/N calculated using Equations K.22 and K.23 tells us, in a sense, what fraction of the *arriving* signal is contributed by the noise. This ratio is given by

$$\left(\frac{S}{N}\right)_{in} = \frac{\dot{C}_s t}{(\dot{C}_s t)^{1/2}} \tag{K.24}$$

or

$$\left(\frac{S}{N}\right)_{in} = (\dot{C}_s t)^{1/2}. \tag{K.25}$$

One hundred total source counts would yield an S/N of 10. Because the noise varies as the square root of the signal, 10,000 total counts would be required to reduce the noise to one percent of the signal or equivalently, to have an S/N of 100.

K.4a Detective Quantum Efficiency

Very few photomultiplier tubes are perfectly efficient. In general, only a fraction, Q, of the incident photons are detected. The parameter Q can be thought of as the efficiency of detection of quanta or photons, or the *quantum efficiency*. Realizing this, the S/N (Equation K.24) for the output signal can be written as:

$$\left(\frac{S}{N}\right)_{out} = \frac{(Q\dot{C}_s t)}{(Q\dot{C}_s t)^{1/2}} \tag{K.26}$$

$$\left(\frac{S}{N}\right)_{out} = (Q\dot{C}_s t)^{1/2}. \tag{K.27}$$

Q is generally not known, but must be measured experimentally. Rearranging,

$$Q = \frac{\left(\dfrac{S}{N}\right)_{out}^{2}}{\dot{C}_s t}$$

$$Q = \frac{\left(\dfrac{S}{N}\right)_{out}^{2}}{\left(\dfrac{S}{N}\right)_{in}^{2}} \tag{K.28}$$

Theoretically, Equation K.28 holds only for the case where the number of detected photons is just a simple linear fraction of the number of incident photons. We can generalize to the more common case by defining

$$\mathrm{DQE} = \frac{\left(\dfrac{S}{N}\right)_{out}^{2}}{\left(\dfrac{S}{N}\right)_{in}^{2}} \tag{K.29}$$

where DQE is the *detective quantum efficiency*, so called because it is a measure of how badly the actual detector deviates from a perfect detector. The actual detector or system performs as if it were an ideal detector with the input signal decreased by the factor DQE (≤ 1). The DQE of a system is always less than the quantum efficiency of the photocathode, as some electrons are lost in the dynodes and different detection systems weight the anode pulses differently. A typical photomultiplier system has a DQE of 2 to 4 percent.

So far we have discussed only shot or Poisson noise, the fluctuations inherent in the source itself. There are several other possible sources of noise, both in the sky and the actual detection system. All do not affect the output S/N in the same manner, and at a given source intensity, one noise component may be dominant or several may contribute. For faint sources, the source signal may be much smaller than the noise.

Listed below are the three major types of noise sources and how they affect the signal-to-noise ratio when they are dominant. Discussions of these sources can be found in other chapters.

1. *Noise dependent on the square root of the signal.*
    a. Photon shot noise, N_s, in the signal.

$$N_s = G(Q\dot{C}_s t)^{1/2}$$

Here Equation K.26 becomes

$$\left(\frac{S}{N}\right)_{out} = \frac{GQ\dot{C}_s t}{G(Q\dot{C}_s t)^{1/2}} = (Q\dot{C}_s t)^{1/2} \quad (K.30)$$

where G is the internal gain of the detector. Note that the S/N is proportional to the square root of the number of counts; increasing counts by a factor of 100 increases the S/N by only a factor of 10. By taking logarithms of both sides,

$$\log\left(\frac{S}{N}\right)_{out} = \tfrac{1}{2} \log(Q\dot{C}_s t) = \tfrac{1}{2} \log(\text{counts}). \quad (K.31)$$

So plotting several readings of log (S/N) versus log (counts) gives a straight line slope ½.

2. *Noise linearly dependent on the signal.*
    a. Scintillation noise from the atmosphere.
      This noise source, N_{sc}, is a fractional modulation, m, of the incident beam, that is

$$N_{sc} = m Q \dot{C}_s t. \quad (K.32)$$

In other words, stars of all magnitudes vary by the same percentage. If a 10^m star is varying by 10 percent, then so is an 18^m star. Equations K.26 and K.31 then become

$$\left(\frac{S}{N}\right)_{out} = \frac{GQ\dot{C}_{sc}t}{GmQ\dot{C}_{sc}t} = \frac{1}{m} \tag{K.33}$$

$$\log\left(\frac{S}{N}\right)_{out} = \text{constant} \tag{K.34}$$

where the slope of the plot is zero.
3. *Noise independent of the signal.*
   a. Amplifier noise.
   Equations K.26 and K.31 become

$$\left(\frac{S}{N}\right)_{out} = \frac{GQ\dot{C}_s t}{N_{am}} \tag{K.35}$$

$$\log\left(\frac{S}{N}\right)_{out} = \log G - \log N_{am} + \log(Q\dot{C}_s t) \tag{K.36}$$

where N_{am} is the amplifier noise in *net equivalent photons*, that is, the number of incident photons that would be required to produce a noise signal the size of the amplifier noise. In this example, the slope of the plot equals 1.
   b. Background shot noise, N_b, such as from the sky.

$$N_b = G(Q\dot{C}_b t)^{1/2}$$

Our equations become:

$$\left(\frac{S}{N}\right)_{out} = \frac{GQ\dot{C}_s t}{G(Q\dot{C}_b t)^{1/2}} \tag{K.37}$$

$$\log\left(\frac{S}{N}\right)_{out} = \log(Q\dot{C}_s t) - \tfrac{1}{2}\log(Q\dot{C}_b t) \tag{K.38}$$

where $\dot{C}_b$ is the background count rate. Again, the slope is equal to 1.
Each of these sources becomes dominant in different count regimes. For instance, bright sources have negligible background, sources near the horizon exhibit much more scintillation, etc. For count rates not dominated by one noise component, all contributors must be accounted for by adding the noise contributions in quadrature, i.e.,

$$N_{tot} = (N_s^2 + N_{sc}^2 + N_{am}^2 + N_b^2 + \cdots)^{1/2} \tag{K.39}$$

370 ASTRONOMICAL PHOTOMETRY

so that

$$\left(\frac{S}{N}\right)_{out} = \frac{GQ\dot{C}_s t}{(N_s^2 + N_{sc}^2 + N_{am}^2 + N_b^2 + \cdots)^{1/2}}. \tag{K.40}$$

A derivation of why uncorrelated noises (or standard deviations) add in quadrature is beyond the scope of this section, but can be described as taking the partial derivative of the total noise count function with respect to each of the contributors (source, sky, etc.) and noting that cross terms vanish (N_s is not dependent on N_b etc.).

A plot of the theoretical signal-to-noise ratio obtainable with $t = 1$ second, $Q = G = 1$, and $\dot{C}_b = 30$ counts per second for various sources (i.e., varying $\dot{C}_s$) is shown in Figure K.4. Note that for high count rates, the slope approaches ½ and for count rates near $\dot{C}_b$ the slope is approximately 1.

K.4b Regimes of Noise Dominance

This section examines three noise regimes as examples of signal-to-noise ratio considerations. These are:

1. *Background.* So far, only those cases where the background and source are uniquely known have been discussed. This is not normally the case, as the background is generally unknown and must be measured during

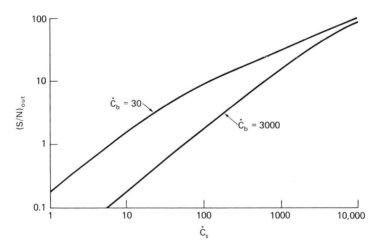

Figure K.4. Signal-to-noise ratio comparisons for differing sources.

part of the total observing time. For this case of first measuring source with sky background and then sky by itself, the noise or standard deviation from the mean is a combination of Poisson noise and background noise:

$$N = \{[\sqrt{(\dot{C}_s + \dot{C}_b)t_s}]^2 + [\frac{t_s}{t_b}\sqrt{\dot{C}_b t_b}]^2\}^{1/2}$$

$$N = \sqrt{\dot{C}_s t_s + \dot{C}_b t_s + \dot{C}_b t_s^2/t_b} \qquad (K.41)$$

where t is the amount of time spent measuring source or background and $\dot{C}_x$ is the count rate for a measurement, with subscript s referring to the source and subscript b the background. Also, the data are represented by:

$$\text{Total source counts} = S \pm N \qquad (K.42)$$

$$\text{Count rate for source (per unit time)} = \frac{S}{t_s} \pm \frac{N}{t_s} \qquad (K.43)$$

2. *Scintillation.* For brighter stars, the dominant noise source is scintillation. But if scintillation and background noises are both important for a source, one should realize that scintillation affects only the source counts and not the sky counts. The S/N for this case is:

$$\left(\frac{S}{N}\right)_{out} = \frac{(\dot{C}_b + \dot{C}_s)t}{\sqrt{\dot{C}_s t + \dot{C}_b t + m^2(\dot{C}_s t)^2}}. \qquad (K.44)$$

As long as the source counts are low or the scintillation modulation, m, is small, the S/N takes on its previous appearance, Equation K.41.

3. *Amplifier noise.* In many cases, such as astronomical TV systems, both amplifier and photon shot noise are important. The S/N for this case is:

$$\left(\frac{S}{N}\right)_{out} = \frac{GQ\dot{C}_s t}{\{[G(\dot{C}_s t)^{1/2}]^2 + N_{am}^2\}^{1/2}}$$

$$= \frac{GQ\dot{C}_s t}{(G^2 Q \dot{C}_s t + N_{am}^2)^{1/2}}$$

$$\left(\frac{S}{N}\right)_{out} = \frac{Q\dot{C}_s t}{\left(Q\dot{C}_s t + \frac{N_{am}^2}{G^2}\right)^{1/2}} \qquad (K.45)$$

Note that the effect of the amplifier noise is decreased by the internal gain of the detector. The ratio is then added, much as a source of photons.

This noise-to-gain ratio can be thought of as the number of equivalent incident photons on the detector that would give shot noise of the same amount. In other words,

$$\frac{N_{am}}{G} = \text{amplifier noise referred to the input.}$$

We then see why amplifier noise, DC amplifier or pulse preamp, is not important for photomultiplier systems, because the noise-free gain in the photomultiplier tube is 10^6, which reduces the effect of amplifier noise to negligible levels.

K.5 THEORETICAL DIFFERENCES BETWEEN DC AND PULSE-COUNTING TECHNIQUES

There are two basic methods of detecting the output from a photomultiplier tube. DC techniques measure the feeble current generated by the incident photons while pulse-counting techniques measure the number of photons or pulses directly. Proponents of both sides have been arguing for decades over which technique, if either, is superior. The major area in which one may have an advantage over the other is in the signal-to-noise ratio. For this reason, the comparison between DC and pulse-counting methods is included in this chapter.

K.5a Pulse Height Distribution

Each dynode of a 1P21 photomultiplier tube has an average gain of about 5. However, statistically one dynode might have a gain of 4 one time, and 6 or even 10 the next. This means that the number of electrons collected at the anode is not constant, depending on which dynode deviates from the mean. This was shown earlier in Figure 7.2 for a typical photomultiplier tube output. The noise is given by N_{am}/G and is small for a photomultiplier tube but not negligible on other devices nor for larger dark currents. Typically the distribution of pulse sizes is called the *pulse height distribution,* shown for an idealized source in Figure 7.3.

The general features of such a distribution are readily seen. As x approaches zero, all pulses are a result of dark current or amplifier noise and the number of such pulses approaches infinity. After this initial peak, the next larger peak is a result of a photon liberating one photoelectron at the cathode. Other peaks occur at larger heights because more than one photoelectron is released by the photon or by cosmic ray induced electrons. Many photoelectrons may be produced if the cosmic ray strikes the photocathode at near grazing incidence. The

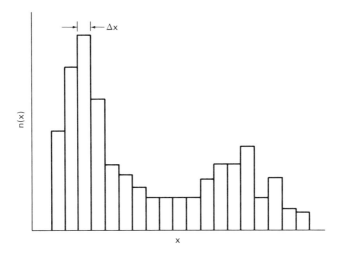

Figure K.5. Schematic pulse height distribution.

probability of more than one photoelectron being liberated by a photon is very small, and most of these higher energy events are caused by the extraneous noise sources such as cosmic rays at a rate of about two per square centimeter of cathode per minute. The shape of the distribution is not constant, but varies according to dynode voltages, temperature, wavelength, and cathode area.

K.5b Effect of Weighting Events on the DQE

The primary theoretical difference between the two detection methods is in their weighting of photomultiplier pulses. The DC method gives the *average* current from the photomultiplier tube; therefore, a large pulse gives more current and is weighted more heavily, even though it signifies the arrival of only one photon just as a smaller pulse does. Pulse-counting techniques treat all pulses equally.

A schematic form of a pulse height distribution is shown in Figure K.5, where $n(x)$ is the number of detected pulses and x is the size of the pulse. Without weights, the detected signal is

$$S = Q\, n(x_1)\, \Delta x_1 + Q\, n(x_2)\, \Delta x_2 + \cdots \qquad (K.46)$$

or, because $\Delta x_1 = \Delta x_2 = \Delta x_3$, and so forth,

$$S = \sum_{i=1}^{N} Q\, n(x_i)\, \Delta x. \qquad (K.47)$$

In integral form ($\Delta x \to 0$),

$$S = \int_0^\infty Q\, n(x)\, dx. \tag{K.48}$$

Incorporating weighting factors, the equations become

$$S = w(x_1)\, Q\, n(x_1)\, \Delta x + w(x_2)\, Q\, n(x_2)\, \Delta x + \cdots \tag{K.49}$$

$$S = \sum_{i=1}^{N} w(x_i)\, Q\, n(x_i)\, \Delta x \tag{K.50}$$

$$S = \int_0^\infty w(x)\, Q\, n(x)\, dx. \tag{K.51}$$

The noise for unweighted signals is just

$$N = \sqrt{S}. \tag{K.52}$$

With weights, the noise equation is much more complicated. Because each bin in Figure K.5 can be thought of as a separate event, the total noise is given by adding in quadrature:

$$N = [\{w(x_1)[Q\, n(x_1)\, \Delta x]^{1/2}\}^2 + \{w(x_2)[Q\, n(x_2)\Delta x]^{1/2}\}^2 + \cdots]^{1/2} \tag{K.53}$$

In integral form,

$$N = \left[\int_0^\infty w(x)^2\, Q\, n(x)\, dx\right]^{1/2}. \tag{K.54}$$

Therefore, the output S/N for weighted events is given by

$$\left(\frac{S}{N}\right)_{out} = \frac{\int_0^\infty w(x)\, Q\, n(x)\, dx}{\left[\int_0^\infty w^2(x)\, Q\, n(x)\, dx\right]^{1/2}}. \tag{K.55}$$

From our definition of DQE, we have

ADVANCED STATISTICS 375

$$\text{DQE} = \frac{\left(\dfrac{S}{N}\right)^2_{\text{out}}}{\left(\dfrac{S}{N}\right)^2_{\text{in}}} = Q \left\{ \frac{[\int w(x)\, n(x)\, dx]^2}{[\int w^2(x)\, n(x)\, dx]\, [\int n(x)\, dx]} \right\} \quad \text{(K.56)}$$

or

$$\text{DQE} = Q f \quad \text{(K.57)}$$

where

$$f = \frac{[\int w(x)\, n(x)\, dx]^2}{[\int w^2(x)\, n(x)\, dx]\, [\int n(x)\, dx]} . \quad \text{(K.58)}$$

So f is the factor that the unweighted DQE is modified by when weights are included. Two cases are involved:

1. *Weights equal to one, as in pulse counting* ($w(x) = 1$). Then

$$\text{DQE} = Q \frac{[\int n(x)\, dx]^2}{[\int n(x)\, dx]\, [\int n(x)\, dx]}$$

This reduces to

$$\text{DQE} = Q \quad \text{(K.59)}$$

for an ideal photomultiplier.

2. *Weights proportional to the size of the pulse, as in DC photometry* ($w(x) = x$).
Then

$$\text{DQE} = \frac{Q\, [\int x n(x)\, dx]^2}{[\int x^2 n(x)\, dx]\, [\int n(x)\, dx]} \quad \text{(K.60)}$$

Some assumptions need to be made about the pulse height distribution before this equation can be solved. Typical trials include $n(x) = $ constant over a window, $n(x) = e^{-x}$, or $n(x) = $ Poisson distribution. These all lead to complicated reductions. Rather than show all three, we use the window function as an example. The pulse height distribution can be approximated by a box, i.e.,

$$n(x) = n_0, \quad a < x < b$$
$$n(x) = 0, \quad x < a \text{ or } x > b,$$

as shown in Figure K.6. Then

$$\text{DQE} = \frac{Q \left[n_0 \int_a^b x \, dx \right]^2}{\left[n_0 \int_a^b x^2 \, dx \right]\left[n_0 \int_a^b dx \right]}$$

$$= \frac{Q n_0^2}{n_0^2} \left\{ \frac{\left[\frac{x^2}{2} \Big|_a^b \right]^2}{\left[\frac{x^3}{3} \Big|_a^b \right]\left[x \Big|_a^b \right]} \right\}$$

$$= Q \left\{ \frac{\left[\frac{b^2 - a^2}{2} \right]^2}{\left[\frac{b^3 - a^3}{3} \right](b - a)} \right\}$$

$$= \frac{3}{4} Q \left\{ \frac{(b + a)^2 (b - a)^2}{(b^2 + ab + a^2)(b - a)(b - a)} \right\}$$

$$\text{DQE} = \frac{3}{4} Q \left\{ \frac{b^2 + 2ab + a^2}{b^2 + ab + a^2} \right\} \qquad (K.61)$$

If the lower limit a is set to zero, then

$$\text{DQE} = \tfrac{3}{4} Q \qquad (K.62)$$

or the DQE for DC photometry is about 25 percent less than that for pulse counting.

Other functions and experimental results give

$$0.5 \, Q \leq \text{DQE} \leq 0.8 \, Q$$

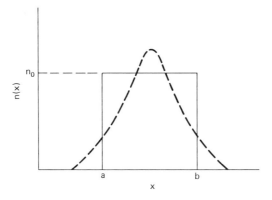

Figure K.6. Box window function.

for DC work. That is, pulse counting gives a better DQE in all cases, by as much as a factor of two. This advantage is greatest for weak signals and becomes much less for strong signals. In general, if the signal is larger than the noise, DC and pulse-counting methods are roughly equivalent, with perhaps 20 percent difference between the two techniques under good conditions. Because of a relatively narrow pulse height distribution, the 1P21 photomultiplier tube is slightly better for DC work than other types, such as the venetian blind tubes.

K.6 PRACTICAL PULSE–DC COMPARISON

In addition to the theoretical signal-to-noise comparison shown above, there are several practical comparisons that must be made when deciding which technique to use. It seems appropriate that, because we have discussed the theoretical differences, we list these observational considerations.

1. Pulse counting is relatively insensitive to amplifier drifts.
2. When high precision (less than 1 percent) is needed, pulse counting is better unless voltage-to-frequency conversion techniques are used with the DC amplifier.
3. When several measurements are to be added or a large number of observations are to be reduced, the data are much easier to handle when in digital form, obtained directly with pulse counting.
4. A digital system has a symmetric and sharp filter function, whereas DC methods have an RC filter, which tends to partially correlate readings, if a long time constant is used.

5. An analog system gives good real-time indication of sky conditions, difficult to obtain from digital methods.
6. Pulse counting provides discrimination against dark current. However, at the same time, some primary photoelectrons must be rejected. A better method to reduce dark current is to cool the photomultiplier tube.
7. Pulse counting is insensitive to leakage currents, but more sensitive to RF fields, such as generated by motors and relays in the telescope's environment.
8. Pulse counting can be performed at higher speed, DC is limited by the pen or meter movement. This advantage is negated if the DC signal is recorded on an instrumentation tape recorder and played back at a slower speed.
9. Pulse counting requires dead-time corrections that can be very significant for bright stars.
10. Complexity of instrumentation for the two methods is approximately the same.

K.7 THEORETICAL S/N COMPARISON OF A PHOTODIODE AND A PHOTOMULTIPLIER TUBE

This section is designed as an example of how two detectors can be compared on paper by their S/N characteristics. This calculation is also intended to be a fairly realistic comparison to help potential photometer builders decide between a photodiode or a photomultiplier as a detector. We have adopted the noise and quantum efficiency characteristics of the 1P21, and typical photodiode characteristics from the EG & G Electro-Optics catalog.[8]

Until now, the S/N discussion has assumed a pulse-counting photometer. However, a photodiode cannot be used for pulse counting because it lacks internal gain. The output of a photomultiplier consists of pulses, as seen in Figure 7.2, because each electron released at the photocathode is amplified by the dynode string by a factor of 10^6. Each detected photon produces a pulse that can be readily counted. Many of the noise sources within the photomultiplier produce smaller pulses at the output because they result from amplification by only part of the dynode string. It is easy to discriminate against much of the noise.

This is not the case for the photodiode because each detected photon contributes just one electron to the output current. There is no strong burst of electrons at the output when a photon is detected. The current produced by the photons looks identical to that produced by noise within the photodiode. This comparison assumes that both detectors are used with excellent quality DC amplifiers.

There are some minor modifications to be made to the S/N treatment for the DC case. Shot noise, whether for photons or electrical current, is a "white" noise. This means it is equally strong at all frequencies. The amount of noise depends on the bandpass of the amplifier. An amplifier that measures a frequency range from 1000 to 10,000 Hz detects much more noise than one that spans 1000 to 2000 Hz. Any equation for shot noise must include the width of the amplifier bandpass. For instance, photon shot noise is written as

$$N_s = G(2Q\dot{C}_s B)^{1/2} t \qquad (K.63)$$

where B is the bandwidth (Hz). For pulse-counting photometry, the bandwidth has a very simple relationship with the integration time, t, namely,

$$B = \frac{1}{2t}. \qquad (K.64)$$

For DC photometry, this relation is

$$B = \frac{1}{4\tau} \qquad (K.65)$$

where τ is the RC time constant of the amplifier. Note that if Equation K.64 is substituted into Equation K.63, we get the same expression for photon shot noise used earlier. Also note that Equations K.64 and K.65 imply that

$$t = 2\tau.$$

Photon shot noise for the DC case is given by

$$N_s = G(2Q\dot{C}_s \tau)^{1/2}. \qquad (K.66)$$

The noise sources that must be considered are the photon shot noise, shot noise from the sky background, amplifier noise, and detector noise. We ignore noise due to atmospheric scintillation because this is highly variable and a function of observatory location. Then by Equation K.40, we have

$$\left(\frac{S}{N}\right)_{out} = \frac{Q\dot{C}_s t}{\left[\left(\frac{N_s}{G}\right)^2 + \left(\frac{N_b}{G}\right)^2 + \left(\frac{N_{am}}{G}\right)^2 + \left(\frac{N_{det}}{G}\right)^2\right]^{1/2}} \qquad (K.67)$$

where N_{det} is the detector noise. For a photomultiplier, this is given by the shot noise of the dark current. For the 1P21 at room temperature, N_{det} is about $2 \times 10^{-11} B^{1/2}$ amps or $1.3 \times 10^8 B^{1/2}$ electrons per second. If the tube is cooled to dry-ice temperature, N_{det} drops to 1.3×10^6 electrons per second. Persha[9] has recommended that photodiodes be used in a photovoltaic mode for astronomical photometry. In this case, there is no dark current but there is a thermal noise generated at the p-n junction given by

$$I_n = \left(\frac{4kTB}{R}\right)^{1/2} \tag{K.68}$$

where k is Boltzmann's constant, T is the temperature (Kelvin), and R is the shunt resistance (ohms) of the photodiode. This resistance is also a function of temperature. According to the EG & G catalog, R approximately doubles for each 5 Celsius degree temperature drop. The largest shunt resistance now available is about 10^9 ohms at room temperature. The noise current for a photodiode is given by

$$I_n = 2.5 \times 10^4 \sqrt{B} \text{ electrons/second} \tag{K.69}$$

at room temperature and

$$I_n = 413 \sqrt{B} \text{ electrons/second} \tag{K.70}$$

at dry-ice temperature.

The amplifier noise for both detectors, referred to the amplifier input, is assumed to be 2500 $B^{1/2}$ electrons per second. This is about the best that can be expected with present FET input amplifiers. We adopt an amplifier time constant of 1 second which fixes B at 0.25 Hz. The internal gain, G, for the photomultiplier is about 10^6 while for the photodiode, which lacks internal amplification, it is 1.0.

Using these various parameters, Equation K.67 becomes for the photomultiplier,

$$\left(\frac{S}{N}\right)_{out} = \frac{2Q\dot{C}_s \tau}{\left[2Q(\dot{C}_s + \dot{C}_b)\tau + \left(\frac{2500}{10^6}\right)^2 \frac{1}{4\tau} + \left(\frac{N_{det}}{10^6}\right)^2\right]^{1/2}} \tag{K.71}$$

and for the photodiode,

$$\left(\frac{S}{N}\right)_{out} = \frac{2Q\dot{C}_s \tau}{\left[2Q(\dot{C}_s + \dot{C}_b)\tau + \frac{(2500)^2}{4\tau} + \left(N_{det}^2\right)\right]^{1/2}} \tag{K.72}$$

The equations have been left in this form to illustrate how a high gain detector minimizes the effects of N_{am} and N_{det}. This advantage over the photodiode becomes evident in the calculations to follow.

The photon arrival rates, $\dot{C}_s$ and $\dot{C}_b$, are calculated for the B filter for various apparent magnitudes using Equation 2.33 with approximate corrections for the transmission of the atmosphere, telescope optics, and filter. A 20-centimeter (8-inch) diameter telescope is assumed. A sky background of twelfth magnitude was established from the known background at a dark site and scaled to a 0.5-millimeter diaphragm. This is a typical size for an active area of a photodiode. The quantum efficiencies at 4400 Å for the photomultiplier and photodiode are 0.10 and 0.60, respectively.

Figure K.7 shows the S/N, calculated from Equations K.71 and K.72 versus B magnitude for uncooled detectors. For very bright stars, the two detectors give comparable results. The reason for this can be seen in Equations K.71 and K.72. For very high photon rates, the S/N approaches a value of $(Q\dot{C}_s)^{1/2}$. Because $\dot{C}_s$ is the same for both detectors, the difference in S/N is set by their quantum efficiencies. In the high photon rate limit, the S/N of the photodiode should exceed that of a photomultiplier by a ratio of $(0.6/0.1)^{1/2}$ or 2.5. For all but the very brightest stars, the photomultiplier is clearly superior. This is a result of the amplifier and detector noise terms becoming important at low photon rates. These terms are not nearly as important for the photomultiplier because of its high internal gain. This more than compensates for the lower quantum efficiency of the photomultiplier.

The S/N curves become a little more meaningful if we consider an example. An 8.4 magnitude star produces a S/N of 100 with a photomultiplier tube and an amplifier time constant of 1 second. With this same time constant, this star would only produce a S/N of 4.5 with a photodiode detector. Because the S/N improves with the square root of the observing time, the observation with the photodiode would have to be 500 times the duration of the observation made with the photomultiplier to obtain the same accuracy. Finally, if we consider a S/N of 1 to be the detection limit, Figure K.7 shows that the photomultiplier can go almost four magnitudes fainter than the photodiode.

The situation improves if the photodiode is cooled to dry-ice temperature. The calculation proceeds as before except that the detector noise terms are reduced as discussed earlier. Figure K.8 shows an uncooled photomultiplier compared to a cooled photodiode. A cooled photodiode comes much closer to matching the performance of an uncooled photomultiplier tube. While the photodiode still appears to be inferior the uncertainties in the calculation are such that it is safest to say that they are roughly comparable. Figure K.9 compares detectors when both are cooled to dry-ice temperature. The photomultiplier has a superior S/N for stars fainter than fifth magnitude while the photodiode is

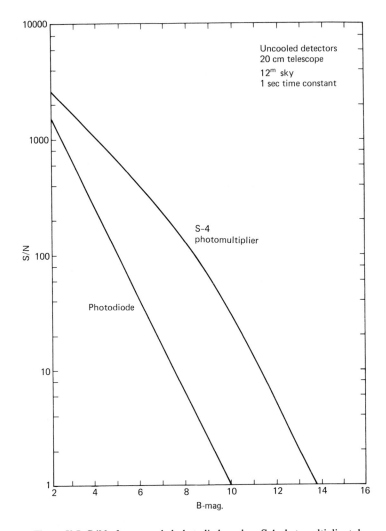

Figure K.7. S/N of an uncooled photodiode and an S-4 photomultiplier tube.

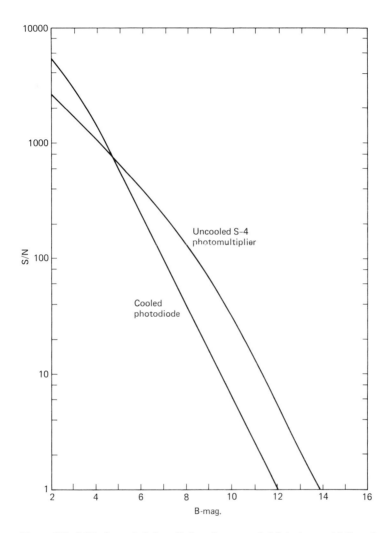

Figure K.8. S/N of a cooled photodiode and an uncooled S-4 photomultiplier tube.

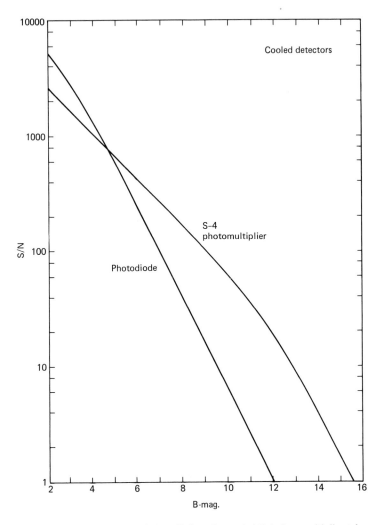

Figure K.9. S/N of a cooled photodiode and a cooled S-4 photomultiplier tube.

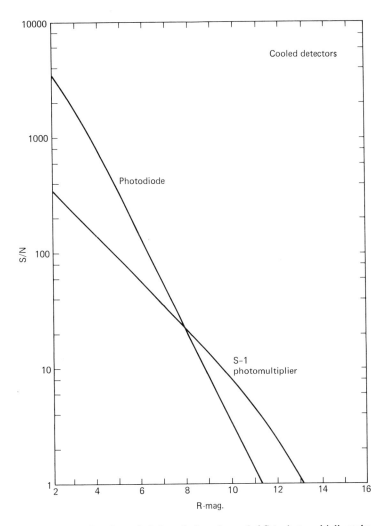

Figure K.10. S/N of a cooled photodiode and a cooled S-1 photomultiplier tube.

better for stars brighter than this. The photomultiplier has a detection limit about 3.5 magnitudes fainter than the photodiode.

Unlike the S-4 response of the 1P21, the photodiode has very high quantum efficiency in the near-infrared. The above calculation was repeated for the R bandpass (7000 Å) comparing the photodiode to an S-1 photomultiplier. Both detectors were assumed to be cooled. The superior quantum efficiency of the photodiode makes a significant difference in the infrared. An S-1 surface has a quantum efficiency of only 0.4 percent compared to the photodiode's 83 percent at 7000 Å. Figure K.10 shows that this difference makes the photodiode a superior detector down to the eighth magnitude, despite the higher internal gain of the photomultiplier. The curves would look much the same in the I bandpass (9000 Å) because the quantum efficiencies are nearly the same.

The above calculations are intended only as a rough comparison of these two detectors. The S/N curve for the photomultiplier has been checked at a few points with observational data and has compared well. Unfortunately, the authors have not had access to comparable empirical data for a photodiode photometer. However, we suspect that the theoretical calculation is not in large error. Figures K.7, K.8, K.9, and K.10 can be used to draw some general conclusions about detector selection. First, if you intend to use an uncooled detector, the photomultiplier tube is clearly the better detector. If you can design and build a cooling system, the photodiode will be only slightly inferior to an uncooled photomultiplier. A cooled photodiode offers the convenience of a single detector with sensitivity from the ultraviolet to the infrared. However, in general, the high internal gain of the photomultiplier makes it a superior detector overall. Finally, it should be noted that the photomultiplier has a potential S/N gain over the photodiode, which has not been included in these calculations. Because of its high internal gain, the photomultiplier can be used to photon count. By the results of Section K.5, this means an additional increase of as much as a factor of $\sqrt{2}$ in the S/N compared to the photodiode, which must use DC measuring techniques.

REFERENCES

1. Young, H. D. 1962. *Statistical Treatment of Experimental Data.* New York: McGraw-Hill.
2. Harnell, D. L. 1975. *Introduction to Statistical Methods.* 2d ed. Reading, Mass.: Addison-Wesley.
3. Bevington, P. R. 1969. *Data Reduction and Error Analysis for the Physical Sciences.* New York: McGraw-Hill.
4. Young, A. T. 1974. In *Methods of Experimental Physics: Astrophysics,* Edited by N. Carleton. New York: Academic Press, vol. **12A**.

5. Golay, M. 1974. *Introduction to Astronomical Photometry.* Boston: D. Reidel.
6. Meaburn, J. 1976. *Detection and Spectrometry of Faint Light.* Boston: D. Reidel.
7. Arfken, G. 1970. *Mathematical Methods for Physicists.* New York: Academic Press.
8. EG & G Electro-Optics Division, 35 Congress St., Salem, Mass. 01970.
9. Persha, G. 1980. IAPPP Com. **2**, 11.

INDEX

AAVSO, 31, 273
Airglow, 230
Air mass, 29, 53, 86, 324, 330
Altitude, 119
Amplifier,
    DC, 188ff
    meter, 193
    noise, 369ff
    operational, 185
    pulse, 167ff
Aperture stops, 83
Asteroids, 271
Atlases
    photographic, 204
    positional, 203
Atmosphere, 28
Aurora, 231
Azimuth, 119

Background
    noise, 369ff
    sky, 10
Balmer discontinuity, 40
Batteries, 149
Binary stars
    Algol, 257
    eclipsing, 2, 256ff
    β Lyrae, 257
    RS CVn, 259
    W UMa, 259
Blackbody, 349

Calculators, 101ff
Calibration
    absolute, 50
    DC, 197
    diaphragm, 227

Cells
    photoconductive, 7
    photoelectric, 7
Chart recorders, 194
Check star, 24
CHU, 109
Clock, sidereal rate, 112
Cold box, 133
Color-color diagram, 45
Color excess, 46
Color index, 27
Colors, instrumental, 25
Comparison stars, 20, 53, 207ff
    selection, 208
    use, 209
Computers, 101ff, 178, 247
Construction, electronic, 147
Coordinates, precesion, 116, 337
Current
    constant sources, 196
    input bias, 190

Date
    Heliocentric Julian, 113
    Julian, 112, 336
Day
    sidereal, 110
    solar, 108
Dead time, 81, 308, 328, 336
    correcting, 82
Depletion region, 20
Detective quantum efficiency, 367ff
Diaphragm, 9, 84, 138ff
    background removal, 226
    calibration, 227
    selection, 220, 224

Diode, PIN, 18, 378
Direct current (DC) methods, 16
Distribution
  Gaussian, 66, 358
  Normal, 66
  Poisson, 360
  probability, 358
Dynode, 14

Eggen, O. J., 254
Einstein, A., 13
Electronics
  DC, 184ff
  pulse, 167ff
Emission
  field, 14
  secondary, 14
  thermionic, 14
Epoch, 264
Equation
  normal, 70
  of condition, 69
Error
  illegitimate, 60
  probable, 66
  propagation, 361
  random, 61
  standard, 71
  systematic, 61
Extinction
  atmospheric, 28, 228
  correction, 325, 331
  first order, 28, 88, 279, 311ff
  second order, 28, 90, 286, 320
  theory, 351ff
Extrapolation, 77

Fabrey lens, 10, 125
Filters
  neutral density, 84
  slide, 136
  UBV, 35, 134
  wheel, 137
Finding charts, 202ff
  preparation, 205
  published, 206
Flare stars, 240
Flip mirror, 141
Flux, 25, 344
Frequency counter, 167

Galaxies, 272, 347
GCVS, 248
Goodness of fit, 71

Hall, D., 259
Harvard spectral classification, 41
Head, photometer, 9, 124ff
Henry Draper catalog, 38
High voltage power supply, 149ff
History, 5
H-R diagram, 41
Hβ photometry, 57

IAPPP, 31, 273
IC4665, 302
Infrared, 55
Intensity, 343
Interpolation, 73ff
  exact, 74
  smoothed, 76
Interstellar
  absorption, 48
  reddening, 48

Johnson, H. L. 34ff, 290ff
Johnson noise, 190
Julian Date, 112, 336

Keenan, P. C. 38ff
Kron, G. E. 8

Least squares
  linear, 68, 338, 339
  multivariate, 363
Light
  night sky, 229
  zodiacal, 229
Light elements, 264
Line blanketing, 55
Luminosity, 349

Magnitude
  difference, 5, 24
  extra atmospheric, 86
  instrumental, 25, 85, 356
  photographic, 6
  photovisual, 6
  standard, 29, 92
Mean, sample, 61
Median, sample, 63
Metals, 40

Meter, amplifier, 193
Minimum light, 263, 266
Mirrors, coating, 12
M-K spectral classification, 34, 38ff
Morgan, W. W., 34ff, 290ff

Noise
  amplifier, 369
  background, 369
  scintillation, 368
  shot, 368
  thermal, 190
Novae, 263

Observing
  applications, 238ff
  diaphragm selection, 224
  faint sources, 218
  first night, 231ff
  optimizing time, 212
Occultation photometry, 245

Period, 264
Photocathode, 14
Photodiodes PIN, 18
Photoelectric effect, 13
Photography, 6
Photometer
  chopping, 159
  dual-beam, 161
  head, 9, 124ff
  multifilter, 163
  single beam, 157
Photometric sequences, 238
Photometry
  all sky, 94
  cluster, 94
  differential, 23, 52ff, 95ff, 216ff
  occultation, 245
  references, 30
Photomultiplier tube (PMT), 11, 13ff, 128ff
  1P21, 13ff, 35, 128
  931A, 129
  cathode types, 16
  dynode, 14
  end-on, 17
  fatigue, 132
  housing, 133
  PIN comparison, 378ff
Photon counting, 16

Photons, 13
Planck's law, 349
Planets, 272
Pleiades, 298
Pogson scale, 5
Point source, 220
Praesepe, 298
Preamp, 168
Precession, 116, 337
Publication of data, 273
Pulse counter, 172ff
  design, 173
  microprocessor, 178
Pulse counting
  DC comparison, 215, 372ff
  operation, 16, 182
Pulse generator, 181
Pulse height distribution, 169, 372

Reference light sources, 155
Refraction
  atmospheric, 104
  calculation, 104
  differential, 107
  effect on air mass, 106
Regression analysis, 68
Rejection of data, 67
RF oscillator power supply, 153

Satellites, 271
Semiconductor, 19
Signal-to-noise (S/N) ratio, 77, 210, 366ff
Standard stars, 33, 290, 297
Standard system, transforming, 29, 92, 355ff
Star
  DC measurements, 213
  profile, 223
  pulse measurements, 210
Statistics, 60ff, 358ff
  sources, 78
Stebbins, J., 7
Stefan's law, 350
Stromgren 4-color system, 55
Sun, rectangular coordinates, 114, 341

Telescope, 11
Time
  sidereal, 110, 340
  solar, 108
  universal, 109

Transformation coefficients
  DC example, 322
  pulse example, 327
Twilight emission lines, 230

U-B problem, 98
UBV
  filters, 35
  IR extension, 55
  standard cluster stars, 297
  standard field stars, 290
  system, 34
  transformation coefficients, 38, 93
  transformation equations, 37
  zero point constraints, 38, 91

Variable stars
  catacylsmic, 263
  cepheids, 3, 249, 250
  dwarf cepheids, 249
  intrinsic, 248ff
  Mira, 254
  RR Lyrae, 249
  RV Tauri, 250
  semiregular, 254
  UV Ceti, 241
  δ Scuti, 249
Variance, 65
Voltage-to-frequency converters, 195

Wien's displacement law, 350

Zero point constants, 38, 91